Engineering Materials

Springer

Berlin
Heidelberg
New York
Barcelona
Hong Kong
London
Milan
Paris
Tokyo

L. Ratke, P. W. Voorhees

Growth and Coarsening

Ostwald Ripening in Material Processing

With 99 Figures

Springer

Professor Dr. Lorenz Ratke
German Aerospace Center DLR
Institute for Space Simulation
D-51170 Cologne
e-mail: lorenz.ratke@dlr.de

Professor Peter W. Voorhees
Northwestern University
Dept. of Materials Science and Engineering
2225 North Campus Drive
Evanston, Illinois 60208-3108
USA
e-mail: p-voorhees@northwestern.edu

ISBN 3-540-42563-2 Springer-Verlag Berlin Heidelberg NewYork

Library of Congress Cataloging-in-Publication Data

Ratke, Lorenz:
Growth and Coarsening: Ostwald Ripening in Materials Processing / L. Ratke; P.W. Voorhees. –
Berlin; Heidelberg; New York; Barcelona; Hongkong; London; Milan; Paris; Tokyo: Springer, 2002
(Engineering Materials)
ISBN 3-540-42563-2

Springer-Verlag is a company in the BertelsmannSpringer publishing group
http://www.springer.de

Printed in Germany

Typesetting: Camera ready by authors
Cover-Design: dé'blik, Berlin
Printed on acid free paper SPIN: 10850790 62/3020/kk - 5 4 3 2 1 0

To Sylvia and Maria

Preface

Whenever a single phase system is quenched into a metastable state nucleation of a second phase lowers the free energy of the system. Following this nucleation event, in solid state transformations, the second phase typically grows by the diffusion of mass either into or from the matrix. When the composition of the matrix is nearly at its equilibrium value coarsening occurs. In this process the matrix is no longer a source or sink of solute, but the majority of the mass transport is from small to large particles. This process is driven by the decrease in interfacial energy per unit volume and results in an increase in the size scale of the two-phase mixture. The first to observe this coarsening process was Ostwald in 1900 who reported that the solubility of small HgO-particles is a function of the radius of the particle. The phenomenon has come to bear his name: Ostwald ripening. Ostwald ripening can occur in virtually any two-phase mixture in which there is significant diffusion. It is thus observed in two-phase mixtures ranging from liquid-vapor mixtures, such as droplets in clouds, to solid-liquid mixtures both during and following solidification. Understanding the dynamics of this ripening process and its surprising self-organizational nature has been at the center of the field since the seminal papers of Lifshitz and Slyozov, and Wagner appeared in the early 60's. From a more applied view, the Ostwald ripening process has a significant impact on manner in which materials are processed, e.g. age hardening of metallic alloys, grain refining, grain growth during recrystallisation, liquid and solid state sintering, and the deoxidation of steel melts. Given the ubiquitous nature of the ripening process and its clear technological importance, we feel that this subject deserves a unified development that is only possible in a monograph.

The main idea behind this monograph is to present students and researchers all the bits and pieces of coarsening theories such that they can understand the main issues and the underlying mathematics of self-similar coarsening of dispersed systems. Growth and coarsening are usually side notes in standard textbooks on physical metallurgy or materials science and as such they are treated in a very simplified fashion. Therefore, the monograph contains all of the background material necessary to understand growth and coarsening of spherical particles or droplets in a liquid or solid matrix. We do not refer the reader to any of the publications given in the bibliography for a derivation of a central result used in the monograph, but try to be self-consistent and comprehensive. Some basic knowledge of heat and mass transfer, thermodynamics, and differential equations would be helpful, but is not necessary as all the concepts required are presented, at least briefly. The monograph is suitable for advanced undergraduate students, graduate students, and researchers. The monograph is not a comprehensive review of coarsening and growth nor is it intended to give a complete survey of both topics. For example, we do not discuss the many mathematical approaches to the theory or the vast experimental literature. Rather, the monograph is intended to present a careful derivation of existing results and to place these results in some prespective.

Cologne and Evanston, July 2001

Lorenz Ratke and Peter W. Voorhees

Table of Contents

1. Introduction

Whenever a single phase system is put into a two-phase metastable state, for example by quenching from a high temperature or changing pressure, a second phase nucleates, grows, and coarsens. Nucleation of the second phase occurs since the energy of the single phase system can be reduced by forming regions of the second phase. Nanometer-sized precipitates of the new phase thus appear. Growth of nuclei then proceeds due to heat diffusion away from the nucleus into the matrix or mass diffusion from the matrix to the nuclei. This results in a dispersion of second phase particles in a matrix. Coarsening, also called Ostwald ripening or competitive growth, then occurs where large particles grow at the expense of small particles. The nucleation, growth, and coarsening is very common, occurring in systems ranging from solid alloys to the precipitation of drops from clouds [246].

Provided there is sufficient atomic mobility, a two-phase dispersion will coarsen by transfer of matter from small to large particles, thus reducing the overall free energy associated with the particle-matrix interfacial area. In 1900, Ostwald reported the first systematic study noting the dependence of the solubility of small HgO-particles on their radii [227]. The phenomenon has come to bear his name. The basic theory of particle coarsening, however, was developed sixty years later by Lifshitz and Slyozov [194] and Wagner [295]. Given that a two-phase mixture always possesses excess energy due to the presence of significant interfacial area, this Ostwald ripening process is observed in the late stages of nearly all phase transformation processes.

Some technically important situations in which nucleation, growth, and coarsening play an essential role in microstructural evolution are:

Quenching a multicomponent alloy from a temperature in a single phase field of the phase diagram rapidly to room temperature generally produces a supersaturated matrix. If such a supersaturated solid solution alloy is then annealed at a higher temperature, precipitates are nucleated and grow by diffusion of solute from the matrix towards the precipitates. After long annealing times the initially large supersaturation of solute has decreased enough for interfacial curvature effects to play an important role: the spectrum of precipitate sizes leads to curvature induced solubility differences in the matrix such that the dispersed two-phase system (precipitates and matrix) coarsens, i.e. the big particles grow on the expense of the smaller ones, see e.g. figure 1.1. This is the basis of all age-hardening alloys (e.g. AlCu, AlAg, AlZn, AlMgSi, AlCuMg, AlZnMg, FeCu, CuCo, NiAl) [207]. During the nucleation and growth stages the yield strength and hardness generally increase, whereas coarsening usually leads to *overaging* and a degradation of the mechanical properties. The processes of nucleation, growth, and coarsening do generally not occur in a strict sequence but overlap.

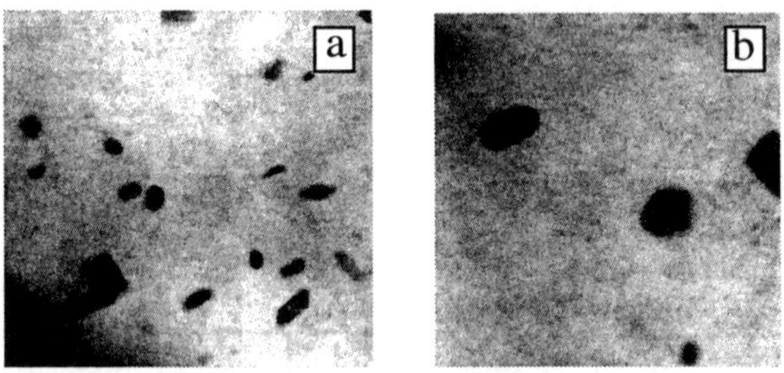

Fig. 1.1. Annealing of an age hardenable AlSiMg1 alloy at 570°C leads to a growth of the equilibrium Mg_2Si precipitates and a reduction of their number per volume after [252]. Figure a) shows the microstructure after 12 h and figure b) the same alloy after an annealing treatment of 48h. Both images were taken using a transmission electron microscope (TEM).

Similarly one can quench an alloy from a single phase liquid state into a two phase liquid state if the phase diagram exhibits a miscibility gap in the liquid state. Then droplets of the second phase will nucleate, grow, and coarsen as in solid solutions, see e.g. figure 1.3, but significantly faster since mass transport in liquids is orders of magnitude faster. However, liquids of different composition usually have different densities. Therefore, they show a rapid spatial phase separation: the denser liquid settles to the bottom of a crucible similar to the separation of oil and water. Under reduced gravity conditions the coarsening of liquid-liquid dispersions can be investigated while avoid-

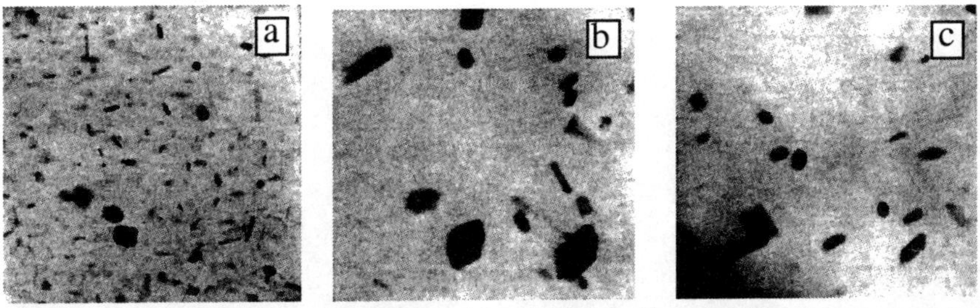

Fig. 1.2. Annealing of an age hardenable AlSiMg1 alloy at different temperatures leads to a growth of the precipitates and a reduction of their number per volume. Figure a) shows the microstructure in a transmission microscope after annealing at 470°C, figure b) at 520°C and figure c) at 570°C. Annealing time was always 12h [252].

ing this problem. Coarsening similar to that of precipitates in solid solutions can then be observed.

Solidification into an undercooled melt frequently results in the formation of solid dendrites that grow due to the transport of latent heat released at the interface to the supercooled melt. During growth the dendrites develop sidebranches that that sufficiently far from the dendrite tips exhibit coarsening due to variations in curvatures along the dendrite surface. Similar to solid solutions, where the solubility is increased when the precipitate size becomes smaller, the melting temperature is a function of the local interfacial curvature. Thus, the morphology of a cast microstructure depends on growth and coarsening of the network of primary dendrite stems and their secondary arms, see e.g. figure 1.4.

Grain refining of cast microstructures, especially of aluminum alloys, is performed by establishing a dispersion of suitable heterogeneous nucleation sites in the melt (e.g. Ti-Boride particles in aluminum) such that solidification begins at these sites. Growth and coarsening of such heterogeneously nucleated dendrites are driven by same processes as growth and coarsening of precipitates: Interface curvature variation driven mass and heat transport via diffusion and convection. Thus, this process can be described using the theoretical framework given in these lectures.

It is also typical in metallurgical processing to heat treat cast ingots to reduce micro- and macrosegregation of alloying elements, which are a natural result of solidification, and to achieve morphological changes of the as-cast microstructure. For example, in cast Aluminum Silicon

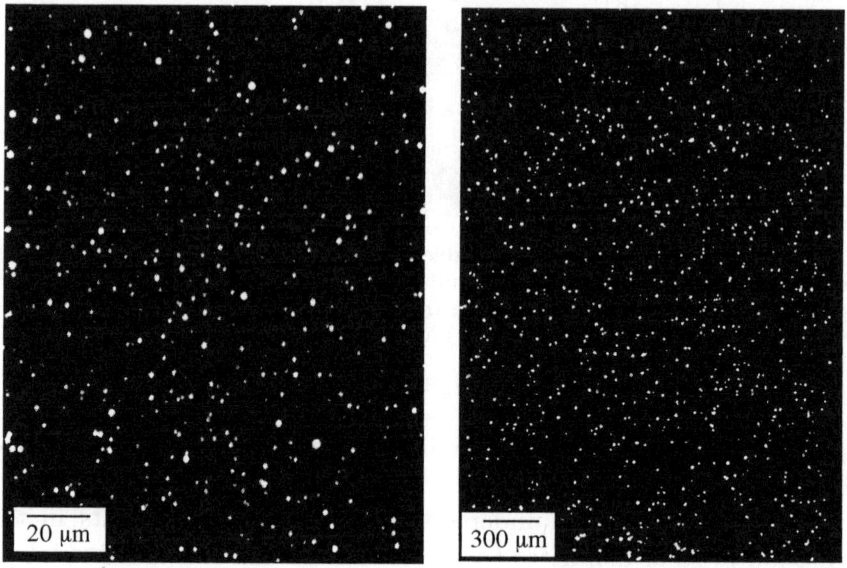

Fig. 1.3. Zn and Pb exhibit a miscibility gap in the liquid state. Annealing them at high temperatures when both components are miscible and quenching them rapidly into the solid state leads to the microstructure shown in the left figure. Annealing them in the miscibility gap for 1h under reduced gravity conditions leads to a considerable coarser microstructure (right figure) as can be seen from the ruler at the top of both figures [236, 237].

Fig. 1.4. A dendrite in a dilute AlCr alloy. From the tip to the bottom of the dendrite the sidebranches or secondary and tertiary dendrites grow and coarsen, i.e. the initial number of secondary arms decreases with increasing distance from the tip (increasing time for coarsening).

alloys the eutectic is of flake-like type. The silicon platelets act as nuclei for internal cracks, they reduce especially the fatigue strength, etc. Heat treating the as-cast alloy at high temperature, just below the eutectic temperature, leads to a morphological change, see figure 1.5. Flakes are transformed into spherical particles as well as growing in size due to the reduction of the total interfacial energy.

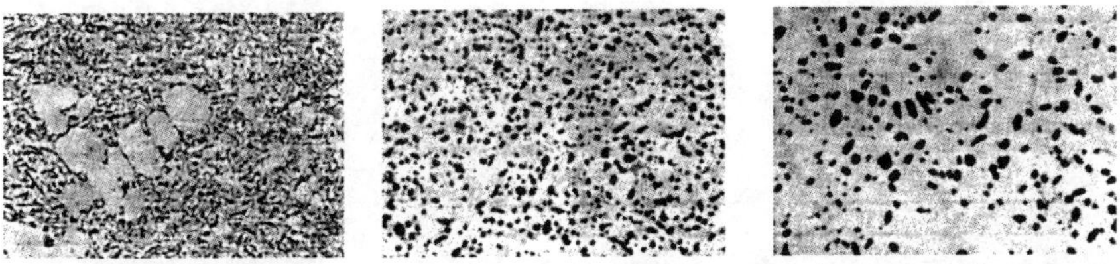

Fig. 1.5. The as-cast microstructure of an hypoeutectic AlSi alloy changes dramatically during heat treatment at different temperatures. The left figure shows the as-cast microstructure, the middle and right figure show the microstructure after 12h annealing at 470°C and 520°C after [252].

Growth and coarsening also occur in other metallurgical processes. For example deoxidation of steel melts is achieved by the addition of suitable deoxidants like FeAl and FeSi leading to the formation of Aluminiumoxide and Silicon-Oxide inclusions in the melt [238]. The solid or liquid inclusions grow rapidly by mass diffusion of oxygen in the melt.

Growth and coarsening also play an important role during liquid phase sintering. It is widely used for the consolidation of ceramics and metal powders into final shapes. Typical materials sintered to full density are heavy metal alloys, hard metals, and structural parts for automotive applications made from Fe-Cu alloys, W-Cu or Ag-Ni for electrical contact applications, and Fe-Pb for bearings [109, 253]. Materials that are produced routinely by liquid phase sintering are the hard metals, e.g. tungsten-carbide(WC)-Co shown in figure 1.6, which cannot be processed via casting routes since the hard and brittle phase W_2C would appear during solidification instead of the hard and tough carbide phase WC. Liquid phase sintering can be divided into different stages: After heating one phase becomes liquid and viscous flow of the solid-liquid mixture leads to a local and global rearrangement of the liquid and solid constituents. This process is followed by dissolution due to alloying and a so-called 'solution and precipitation' stage, in which dissolution of smaller solid phase particles occurs together with reprecipitation of mass onto the larger solid phase particles. This process leads to significant shape accommodation. This stage is quite often modeled using the coarsening models explained in some detail in these lectures.

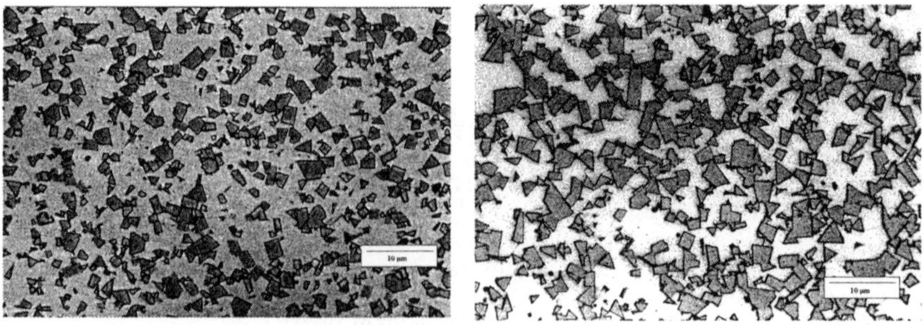

Fig. 1.6. Coarsening in a liquid phase sintered hard metal consisting of solid WC particles of polyhedral shape and a Co(WC) matrix. The sintering temperature was 1400°C. The left figure shows the microstructure after 6 minutes and the right after a sintering treatment of 1400 minutes. At the sintering temperature the volume content of WC is 30vol.%.

Thus, growth and coarsening are common features of all structural transformations in two- or multi-phase mixtures. The principles of growth and coarsening are described in this book.

2. Thermodynamics of alloys

In this chapter we will present briefly some basic thermodynamic relations that are useful in understanding the fundamentals of growth and coarsening of particles embedded in a matrix. The reader who is already familiar with the concepts of free energy, chemical potential, equilibrium of heterogeneous alloys, the Gibbs-Thomson effect etc. may skip this chapter.

2.1 Basic definitions

Experience tells that under given conditions a material is in a definitive *state*. This state is determined exactly by certain properties. For example at 200 K the compound H_2O is solid, at 300 K it is a liquid and at 400 K it is a gas. Changing temperature changes the state and the properties of the material. All parameters defining a state are commonly called *state variables*. For a fluid, such parameters are temperature T, pressure P and composition of species x_i, where $x_i = n_i/n$. n_i is the number of moles of component i and n is the total number of moles in the alloy. The key issue in thermodynamics is to determine the conditions and relationships between variables at *equilibrium*. Equilibrium is attained when the energy of the system is an extremum. If the energy is maximized, the system is in a state of unstable equilibrium. If the energy of the system is at a local minimum, the system is in a metastable state. Finally, if the energy of the system is in a global energy minimum, the system is in stable equilibrium. All three states of equilibrium have important consequences in phase transformation processes. Equilibrium is independent of the way it is achieved: one can produce ice by condensation of water vapor or freezing of liquid water. Ice can be directly converted into vapor without going through the liquid state (sublimation) and vapor can be achieved by heating water above 373 K. Thermodynamics provides no information on the rate of the transformation, only on whether the transformation can occur.

The appearance or condition of a material (e.g. solid, liquid, vapor) is called a *phase.* A phase is macroscopically homogeneous. A phases can be differentiated from one another by their atomic structure, e.g. liquid, or crystal. In crystals, materials can have different crystal structures, e.g. iron is at low temperatures body centered cubic and at high temperatures face centered cubic. However, it is possible to have two phases with the same crystal structure in one material. In this cases, for example, the two phases differ in composition. If more than one phase is present, e.g. solid and liquid, they contact each other at a *phase boundary* or *interface.* For example, in a glass of water there is a phase boundary between the liquid water and the gas phase above it. The totality of all phases is called a *system.* In a system it is possible that under given external conditions two or more phases are in equilibrium. A system consisting of only one chemical component (like water) is called a *one-component system.* In materials science one is usually concerned with systems having two and more components, or *multi-component systems.*

The definition of equilibrium follows from the first and second laws of thermodynamics. These are that the internal energy in a closed system is conserved and that the entropy of a system must be maximized in a closed system. Assuming that the internal energy can be written as a function of the entropy, volume, and the number of moles of each chemical component, these two laws state that equilibrium exists when the internal energy of the system is extremized under the constraints of constant entropy, volume and amount of each chemical component. Extremizing the total energy under these constraints yields in a single phase system yields the conditions for an energy extremum,

$$\begin{aligned} &T \quad \text{constant} \\ &P \quad \text{constant} \\ &\mu_i \quad \text{constant for } i = 1 \dots N \end{aligned} \tag{2.1}$$

where T is the temperature, P is the pressure, μ_i is the chemical potential of each component, and N is the number of chemical components in the system. Given that these quantities are constants, the system is uniform at equilibrium. A single phase is homogeneous at equilibrium: the temperature, pressure and chemical potential are constants. In particular, since the chemical potential of a component in a single phase system is an invertible function of composition and the chemical potential is uniform, this implies that the composition must also be uniform throughout the phase.

In two phase systems where the two phases are separated by a planar interface, the same equilibrium conditions apply: temperature, pressure and chemical potential are constant. Thus, at the interface between two phases, α and β,

$$T^{\alpha} = T^{\beta} \tag{2.2}$$

$$P^{\alpha} = P^{\beta} \tag{2.3}$$

$$\mu_i^{\alpha}\left(T, P, x_i^{\alpha}\right) = \mu_i^{\beta}\left(T, P, x_i^{\beta}\right) \quad \text{for } i = 1 \ldots N \tag{2.4}$$

However, as we will see the chemical potentials of the two phases are different functions of composition[1]. Thus, while the chemical potentials of the two phases are equal at the interface, this only occurs if the compositions of the two phases are different.

It is possible to define other energy functions which are extremized under different sets of constraints. A particularly convenient energy is the *Gibbs free energy* G, which is defined as

$$G = H - TS, \tag{2.5}$$

or in molar terms

$$G_m = H_m - TS_m. \tag{2.6}$$

This quantity is also called free energy. H is the heat content or *enthalpy* and S is the *entropy*. H_m is enthalpy per mole and S_m is the entropy per mole. The widespread use of the Gibbs free energy lies in the fact that it is an extremum under the constraints of constant T, P, quantities that are particularly easy to control experimentally. At constant T, P this implies that a system will evolve to minimize the Gibbs free energy until it reaches a local or global minimum.

Let us now further define and discuss the meaning of H and S. The enthalpy H of a system is related to its internal energy E, pressure P, and volume V by

$$H = E + PV, \tag{2.7}$$

[1] For a more detailed discussion of the chemical potential and its meaning will be given in section 2.3

or again in molar terms

$$H_m = E_m + PV_m. \tag{2.8}$$

The *internal energy* E of a system is the sum of the kinetic and potential energies of all parts of the system, for example it can be computed with knowledge of the bond energy between atoms. The *heat capacity* or *specific heat* of a substance is the amount of heat required to raise the temperature of one mole substance by one degree. It is related to the molar enthalpy H_m by the relation

$$c_P = \left(\frac{\partial H_m}{\partial T}\right)_P, \tag{2.9}$$

or at constant volume to the molar internal energy E_m:

$$c_V = \left(\frac{\partial E_m}{\partial T}\right)_V. \tag{2.10}$$

Often the heat capacity is given in terms of a unit weight rather than a mole of substance. The heat capacity C_P or C_V is then defined as

$$C_P = \frac{1}{V\rho}\left(\frac{\partial H}{\partial T}\right)_P, \tag{2.11}$$

and

$$C_V = \frac{1}{V\rho}\left(\frac{\partial E}{\partial T}\right)_V. \tag{2.12}$$

with ρ the density of the phase. We use uppercase letters to distinguish these from the molar quantities. In alloys one usually considers systems at constant pressure so that enthalpy changes are more important than internal energy changes.

The entropy used in the equation above can best be considered as a measure for the probability of a given state. In a sense it is a measure of the order of a system, for example for water at 300K the liquid state is less ordered than the crystalline state. Calculating the probability that all molecules are arranged on a crystal lattice and comparing it with the probability that they are arranged randomly in space shows that the crystal state at 300K is less probable than the liquid state. Thus the liquid state has a higher entropy than the crystal state and is therefore attained at 300K. Statistical mechanics is used to calculate the probability of a given state. A more phenomenological description of entropy can be given using the second law of thermodynamics. At constant pressure this can be expressed as

$$dS = \left.\frac{dH}{T}\right|_P . \tag{2.13}$$

A system is in equilibrium if the entropy attains a maximum value under the constraint of constant energy, volume and total number of moles of each chemical component. Thus under these conditions, it exists in a state of maximum disorder. Any spontaneous reaction occurring in an isolated system implies the relation $dS > 0$ or $dS/dt > 0$. The rate of entropy production always is positive. The principle of the increase in entropy of a system refers only to isolated systems. In contrast, if a system can exchange heat with its surroundings there can be either an increase or decrease in the entropy of the system, but this is counterbalanced by a decrease or increase in entropy of the surrounding.

2.2 Single component systems

We will first calculate the free energy of a single component system as a function of temperature. We will show that the Gibbs free energy decreases in a non-linear fashion with temperature and then use this knowledge to calculate the equilibrium melting temperature of a pure, single component substance. The Gibbs free energy G of a single component material per mole at constant pressure is only a function of temperature, see e.g. Eq.(2.6). The change in free energy of the system, dG, with increase in temperature dT is calculated by taking the total derivative of this equation and using eq.(2.13):

$$dG_m = dH_m - TdS_m - S_m dT = dH_m - T\frac{dH_m}{T} - S_m dT = -S_m dT \tag{2.14}$$

Integrating this equation from zero K to a temperature T leads to

$$G_m(T) = G_m(0) - \int_0^T S_m dT, \tag{2.15}$$

where $G_m(0)$ is the Gibbs free energy at the absolute zero. Since per definition (or third law of thermodynamics) we have $S(T = 0) = 0$ it follows $G_m(0) = H_m(0)$. The entropy at a given temperature can be obtained integrating eq.(2.13) from zero K to T at constant pressure

$$S_m = S_m(0) + \int_0^T dS_m = \int_0^T \frac{1}{T} dH_m \tag{2.16}$$

since $S_m(0) = 0$. If $H_m = H_m(T, P)$ at constant pressure P,

$$dH_m = \left.\frac{\partial H_m}{\partial T}\right|_P dT. \tag{2.17}$$

Using the relation eq.(2.9) this equation becomes

$$dH_m = c_p dT. \tag{2.18}$$

Using eq.(2.18) in eq.(2.16) we get

$$S_m = \int_0^T \frac{c_p}{T} dT. \tag{2.19}$$

Therefore, we obtain for the Gibbs free energy change of a single component system on changing the temperature

$$G_m(T) = H_m(0) + \int_0^T \left(\int_0^{T''} \frac{c_P}{T'} dT' \right) dT''. \tag{2.20}$$

This equation cannot be integrated without the knowledge of the manner in which the heat capacity, c_P, changes with temperature.

To simplify, instead of integrating from the absolute zero we integrate from an arbitrary reference temperature T_0 and define the state of the material at this temperature as a reference state (usually atmospheric pressure and temperature of 298K). We also assume that the specific heat to be a constant in the temperature range under consideration and define it as c_{P0}. Then, the inner integration can be performed leading to

$$\int_{T_0}^T \frac{c_P}{T'} dT' = \int_{T_0}^T \frac{c_{P0}}{T'} dT' = c_{P0}(\ln T/T_0). \tag{2.21}$$

Using this in Eq.(2.20) and performing the second integration gives

$$G_m(T) = H_m(T_0) + c_{P0} T \left(\ln(\frac{T}{T_0}) - 1 \right) - c_{P0} T_0. \tag{2.22}$$

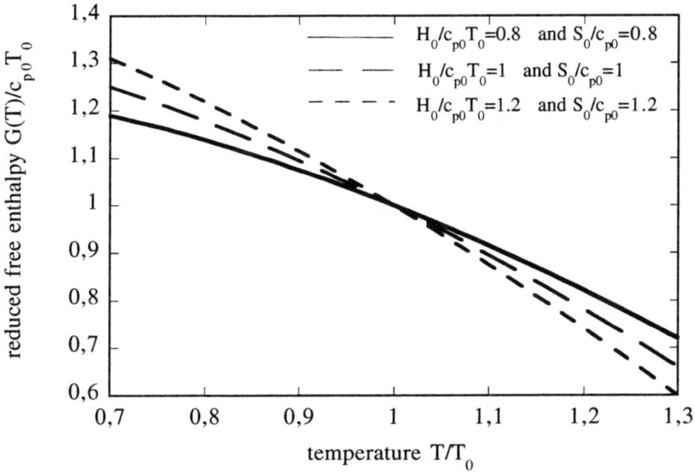

Fig. 2.1. Dependence of Gibbs free energy as a function of temperature for different parameters with reference to eq.(2.22) - valid for a pure substance.

The behavior of this Gibbs free energy as a function of reduced temperature T/T_0 is shown in Fig. 2.1. With increasing temperature the Gibbs free energy decreases in a non-linear fashion.

Now assume that the solid and liquid state have different specific heats $c_{p0}^L > c_{p0}^s$ (the greater specific heat of the liquid compared to the solid corresponds to the greater entropy of the liquid state). Therefore their free energies differ as shown in Fig. 2.2. From the curves shown, the solid is the equilibrium at temperatures below the crossing point and above the liquid state is in equilibrium. The point where both free energies cross and thus are equal defines the melting point T_m.

Let us calculate the melting point. The Gibbs free energy of the solid and melt can be written as

$$G_m^{sol.} = H_m^{sol.} - TS_m^{sol.}, \tag{2.23}$$
$$G_m^{liq.} = H_m^{liq.} - TS_m^{liq.}. \tag{2.24}$$

The temperature, at which both free energies are equal, defines the equilibrium melting temperature T_m. This is the equivalent of the equality of the chemical potential condition in eq.(2.4) for a pure material.

$$\begin{aligned} G_m^{sol.}(T_m) &= G_m^{liq.}(T_m) \\ H_m^{liq.} - H_m^{sol.} &= -T_m S_m^{liq.} + T_m S_m^{sol.} \end{aligned} \tag{2.25}$$

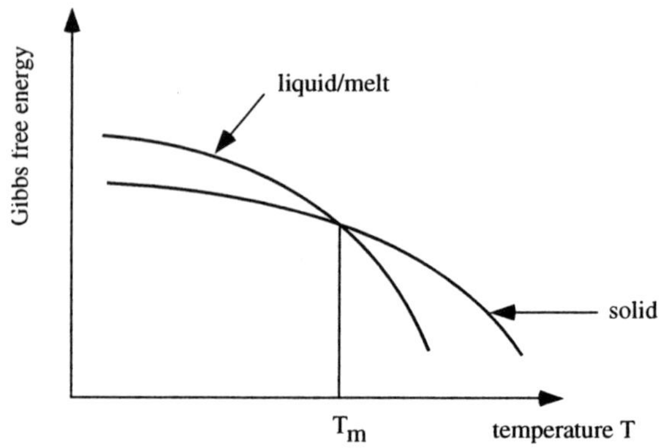

Fig. 2.2. Dependence of Gibbs free energy as a function of temperature for a liquid and its solid.

The enthalpies and entropies are evaluated at T_m. Let us define the heat of fusion or latent heat per mole L_m, measured in J/mole, as the difference in enthalpy between the liquid and solid,

$$L_m = H_m^{liq.} - H_m^{sol.}. \tag{2.26}$$

Often the latent heat L can be expressed in units of energy/mass, or J/kg where $L = L_m\rho/V_m$. The entropy of fusion per mole is defined as

$$\Delta S_m^f = S_m^{liq.} - S_m^{sol.}. \tag{2.27}$$

With these expressions we obtain from eq.(2.25)

$$\begin{aligned} L_m &= T_m \Delta S_m^f, \\ \Delta S_m^f &= \frac{L_m}{T_m}, \end{aligned} \tag{2.28}$$

and

$$T_m = \frac{L_m}{\Delta S_m^f}. \tag{2.29}$$

The latent heat or the difference in enthalpy between solid and liquid is related to the binding energy difference between the solid and the liquid state. The larger this difference the higher the melting point. The entropy of fusion plays a similar role: the smaller it is the higher the melting point. Generally all metals have a similar entropy of melting and, thus, different melting points can in general be accounted to binding energy differences. Semiconductors like Si and Ge have higher entropies of fusion than metals and accordingly high melting temperatures.

2.3 Binary and multicomponent systems

If a material or substance consists of more than one component the Gibbs free energy is not only a function of temperature and pressure but also of the number of atoms, or moles n_i of each chemical component. The Gibbs free energy is a function of T, P, n_i and, thus,

$$dG = \left.\frac{\partial G}{\partial T}\right|_{P,n_i} dT + \left.\frac{\partial G}{\partial P}\right|_{T,n_i} dP + \sum_i^N \left.\frac{\partial G}{\partial n_i}\right|_{T,P,n_i \neq n_j} dn_i. \qquad (2.30)$$

Looking at the definition of the Gibbs free energy we can see that

$$\left.\frac{\partial G}{\partial T}\right|_{P,n_i} = -S \qquad (2.31)$$

and

$$\left.\frac{\partial G}{\partial P}\right|_{T,n_i} = V. \qquad (2.32)$$

There is an additional, and very important, thermodynamic variable, the *chemical potential* of component i, which is defined as

$$\mu_i = \left.\frac{\partial G}{\partial n_i}\right|_{T,P,n_i \neq n_j}. \qquad (2.33)$$

Using the definitions of the partial derivatives we obtain

$$dG = -SdT + VdP + \sum_i^N \mu_i dn_i. \qquad (2.34)$$

The chemical potential measures the change in the Gibbs free energy upon the addition of a small amount of component i to an alloy composed of N chemical components and has units of an energy/mole.

The existence of a uniform system at equilibrium allows us to derive a very useful result relating the Gibbs free energy and chemical potential. Assume that we have a system that is at equilibrium. If the size of the system is doubled by material with the same temperature, pressure, and composition, the Gibbs free energy of the system must double. Thus, if we increase the amount of each element n_i by a factor of λ, the free energy must increase by a factor of λ as well: $G(T, P, \lambda n_1, \lambda n_2 \ldots \lambda n_N) = \lambda G(T, P, n_1, n_2 \ldots n_N)$. Thus, the free energy is a homogeneous function of degree 1. A homogeneous function of degree α is defined as, for a function of two variables for example, $f(\lambda x, \lambda y) = \lambda^\alpha f(x,y)$. The derivatives of this function $f(x,y)$ satisfy Euler's theorem for homogeneous functions of degree α:

$$x \left.\frac{\partial f}{\partial x}\right|_y + y \left.\frac{\partial f}{\partial y}\right|_x = \alpha f(x,y) \tag{2.35}$$

This is *not* the first term of a Taylor's expansion, it is exact! Replacing $x, y \ldots$ with n_1, n_2 etc. and setting $\alpha = 1$ in eq.(2.35) yields

$$\begin{aligned} G(T,P,n_i) = n_1 \left.\frac{\partial G}{\partial n_1}\right|_{T,P,n_2\ldots n_n} &+ n_2 \left.\frac{\partial G}{\partial n_2}\right|_{T,P,n_1,n_3\ldots n_n} + \ldots \\ &+ n_N \left.\frac{\partial G}{\partial n_N}\right|_{T,P,n_1,\ldots n_{n-1}} . \end{aligned} \tag{2.36}$$

Using the definition of the chemical potential we find

$$G(T,P,n_i) = \sum_i^N \mu_i (T,p,x_i)\, n_i. \tag{2.37}$$

Thus, the Gibbs free energy can be represented in terms of the chemical potentials of the chemical components. It should be noted that this is in perfect agreement with the explanation of the chemical potential given above: The chemical potential describes the change in G upon addition of small amounts of the components. Thus, the total Gibbs free energy can be constructed by adding up the small changes in composition from zero to the final system size as described by equation eq.(2.37).

We now simplify the results to binary alloys and thus $N = 2$. We also let $x_1 = x_A$ and $x_2 = x_B$. Taking the total derivative of eq.(2.37),

$$dG = n_A d\mu_A + \mu_A dn_A + n_B d\mu_B + \mu_B dn_B, \tag{2.38}$$

and using eq.(2.34) yields the Gibbs-Duhem equation

$$0 = -SdT + VdP + n_A d\mu_A + n_B d\mu_B. \tag{2.39}$$

This implies that at constant T, P the chemical potentials of the two chemical components in a binary alloy are not independent.

It is convenient to employ the Gibbs free energy per mole in many cases as this energy is independent of the size of the system. The Gibbs free energy per mole $G_m = G/n$ where n is the total number of moles of atoms in the system, $n = n_A + n_B$. As noted above $G = G(n_A, n_B)$. However, the molar free energy cannot be a function of both chemical components, it can only be a function of the mole fraction of one component, $G_m = G_m(x_A)$, since G_m is not a function of the total amount of material in the system. We could equally have chosen x_B as our independent variable since $x_A + x_B = 1$. We can use these results to determine the relationship between the molar Gibbs free energy and the chemical potential as follows. Using the definition of the chemical potential eq.(2.33) and the relationship between G and G_m we get

$$\mu_A(x_A) = \left.\frac{\partial (nG_m(x_A))}{\partial n_A}\right|_{n_B}, \tag{2.40}$$

where we have suppressed the dependence of G_m and μ_A on T, P to simplify notation. Since $\partial n/\partial n_A = 1$

$$\mu_A(x_A) = G_m(x_A) + n \left.\frac{\partial G_m(x_A)}{\partial n_A}\right|_{n_B}. \tag{2.41}$$

Thus,

$$\mu_A(x_A) = G_m(x_A) + n \frac{dG_m(x_A)}{dx_A} \left.\frac{\partial x_A}{\partial n_A}\right|_{n_B}, \tag{2.42}$$

since G_m is only a function of x_A. Using the definition of x_A we find

$$\mu_A(x_A) = G_m(x_A) + n \frac{dG_m(x_A)}{dx_A} \left[\frac{\partial}{\partial n_A}\left(\frac{n_A}{n}\right)\right]_{n_B}. \tag{2.43}$$

Performing the differentiation and using the result, $(n - n_A)/n^2 = x_B/n$ yields

$$\mu_A(x_A) = G_m(x_A) + x_B \frac{dG_m(x_A)}{dx_A}. \tag{2.44}$$

Since $x_A = 1 - x_B$ we can equally view G_m as a function of x_B and obtain

$$\frac{dG_m(x_B)}{dx_A} = -\frac{dG_m(x_B)}{dx_B}. \tag{2.45}$$

Therefore,

$$\mu_A(x_B) = G_m(x_B) - x_B \frac{dG_m}{dx_B}, \tag{2.46}$$

where the derivative of G_m is evaluated at x_B. Following a similar procedure for μ_B yields

$$\mu_B(x_B) = G_m(x_B) + (1 - x_B)\frac{dG_m}{dx_B}. \tag{2.47}$$

We shall use these relationships extensively in what follows as they can be used to develop very useful graphical constructions.

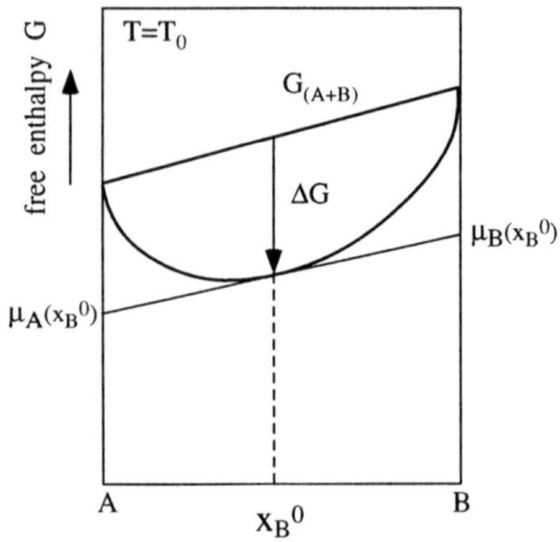

Fig. 2.3. Change of molar Gibbs free energy on alloy formation.

It is possible to use a molar Gibbs free energy plot, such as the one shown in Fig. 2.3, to determine the chemical potential of an alloy of a certain composition. This construction follows from eq.(2.46) and eq.(2.47). The chemical potential of component A in an alloy of composition x_B^o is by eq.(2.46)

$$\mu_A(x_B^o) = G_m(x_B^o) - x_B^o \left.\frac{dG_m}{dx_B}\right|_{x_B = x_B^o} \tag{2.48}$$

and the chemical potential of component B by eq.(2.47)

$$\mu_B(x_B^o) = G_m(x_B^o) + (1 - x_B^o)\left.\frac{dG_m}{dx_B}\right|_{x_B = x_B^o}, \tag{2.49}$$

where the derivative in both equations is evaluated at $x_B = x_B^o$. We can now interpret each of the terms on the RHS of eq.(2.48) and eq.(2.49) graphically. The equation of a line $y(x_B)$ drawn tangent to the Gibbs free energy curve at x_B^o is

$$y(x_B) = G_m(x_B^o) + (x_b - x_b^o)\left.\frac{dG_m}{dx_B}\right|_{x_b = x_b^o}. \tag{2.50}$$

Thus, when $x_B = 0$, $y(0) = \mu_A(x_B^o)$. $\mu_A(x_B^o)$ is the chemical potential of atoms of component A in a mixture at composition x_B^o. This is where the line drawn tangent to the free energy curve intersects the $x_B = 0$ axis. When $x_B = 1$, $y(1) = \mu_B(x_B^o)$. $\mu_B(x_B^o)$ is the chemical potential of atoms of component B in a mixture at composition x_B^o. This is where the line drawn tangent to the free energy curve intersects the $x_B = 1$ axis. Thus, it is possible to determine the chemical potentials of each component simply by drawing a tangent to the molar free energy curve and determining where this line intersects the $x_B = 0$ or $x_B = 1$.

Many processes require knowledge of the energy change associated with moving or forming small amounts of material of a certain composition from another material of different composition. We shall determine the energy change when an infinitesimal amount of material of composition x'_B is formed from a solution of composition x_B^i. This change in energy can be calculation from a knowledge of: (1) the energy change on removing a small amount of material of composition x'_B from a large amount of material of composition x_B^i and (2) the energy on forming the same number of moles of material of composition x'_B. The energy of (1) is not equal to the energy of (2) because the system is closed, the total number of atoms in the system remains fixed. The free energy change of removing material, *per mole of material removed* at constant T, P can be found as follows: In the initial state we have

$$G^i = G\left(n_A^i, n_B^i\right), \tag{2.51}$$

where G^i is the initial free energy, n_A^i and n_B^i are the initial number of A and B atoms in the system. In the final state, after the material has been removed we have,

$$G^f = G\left(n_A^f, n_B^f\right), \tag{2.52}$$

where G^f is the final free energy, n_A^f and n_B^f are the final number of A and B atoms in the system. Since the amount of material removed is very small, n_A^i and n_A^f differ by an infinitesimal amount, similarly for n_B, we can expand G^f about G^i at constant T, P to first order only,

$$\begin{aligned} G^f\left(n_A^f, n_B^f\right) &= G^i\left(n_A^i, n_B^i\right) + \left.\frac{\partial G}{\partial n_A}\right|_{n_A=n_A^i, n_B=n_B^i}\left(n_A^f - n_A^i\right) \\ &+ \left.\frac{\partial G}{\partial n_B}\right|_{n_A=n_A^i, n_B=n_B^i}\left(n_B^f - n_B^i\right). \end{aligned} \tag{2.53}$$

The free energy change on removing material, ΔG^R is $\Delta G^R = G^f - G^i$, and using the definitions of the chemical potentials,

$$\Delta G^R = \mu_A(x_B^i)\left(n_A^f - n_A^i\right) + \mu_B(x_B^i)\left(n_B^f - n_B^i\right). \tag{2.54}$$

The number of moles of each component removed is $n'_A = n_A^i - n_A^f$, $n'_B = n_B^i - n_B^f$ and thus

$$\Delta G^R = -\mu_A(x_B^i)n'_A - \mu_B(x_B^i)n'_B. \tag{2.55}$$

The total number of moles of atoms removed is $n'_A + n'_B$. The free energy change per mole of material removed is then

$$\Delta G_m^R = \frac{\Delta G^R}{n'_A + n'_B}. \tag{2.56}$$

Using eq.(2.55) in eq.(2.56) we obtain

$$\Delta G_m^R = -\mu_A(x_B^i)x'_A - \mu_B(x_B^i)x'_B. \tag{2.57}$$

The free energy of the infinitesimal amount of material formed is simply the molar free energy of material of composition x'_B which follows from eq.(2.37),

$$G_m = \mu_A(x'_B)x'_A + \mu_B(x'_B)x'_B. \tag{2.58}$$

The total free energy change on forming a small amount of material of composition x'_B from material of composition x_B^i is thus the sum of eq.(2.57) and eq.(2.58),

$$\Delta G_m = \left[\mu_A(x'_B) - \mu_A(x_B^i)\right] x'_A + \left[\mu_B(x'_B) - \mu_B(x_B^i)\right] x'_B. \tag{2.59}$$

This is a very general result in that the difference in composition between that removed and the original alloy can be very different. The only requirement is that the amount of material removed be very small.

The right hand side of eq.(2.59) can be rewritten in terms of Gibbs free energies. This will also allow us to develop a graphical construction for ΔG_m. Regrouping the terms in eq.(2.59),

$$\Delta G_m = \mu_A(x'_B)x'_B + \mu_B(x'_B)x'_A - \mu_A(x_B^i)x'_A - \mu_B(x'_B)x'_B. \tag{2.60}$$

Adding and subtracting $G_m\left(x_B^i\right)$, and writing $G_m\left(x_B^i\right)$ in terms of chemical potentials yields,

$$\Delta G_m = G_m\left(x'_B\right) - G_m\left(x^i_B\right) + \mu_A(x^i_B)(x^i_A - x'_A) + \mu_B(x^i_B)(x^i_B - x'_B). \tag{2.61}$$

Since $x_A - 1 - x_B$,

$$\Delta G_m = G_m\left(x'_B\right) - G_m\left(x^i_B\right) + \left[\mu_A(x^i_B) - \mu_B(x^i_B)\right]\left(x^i_A - x'_A\right). \tag{2.62}$$

To eliminate the difference in chemical potentials, take the total derivative of eq.(2.37), use the Gibbs-Duhem equation at constant T, P, and the definition of a total derivative in terms of partial derivatives to yield,

$$\mu_A - \mu_B = \frac{\partial G_m}{\partial x_A}. \tag{2.63}$$

Using eq.(2.63) in eq.(2.62) yields the desired result,

$$\Delta G_m = G_m\left(x'_B\right) - G_m\left(x^i_B\right) - \left.\frac{\partial G_m}{\partial x_A}\right|_{x_B = x^i_B}\left(x'_A - x^i_A\right). \tag{2.64}$$

This form of the free energy is particularly useful when dealing with thermodynamic quantities expressed in terms of free energies and in deriving graphical constructions to show ΔG_m.

To understand the physics behind the Gibbs free energy and also to extend our ability to calculate the thermodynamic behavior of binary alloys, consider two pure materials, one with n_A moles of component A and one of n_B moles of component B that are going to be mixed at constant T, P. The Gibbs free energy of the system before the two systems with components A and B are mixed, G^b is

$$G^b = n_A G^o_A + n_B G^o_B, \tag{2.65}$$

where G^o_A and G^o_B are the molar Gibbs free energies of pure components A and B, respectively. Since they are the molar free energies of pure materials they are only functions of T and P. Dividing eq.(2.65) by n yields

$$G^b_m = x_A G^o_A + x_B G^o_B. \tag{2.66}$$

This is a straight line on a molar Gibbs free energy plot, see Fig. 2.3. If we now mix components A and B together, the Gibbs free energy of the system after mixing, G^a_m follows from eq.(2.37),

$$G^a_m = x_A \mu_A + x_B \mu_B. \tag{2.67}$$

The molar Gibbs free energy accompanying mixing $\Delta G^M = G_m^a - G_m^b$ is

$$\Delta G^M = x_A \left(\mu_A - G_A^o\right) + x_B \left(\mu_B - G_B^o\right). \tag{2.68}$$

This results in a change of the Gibbs free energy ΔG as shown in Fig. 2.3. In this case the Gibbs free energy decreases and, thus, the two elements intermix.

To proceed further it is necessary to develop analytical expressions for the Gibbs free energy of a solution. From the definition of the molar free energy of mixing $\Delta G^M = G_m^a - G_m^b$,

$$G_m^a = G_m^b + \Delta G^M. \tag{2.69}$$

The free energy after mixing, G_m^a, is the free energy of the solution which we will henceforth denote as simply G_m. Since G_m^b is known from knowledge of the free energies of the pure materials, see eq.(2.66), we only have to choose an appropriate ΔG. This Gibbs free energy of mixing ΔG is related to the molar mixing enthalpy ΔH_m^M and the molar mixing entropy ΔS_m^M via

$$\Delta G^M = \Delta H_m^M - T\Delta S_m^M. \tag{2.70}$$

The shape of the molar free energy as a function of composition depends on the enthalpy and entropy terms in eq.(2.70). The entropy change stems from the change of ordering of the atoms on alloy formation. In the simplest case, the atoms of the constituents are distributed randomly in the volume of the liquid alloy or on the lattice sites of a crystal and, thus, the entropy of mixing is given as

$$\Delta S_m^{M,ideal} = -\mathcal{R}(x_A \ln x_A + x_B \ln x_B), \tag{2.71}$$

where $\mathcal{R}$ the the universal gas constant. If the mixing of atoms occurs without any change in the heat content (no evolution or consumption of energy) $\Delta H = 0$ and only the entropy change on alloying defines the Gibbs free energy change. This defines one extreme situation of alloy formation called the *ideal solution model.* Using the ideal solution entropy and enthalpy of mixing in eq.(2.70), substituting this result in eq.(2.69) and using eq.(2.66) yields the free energy of an ideal solution,

$$G_m = x_A G_A^o + x_B G_B^o + \mathcal{R}T(x_A \ln x_A + x_B \ln x_B). \tag{2.72}$$

The Gibbs free energy as a function of temperature and concentration has a shape depicted in Fig. 2.4. Note that the tangent at $x_B = 0$

and $x_B = 1$ is infinite. This means that in reality no pure substance exists, but any alloying of even the smallest amount decreases the free enthalpy.

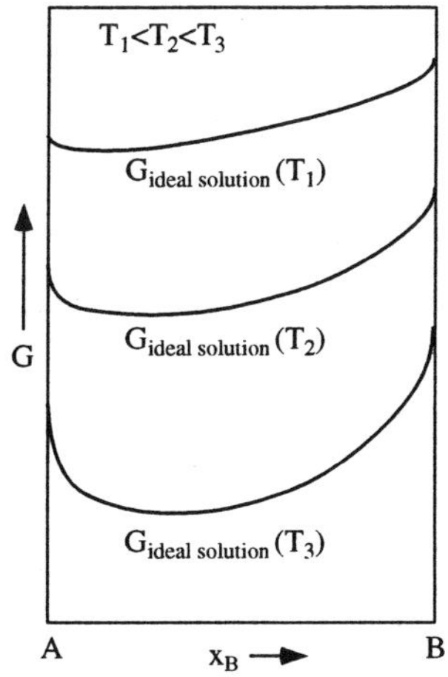

Fig. 2.4. Gibbs free energy in an ideal solution of two components for three different temperatures.

As mentioned, ΔH_m is the latent heat evolving during mixing of the constituents A and B and is negative reflecting the reduction in Gibbs free energy upon mixing. It is related to the interatomic interactions between the A and B atoms, between A and A, and B and B atoms. Considering only nearest neighbor interactions, the so-called *regular solution model*, ΔH_m^M can be written as

$$\Delta H_m^M = \Delta H_m^0 x_A x_B, \tag{2.73}$$

where the constant ΔH_m^0 is related to the interatomic interactions. Depending on whether the average interatomic energy in the alloy is stronger or weaker than in the unmixed components at the given molar fraction x_B, there is a tendency towards compound formation ($\Delta H_m^0 < 0$) or towards phase separation ($\Delta H_m^0 > 0$). Positive ΔH_m^0 results in atoms preferring to be surrounded by atoms of the same species and occurs for alloys with large atomic radii differences of the components and small differences in electronegativity. Negative ΔH_m^0 results in atoms preferring to be surrounded by atoms of the different species and occurs for alloys with small atomic radii differences of the components and large differences in electronegativity. We shall assume that the entropy of mixing is not affected by the interactions between atoms and thus the ideal entropy of mixing is employed. Substituting the regular solution enthalpy of mixing and the ideal solution entropy of mixing into eq.(2.70), substituting this result in eq.(2.69) and using eq.(2.66) yields the free energy of a regular solution,

$$G_m = x_A G_A^o + x_B G_B^o + \mathcal{R}T(x_A \ln x_A + x_B \ln x_B) + \Delta H_m^0 x_A x_B. \quad (2.74)$$

For positive ΔH_m^0 values the $G_m(x_B)$ curve has a shape as shown in

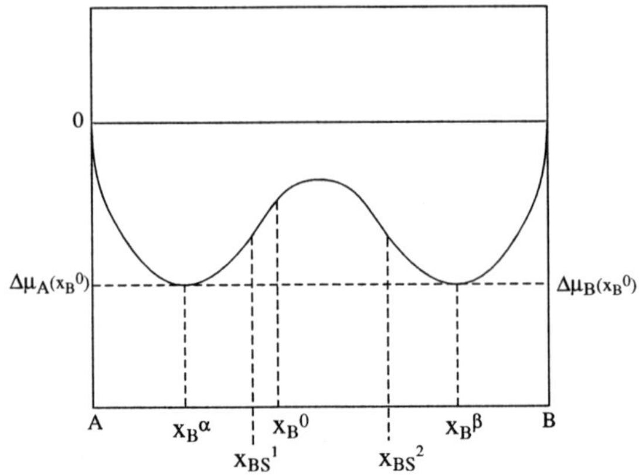

Fig. 2.5. Energetic conditions in a system with limited miscibility

Fig. 2.5 at sufficiently low temperatures.

For each phase the chemical potentials of each component are given by the intersection of the tangent on the Gibbs free energy curve with the $x_B = 0$ and the $x_B = 1$ axes as discussed above with Fig. 2.3. Since in thermodynamic equilibrium the chemical potentials for each component have to be equal in both phases, this can only be accomplished if the tangent lines drawn at the two different compositions on the Gibbs free energy curves for both phases are the same line. Thus, we can obtain the equilibrium compositions x_B^α and x_B^β by drawing the *common tangent* as shown in Fig. 2.5 .

The Gibbs free energy of the mixture of the two phases is given by the weighted average of the Gibbs free energies of the two phases, since each phase is uniform,

$$G_m(x_B) = x^\alpha G_m^\alpha + x^\beta G_m^\beta. \quad (2.75)$$

x^α and x^β are the relative quantities of the α- and β-phases and are determined by the lever-law. $G_m(x_B)$ thus lies on the common tangent connecting the two equilibrium compositions.

For the concentrations of the phases in equilibrium, the chemical potentials of each component are equal in both phases as shown above. For a simple regular solution, with increasing temperature, the compositions x_B^α and x_B^β approach each other until both are equal at the critical temperature T_c . This can be understood by considering eq.(2.70). With increasing T the $-T\Delta S$ term will eventually dominate the ΔH term, which results in the situation described. This fact can be used to calculate a simple relation between temperature and composition of the coexisting phases[2]. Using the regular solution free energy in eq.(2.46) and eq.(2.47) and substituting the results into the equilibrium conditions yields an expression for the equilibrium composition of the phases, or phase boundary, as a function of temperature $T = T(x_B)$,

$$\frac{T}{T_c} = \frac{2(1 - 2x_B)}{\ln\left[(1 - x_B)/x_B)\right]} \tag{2.76}$$

The critical temperature of the miscibility gap in the regular solution model occurs at $x_A = x_B = 1/2$ and, thus,

$$T_c = \frac{\Delta H_m^0}{2\mathcal{R}}. \tag{2.77}$$

An example is shown in Fig. 2.6.

The points of inflection in the $G_m - x$-curve x'_{Bs} and x''_{Bs} in Fig. 2.5 are of special importance: the locus of these points defines the so-called spinodal line, shown in Fig. 2.6 as a dotted line. The spinodal is at those compositions for which

$$\frac{\partial^2 G_m}{\partial x_B^2} = 0, \tag{2.78}$$

leading, after some manipulations, for a regular solution to

$$T_{\text{sp}}(x_B) = \frac{2\Delta H_0}{\mathcal{R}} x_B(1 - x_B) = 4T_c x_B(1 - x_B). \tag{2.79}$$

The region between the binodal line and the spinodal line is metastable against concentration fluctuations (at constant temperature) whereas

[2] In general this is not the case and is simply a result of the representation of the Gibbs free energy in terms of the energy of mixing and the symmetry of the free energy curves of a regular solution.

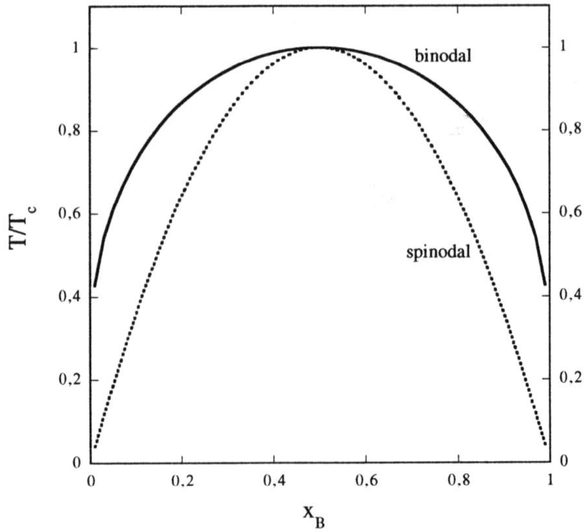

Fig. 2.6. The miscibility gap and spinodal line in a regular solution system.

the region below the spinodal line is unstable against even the smallest concentration fluctuations.

In order to understand this let's look at Fig. 2.7. Shown is the same situation as in Fig. 2.5, but now we look at two compositions: x_B^0 in the unstable region and x_B^0 in the metastable region. Quench the system from some elevated temperature in the single phase region to the temperature where the Gibbs free energy curve assumes the form shown in Fig. 2.7. Assume that at this temperature and composition the system is at equilibrium. The phase will then have at first a total Gibbs free energy that lies on the curve and determined by the composition. While for both compositions the optimal free energy lies on the common tangent of the demixed phases as discussed above, the problem is getting there. The system has to separate into two phases. Let's assume that we have small spontaneous fluctuations in concentration due to thermal excitation. Locally the average concentration varies from $x_B^0 - \Delta x$ to $x_B^0 + \Delta x$ or $x_B^1 - \Delta x$ to $x_B^1 + \Delta x$, depending on whether we are looking at x_B^0 or x_B^1. The Gibbs free energy is again determined by the weighted average of the Gibbs free energy at the two different concentrations and lies on the line connecting the Gibbs free energies of those. This line is shown for fluctuations around x_B^0 as well as around x_B^1. One can easily see that in the first case even the slightest fluctuation in concentration lowers the total Gibbs free energy and thus the phase is unstable. The path to the minimum in Gibbs free

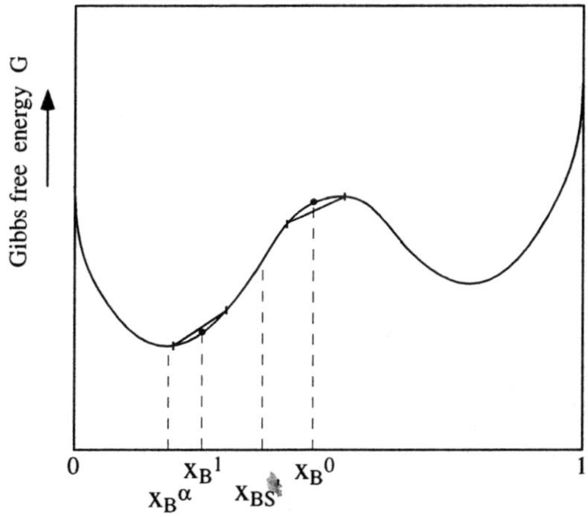

Fig. 2.7. Gibbs free energy as a function of concentration showing that small fluctuations about x_B^0 decrease the Gibbs free energy of the system but small fluctuations about x'_B increase the Gibbs free energy of the system.

energy is one of continuously decreasing G. In the second case small concentration fluctuations *increase* the total Gibbs free energy. In this case an energy barrier exists and the phase is metastable.

The regular solution model always is only a first approximation to describe real alloys and their thermodynamics. Generally, the enthalpy and entropy of mixing depend in a more complicated way on the composition and temperature. Deviations of both from the ideal behavior are generally expressed as so-called excess quantities,

$$\Delta G^{ex} = \Delta G - \Delta G^{id}, \tag{2.80}$$

$$\Delta S^{ex} = \Delta S - \Delta S^{id}, \tag{2.81}$$

where in the ideal case there is no enthalpy of mixing ($\Delta H_0 = 0$) and the ideal entropy of mixing is given by eq.(2.71). The thermodynamic quantities are then expressed by a so-called Redlich-Kister Ansatz ,

$$\Delta H = \sum_{\mathrm{i}=0}^{n} A_{\mathrm{i}} x_A x_B (x_A - x_B)^i, \tag{2.82}$$

$$\Delta S^{ex} = -\sum_{\mathrm{i}=0}^{n} B_{\mathrm{i}} x_A x_B (x_A - x_B)^i, \tag{2.83}$$

$$\Delta G^{ex} = \sum_{\mathrm{i}=0}^{n} (A_{\mathrm{i}} - TB_i) x_A x_B (x_A - x_B)^i \tag{2.84}$$

The coefficients A_i and B_i are determined from measurements of the thermodynamic quantities and optimized to fit as close as possible to the measured shape of the phase boundary. In eq.(2.82) it can be noticed, that the term $i = 0$ represents the regular solution. Within the regular solution the coefficients B_i are all zero as well as all A_i for $i > 0$.

2.4 Applications: Phase diagrams

Let us apply the thermodynamic relations sketched above to two important cases, a system that is completely miscible in the liquid and the solid state and a eutectic system showing limited miscibility in the solid state and the eutectic reaction.

2.4.1 Complete miscibility

Figures 2.8 shows Gibbs free energy curves of the liquid and the solid state at six different temperatures for two components that shall completely be miscible in the solid and liquid state. This occurs if either there is no heat evolution on mixing of the components (ideal solution as explained above) or if the enthalpy change on formation of the alloy is negative $\Delta H_0 < 0$. At the temperatures T_1, T_6 both free enthalpy curves have no crossing point. At the lower temperature the curve of the solid is lower than that of the liquid and at the higher temperature just opposite, defining either the solid or the liquid state as the stable ones. At T_2 and T_5 the curves of the solid and liquid touch just at the melting point of both components defining in the phase diagram (see Fig. 2.9) two points. At the points T_3 and T_4 both free enthalpy curves have an intersecting point. Using the common tangent construction explained above and shown in the figures we obtain four new points, labeled a,b,c,d and shown in the phase diagram. If we would have the exact mathematical descriptions of the G-x curves for both the solid and the liquid it is easy to define the solidus and liquidus lines mathematically as shown schematically in Fig. 2.9.

2.4.2 Eutectic phase diagram

Figures 2.10 shows Gibbs free energy curves at six different temperatures for two components that shall show a eutectic reaction. A eutectic reaction is a three phase equilibrium: at constant temperature

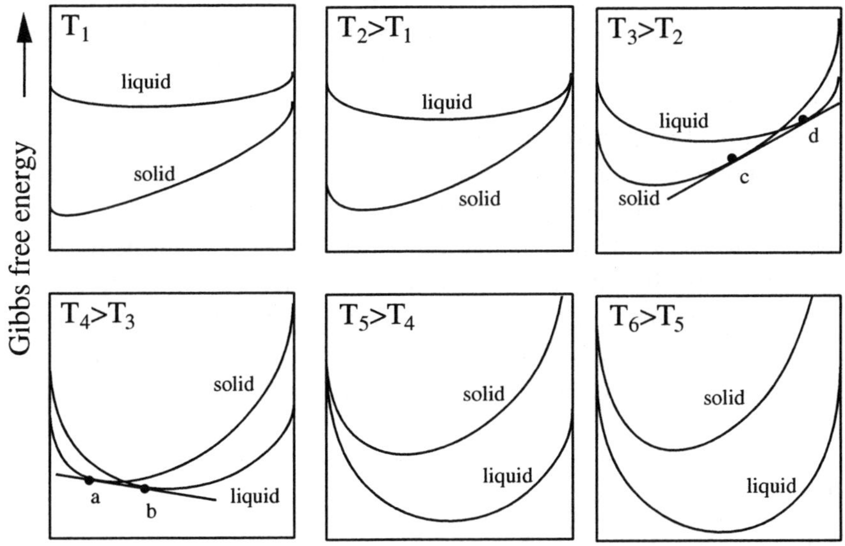

Fig. 2.8. Gibbs free energy at different temperatures in a binary alloy system exhibiting complete miscibility in the solid and liquid state.

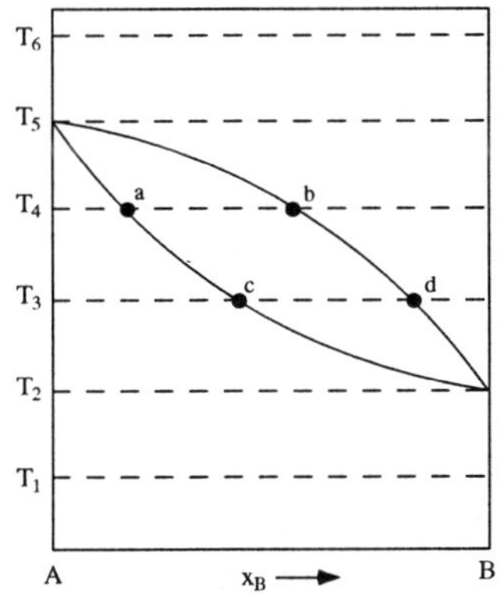

Fig. 2.9. Phase diagram constructed from the free energy curves shown in Fig. 2.8.

a liquid simultaneously decomposes into two different solid phases. At the temperature T_1 we have three Gibbs free energy curves: the liquid curve and two belonging to the solid phases α and β. The latter of course has much higher values than both solid phases. The common tangent at the points n and o define the first two points of the phase diagram shown in Fig. 2.11. Note that we have drawn the Gibbs free energy curves like parabolas, which is due to the fact that in a regular solution model they are very close to parabolas (enthalpy plus entropy term). However, the temperatures are sufficiently high that $T > T_c$ in both phases or $\Delta H_0 < 0$. At the next higher temperature T_2 the construction is similar as before and we obtain two new points l,m. At the temperature T_3, the eutectic temperature, three phases are in equilibrium: the liquid and the two solid phases. The common tangent to all three curves gives the points i,j,k. At the higher temperature T_4

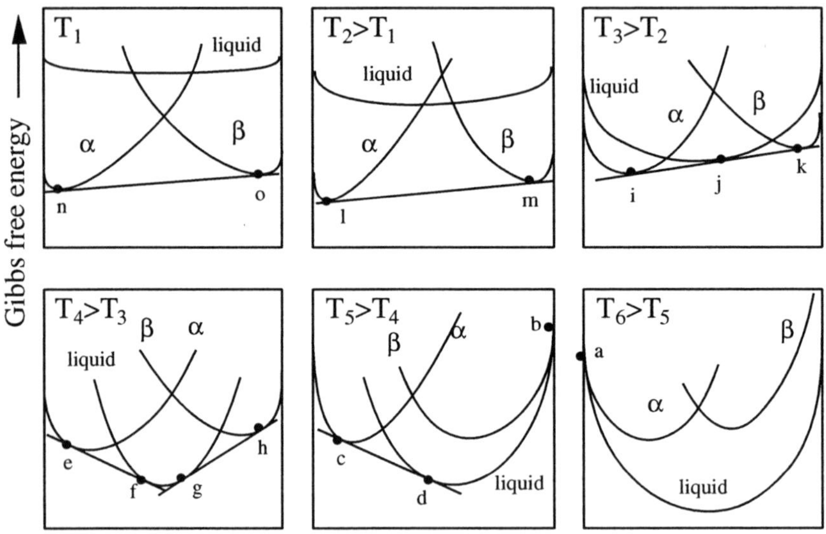

Fig. 2.10. Gibbs free energy at different temperatures in a binary alloy system exhibiting a eutectic reaction.

the two solid phase have no common tangent together but both have one with the liquid phase. This is similar to the situation treated in the case of two completely miscible components above. We obtain four points e,f,g,h. At the temperature T_5 we obtain three points including the melting point of component B and at T_6 we just got the melting point of component A. The phase diagram then looks as shown in

Fig. 2.9. The curves connecting the points n,l,i and o,m,k are called

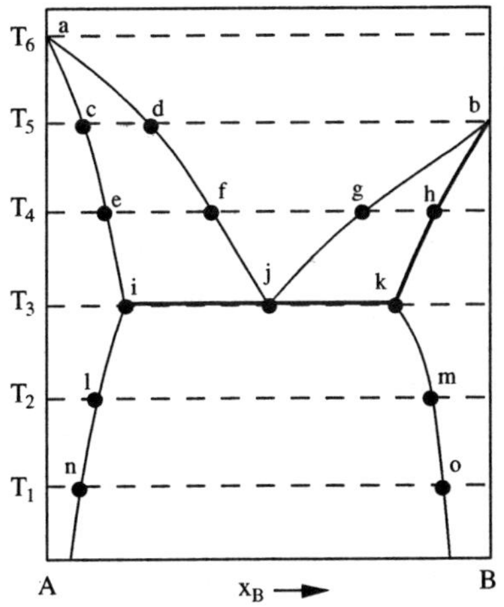

Fig. 2.11. Phase diagram constructed from the free energy curves shown in Fig. 2.10.

solubility lines and the points i and k denote the maximum solubility of the α and β phases for either the B or the A component. The point j is the eutectic point and the line connecting the points i,j,k is called the eutectic tie line. Generally a line connecting two equilibrium state points like e and f or l and m is called a tie line. For a given composition x_B being in between e.g. l,m at temperature T_2 the solid decomposes into two different solid phases of composition given by the end points of the tie line, namely l and m.

2.5 Gibbs-Thomson equation

The origin of Ostwald ripening lies in the dependence of interfacial concentrations or melting temperature on the curvature of the interface. Describing these relationships is thus necessarily the first step in describing Ostwald ripening.

We will start out by assuming a two-phase mixture. We will ignore nonhydrostatic stresses and the effects of interfacial stresses. Equilibrium requires that

$$\begin{aligned} \mu_i^{\beta}(T, P^{\beta}) &= \mu_i^{\alpha}(T, P^{\alpha}) \quad \text{for } i = 1 \ldots N, \\ P^{\beta} &= P^{\alpha} + \sigma\kappa, \end{aligned} \tag{2.85}$$

$$T = \text{constant},$$

where σ is the interfacial energy and κ is the mean curvature of the interface reckoned positive for a spherical particle of β. The mean curvature is related to the two principle radii of curvatures of the interface, R_1 and R_2 as $\kappa = 1/R_1 + 1/R_2$. For a sphere $R_1 = R_2 = R$ and $\kappa = 2/R$, P^α and P^β are the pressures in the α- and β-phase respectively. σ is the excess Grand Canonical free energy due to the presence of an interface, not an excess Gibbs free energy. A Grand canonical free energy, Ω is defined as $\Omega = E - TS - \sum_i \mu_i n_i$.

It should be noted, that the pressure in the two phases does not have to be the same. In fact, as described by eq.(2.85), if the interface between the two phases is curved there will be a pressure difference due to the *interfacial tension*, which under the assumptions given above is identical to the interfacial energy, σ. The same phenomenon occurs in soap bubbles, although here we have a membrane splitting two portions of air (inside the membrane vs. outside the membrane). The membrane is the interface that tries to decrease its total area. As a result the pressure inside the soap bubble is above that of the phase outside. With the pressure increase the volume decreases and the density increases. The soap bubbles fall to the ground. The smaller the soap bubble and thus the higher the curvature, the larger the pressure difference and the faster the bubble falls. For metals a pressure increase will not change volume or density, metals are virtually incompressible. However, the behavior of a pressure shift due to a curved interface, that is very familiar with soap-bubbles, is the same.

2.5.1 Pure Material

The first case we consider is that of a one-component material. Let's assume that solid and liquid phase are in equilibrium, see Fig. 2.12. As discussed earlier (see eq.(2.25) and eq.(2.85)), this implies that

$$\begin{aligned} G_m^{sol.}(T, P^{sol.}) &= G_m^{liq.}(T, P^{liq.}), \\ P^{sol.} &= P^{liq.} + \sigma\kappa, \\ T &= \text{constant}. \end{aligned} \tag{2.86}$$

Since classical thermodynamics deals only with changes in thermodynamic quantities, we will first describe a reference state. The

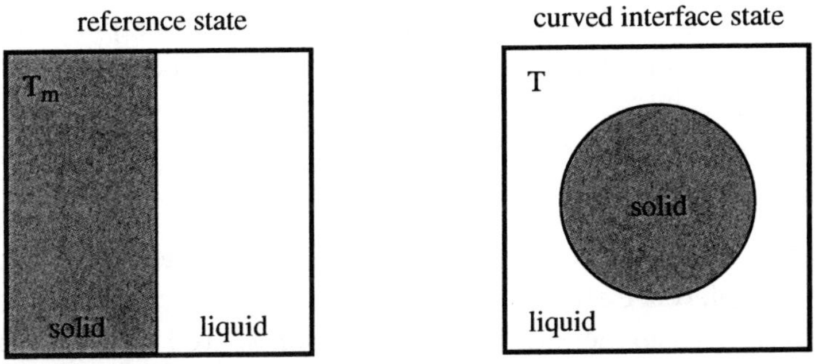

Fig. 2.12. Situation in which the solid phase of a pure substance is in equilibrium with its liquid phase. Shown on the left side is the reference state with a planar interface and at the right side the one with a curved interface

obvious one is a system with a flat interface ($\kappa = 0$) between the solid and liquid phases and, thus,

$$G_m^{sol.}(T, P^{sol.}) = G_m^{liq.}(T, P^{liq.}), \tag{2.87}$$

$$P^{sol.} = P^{liq.}. \tag{2.88}$$

The pressurc in both phases is equal. The temperature at which both phases coexist is commonly termed the bulk melting point, T_m. eq.(2.87) can be rewritten as

$$G_m^{sol.}(T_m, P^{liq.}) = G_m^{liq.}(T_m, P^{liq.}). \tag{2.89}$$

By comparing eq.(2.87) and eq.(2.86) it becomes clear that the temperature at which the two phases are in equilibrium changes with changing pressure and thus depends on the interface curvature. We assume that we have a small solid particle in a liquid. The pressure of the liquid phase is the pressure applied to the system. The pressure in the solid phase is shifted according to the interface curvature. In order to determine the magnitude of the shift in equilibrium temperature due to the curved interface we perform a Taylor expansion of the Gibbs free energy of the solid to first order in the deviation of the temperature and pressure from T_m and $P^{liq.}$,

$$G_m^{sol.}(T, P^{sol.}) = G_m^{sol.}(T_m, P^{liq.}) + \left(\frac{\partial G_m^{sol.}}{\partial T}\right)_{T_m, P^{liq.}} (T - T_m) + \\ + \left(\frac{\partial G_m^{sol.}}{\partial P}\right)_{T_m, P^{liq.}} (P^{sol.} - P^{liq.}), \tag{2.90}$$

and similarly for the liquid but only in terms of temperature as the pressure of the liquid is the independent of the curvature,

$$G_m^{liq.}(T, P^{liq.}) = G_m^{liq.}(T_m, P^{liq.}) + \left(\frac{\partial G_m^{liq.}}{\partial T}\right)_{T_m, P^{liq.}} (T - T_m). \quad (2.91)$$

The partial derivatives appearing in eq.(2.90) and eq.(2.91) are evaluated in the reference state (flat interface); they are constants. Furthermore, these partial derivatives can be represented in terms of other thermodynamic quantities,

$$\left(\frac{\partial G_m^{sol.}}{\partial T}\right)_{T_m, P^{liq.}} = -S_m^s, \quad (2.92)$$

where S_m^s is the molar entropy of the solid evaluated in the reference state, at T_m and $P^{liq.}$. A similar expression holds for the liquid. The derivative of the molar Gibbs free energy of the solid with respect to pressure is

$$\left(\frac{\partial G_m^{sol.}}{\partial P}\right)_{T_m, P^{liq.}} = V_m^s. \quad (2.93)$$

V_m^s is the molar volume of the solid phase evaluated at T_m and $P^{liq.}$. Using eq.(2.93) and eq.(2.92) in eq.(2.90) and eq.(2.91) yields

$$G_m^{sol.}(T, P^{sol.}) = G_m^{sol.}(T_m, P^{liq.}) - S_m^s(T - T_m) + V_m^s(P^{sol.} - P^{liq.}) \quad (2.94)$$

and

$$G_m^{liq.}(T, P^{liq.}) = G_m^{liq.}(T_m, P^{liq.}) - S_m^l(T - T_m). \quad (2.95)$$

These expressions are good approximations for the Gibbs free energy for systems if the changes in pressure or temperature from the comparison state are small.

Using these expansions for $G_m^{sol.}(T, P^{sol.})$ and $G_m^{liq.}(T, P^{liq.})$ in the equilibrium conditions for a system with a curved interface, eq.(2.86), we get

$$\begin{aligned} G_m^{sol.}(T_m, P^{liq.}) - S_m^s(T - T_m) + V_m^s(P^{sol.} - P^{liq.}) = \\ G_m^{liq.}(T_m, P^{liq.}) - S_m^l(T - T_m). \end{aligned} \quad (2.96)$$

Using the equilibrium condition at a planar interface eq.(2.87) and the expression for the difference in pressures between the solid and liquid, eq.(2.86), we get

$$-S_m^s(T - T_m) + V_m^s \sigma \kappa = -S_m^l(T - T_m). \tag{2.97}$$

Solving this equation for the temperature difference yields the *Gibbs-Thomson equation* for a pure material,

$$T - T_m = \frac{V_m^s \sigma \kappa}{S_m^s - S_m^l}. \tag{2.98}$$

For a spherical particle this becomes

$$T - T_m = \frac{2V_m^s \sigma}{(S_m^s - S_m^l)R}. \tag{2.99}$$

Thus, the temperature at the interface between a solid and liquid will vary with the curvature of the interface. This temperature dependence follows from the pressure change in the solid due to the interfacial energy and the pressure and temperature dependence of the molar Gibbs free energy. Since $S_m^l > S_m^s$ the equilibrium temperature in a system with a spherical solid particle must be lower than that of a planar interface.

Since the molar entropies that appear in eq.(2.98) and eq.(2.99) are evaluated at the bulk melting point, these can be replaced using eq.(2.28) to yield the standard form of the Gibbs-Thomson equation for a spherical solid particle,

$$T = T_m - \frac{2T_m V_m^s \sigma}{L_m R} \tag{2.100}$$

or

$$T = T_m - \frac{2\Gamma}{R}, \tag{2.101}$$

where we used the so-called *Gibbs-Thomson coefficient*, Γ, which is defined as

$$\Gamma = \frac{T_m V_m^s \sigma}{L_m}. \tag{2.102}$$

2.6 Binary Alloy

In a binary alloy, where the two phases α and β are in equilibrium, a curved interface will again change the pressure of one phase with regard to the other, see Fig. 2.13. The difference between the two possible states (flat interface and curved interface) can be expressed in terms of a equilibrium temperature change as before. However, we can describe it in terms of concentration changes at a fixed temperature. In order to derive the compositional Gibbs-Thomson equation we begin with the equilibrium conditions for a binary alloy with a spherical particle of β,

$$\begin{aligned} \mu_A^\beta(x_B^\beta, P^\beta) &= \mu_A^\alpha\left(x_B^\alpha, P^\alpha\right), \\ \mu_B^\beta(x_B^\beta, P^\beta) &= \mu_B^\alpha\left(x_B^\alpha, P^\alpha\right), \\ P^\beta &= P^\alpha + 2\sigma/R, \end{aligned} \tag{2.103}$$

where x_B^α and x_B^β are the mole fractions of component B in equilibrium in the α- and β-phase respectively. The last equation is written assuming a system with a particle of β-phase of radius R surrounded by α-phase. Since this is a binary mixture the mole fraction of component A, x_A, is simply $1 - x_B$. The equilibrium conditions give two equations, the two chemical potential conditions, for the two unknown concentrations x_B^β and x_B^α. Thus, in an isothermal binary alloy the concentrations are fixed for a given particle radius.

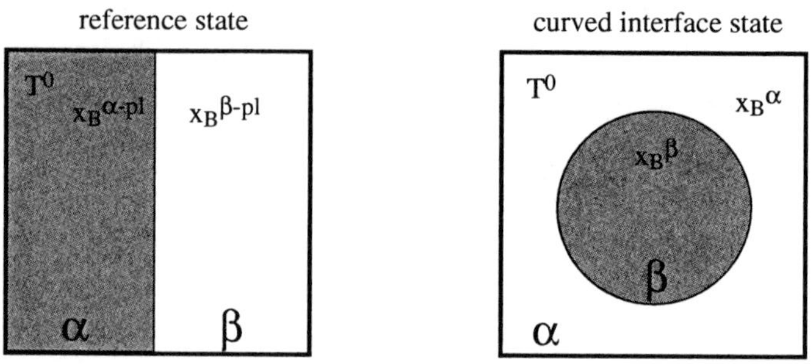

Fig. 2.13. Situation in which the α-phase of a binary substance is in equilibrium with the β-phase. Shown on the left side is the reference state with a planar interface and at the right side the one with a curved interface

Similar to the one-component system we choose as the reference system a system with a planar interface at a pressure of the matrix phase P^α. We again assume that the matrix pressure is the same in both the reference and the spherical particle system. The equilibrium conditions in the reference system are

$$\mu_A^\beta(x_B^{\beta,e}, P^\alpha) = \mu_A^\alpha\left(x_B^{\alpha,e}, P^\alpha\right), \tag{2.104}$$

$$\mu_B^\beta(x_B^{\beta,e}, P^\alpha) = \mu_B^\alpha\left(x_B^{\alpha,e}, P^\alpha\right), \tag{2.105}$$

since $P^\alpha = P^\beta$. $x_B^{\beta,e}$ and $x_B^{\alpha,e}$ are the equilibrium concentrations at a planar interface. These are, thus, the concentrations as given by the tie line on a standard binary phase diagram.

To relate the chemical potentials in the reference and the spherical-particle systems, we perform a Taylor-expansion of the chemical potentials in the system with the spherical particle about that of the reference state to first order in the change in composition and pressure. This yields for the particle phase (β-phase)

$$\mu_i^\beta(x_B^\beta, P^\beta) = \mu_i^\beta(x_B^{\beta,e}, P^\alpha) + \left(\frac{\partial \mu_i^\beta}{\partial x_B}\right)_{x_B^{\beta,e}, P^\alpha} \left(x_B^\beta - x_B^{\beta,e}\right) + \left(\frac{\partial \mu_i^\beta}{\partial P^\beta}\right)_{x_B^{\beta,e}, P^\alpha} \left(P^\beta - P^\alpha\right) \tag{2.106}$$

and for the matrix phase (α-phase)

$$\mu_i^\alpha(x_B^\alpha, P^\alpha) = \mu_i^\alpha(x_B^{\alpha,e}, P^\alpha) + \left(\frac{\partial \mu_i^\alpha}{\partial x_B}\right)_{x_B^{\alpha,e}, P^\alpha} \left(x_B^\alpha - x_B^{\alpha,e}\right), \tag{2.107}$$

where $i = A, B$. As was the case for the pure material the partial derivatives can be replaced by other quantities. The molar Gibbs free energy in a binary alloy is related to the chemical potentials by

$$\mu_A = G_m - x_B \frac{\partial G_m}{\partial x_B} \tag{2.108}$$

and

$$\mu_B = G_m + (1 - x_B) \frac{\partial G_m}{\partial x_B}. \tag{2.109}$$

Thus,

$$\frac{\partial \mu_A}{\partial x_B} = -x_B \frac{\partial^2 G_m}{\partial {x_B}^2} \tag{2.110}$$

and

$$\frac{\partial \mu_B}{\partial x_B} = (1 - x_B) \frac{\partial^2 G_m}{\partial {x_B}^2}. \tag{2.111}$$

Also,

$$\frac{\partial \mu_i}{\partial P} = \overline{V}_i, \tag{2.112}$$

where $\overline{V}_i$ is the partial molar volume of component i. Using eq.(2.110), eq.(2.111), and eq.(2.112) in eq.(2.106) yields

$$\mu_A^\beta(x_B^\beta, P^\beta) = \mu_A^\beta(x_B^{\beta,e}, P^\alpha) - x_B^{\beta,e} G_m''^\beta (x_B^\beta - x_B^{\beta,e}) + \overline{V}_A^\beta (P^\beta - P^\alpha) \tag{2.113}$$

and

$$\mu_B^\beta(x_B^\beta, P^\beta) = \mu_B^\beta(x_B^{\beta,e}, P^\alpha) + (1 - x_B^{\beta,e}) G_m''^\beta (x_B^\beta - x_B^{\beta,e}) + \overline{V}_B^\beta (P^\beta - P^\alpha) \tag{2.114}$$

where

$$G_m''^\beta = \frac{\partial^2 G_m^\beta}{\partial x_B{}^2} \tag{2.115}$$

and the $\overline{V}_i$ are evaluated at $x_B^{\beta,e}, P^\alpha$. For the matrix phase we obtain

$$\mu_A^\alpha(x_B^\alpha, P^\alpha) = \mu_A^\alpha(x_B^{\alpha,e}, P^\alpha) - x_B^{\alpha,e} G_m''^\alpha (x_B^\alpha - x_B^{\alpha,e}) \tag{2.116}$$

and

$$\mu_B^\alpha(x_B^\alpha, P^\alpha) = \mu_B^\alpha(x_B^{\alpha,e}, P^\alpha) + (1 - x_B^{\alpha,e}) G_m''^\alpha (x_B^\alpha - x_B^{\alpha,e}), \tag{2.117}$$

where

$$G_m''^\alpha = \frac{\partial^2 G_m^\alpha}{\partial x_B{}^2} \tag{2.118}$$

is evaluated at $x_B^{\beta,e}, P^\alpha$.

Substituting eq.(2.113), eq.(2.114), eq.(2.116), and eq.(2.117) in the equilibrium conditions for a spherical particle, eq.(2.103), and using the equilibrium conditions at a planar interface eq.(2.106) yields

$$\begin{aligned} -x_B^{\alpha,e} G_m''^\alpha (x_B^\alpha - x_B^{\alpha,e}) &= -x_B^{\beta,e} G_m''^\beta (x_B^\beta - x_B^{\beta,e}) + \overline{V}_A^\beta (P^\beta - P^\alpha), \\ (1 - x_B^{\alpha,e}) G_m''^\alpha (x_B^\alpha - x_B^{\alpha,e}) &= (1 - x_B^{\beta,e}) G_m''^\beta (x_B^\beta - x_B^{\beta,e}) + \overline{V}_B^\beta (P^\beta - P^\alpha), \\ P^\beta - P^\alpha &= 2\sigma/R. \end{aligned} \tag{2.119}$$

After eliminating $P^\beta - P^\alpha$ using the third equation above, this yields two equations in the two unknowns, $x_B^\beta - x_B^{\beta,e}$ and $x_B^\alpha - x_B^{\alpha,e}$. Solving this system and using the relation $V_m^\beta = (1 - x_B^{\beta,e})\overline{V}_A^\beta + x_B^{\beta,e}\overline{V}_B^\beta$ yields the Gibbs-Thomson equation for the matrix phase,

$$x_B^\alpha = x_B^{\alpha,e} + \frac{2V_m^\beta \sigma}{\left(x_B^{\beta,e} - x_B^{\alpha,e}\right) G_m''^\alpha R}, \tag{2.120}$$

and for the particle phase,

$$x_B^\beta = x_B^{\beta,e} + \frac{2\left[(1 - x_B^{\alpha,e}) \overline{V}_A^\beta + x_B^{\alpha,e} \overline{V}_B^\beta\right] \sigma}{\left(x_B^{\beta,e} - x_B^{\alpha,e}\right) G_m''^\beta R}. \tag{2.121}$$

These expressions are valid for a general nonideal nondilute solution as the form of the molar Gibbs free energy has not been specified at this point.

A portion of the quantities appearing in the second terms of both Gibbs-Thomson equations have units of length and are termed *capillary lengths.* The capillary lengths for the two phases, ℓ^β and ℓ^α, are defined as

$$\ell^\alpha = \frac{2V_m^\beta \sigma}{\left(x_B^{\beta,e} - x_B^{\alpha,e}\right) G_m^{\prime\prime\alpha}} \tag{2.122}$$

and

$$\ell^\beta = \frac{2\left[(1 - x_B^{\alpha,e})\,\overline{V}_A^\beta + x_B^{\alpha,e}\overline{V}_B^\beta\right]\sigma}{\left(x_B^{\beta,e} - x_B^{\alpha,e}\right) G_m^{\prime\prime\beta}}. \tag{2.123}$$

Using these definitions, the Gibbs-Thomson equations become

$$x_B^\alpha = x_B^{\alpha,e} + \ell^\alpha / R \tag{2.124}$$

and

$$x_B^\beta = x_B^{\beta,e} + \ell^\beta / R. \tag{2.125}$$

The capillary lengths are quite small, typically on the order of 0.1nm. Thus, unless the particle sizes are on the scale of the critical radius for nucleation, the deviation in the concentration from the equilibrium concentrations is very small. For $x_B^{\beta,e} > x_B^{\alpha,e}$ the composition of both phases at the interface of a spherical particle will be above that of the equilibrium concentration, since G_m'' must be positive for the alloy to be thermodynamically stable.

To determine the dependence of the concentration on particle size requires knowledge of the molar free energy of the phase of interest. This can be represented using CALPHAD methods in the most general cases. Here the Gibbs free energy is given by a complex function of composition and temperature as necessary to reproduce the measured phase diagram and other thermodynamical data. As an illustration, however, we take the molar Gibbs free energy to be an ideal solution, eq.(2.72). Thus,

$$G_m'' = \frac{\mathcal{R}T}{x_B\,(1 - x_B)}. \tag{2.126}$$

Using this in the Gibbs-Thomson equation for the matrix phase we obtain

$$x_B^\alpha = x_B^{\alpha,e} + \frac{2V_m^\beta x_B^{\alpha,e}\left(1 - x_B^{\alpha,e}\right)\sigma}{\left(x_B^{\beta,e} - x_B^{\alpha,e}\right)\mathcal{R}T R}. \tag{2.127}$$

Finally, for a system in which the precipitate phase is nearly pure component B, $x_B^{\beta,e}$ is nearly one, and the matrix phase is dilute in component B, $x_B^{\alpha,e} \ll 1$ this equation becomes

$$x_B^\alpha = x_B^{\alpha,e} + \frac{2V_m^\beta x_B^{\alpha,e}\sigma}{\mathcal{R}T R}. \tag{2.128}$$

We have here derived the Gibbs-Thomson equation that describes the shift in composition in the α- and β-phase as a result of a curved interface while keeping the temperature of the system constant. If the

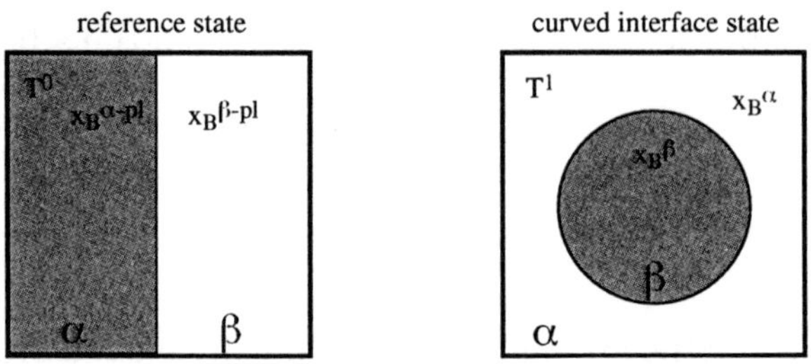

Fig. 2.14. Situation in which the α-phase of a binary substance is in equilibrium with the β-phase. Shown on the left side is the reference state with a planar interface and at the right side the one with a curved interface.

temperature is not held fixed then there are three variables with only two chemical potential equations to fix relationships between these variables. If the solution is dilute, then the phase boundaries are straight lines. As a result, the Gibbs Thomson equation simply results in a vertical shift of these phase boundaries. Thus, the slope of the phase boundary and the distribution coefficient is independent of curvature. In general, however, this is not true, see Fig. 2.15 left side. Here the distribution defined as the ratio of the solidus to the liquidus concentration is not a constant but depends on concentration. This can easily be observed in a system with complete miscibility in the liquid and solid state as shown in Fig. 2.15. It is possible to correctly describe concentrations and temperatures of the system with a curved interface in terms of a temperature shift if we add a second equation describing the relative shift in composition of the two phases or the dependence of the distribution coefficient on curvature [294]. In

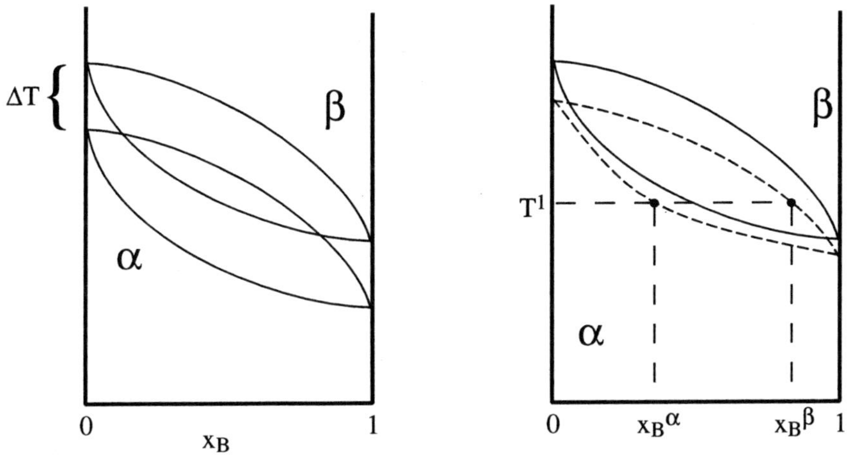

Fig. 2.15. Phase diagram in view of the Gibbs-Thomson equation. The left diagram displays the *wrong* view that a curved interface can be treated as a shift of the liquidus and solidus by ΔT. The right side reflects the correct view: at each given temperature we obtain two new equilibrium concentrations x_B^α and x_B^β.

general, we have three options for describing the new situation. This can be easily understood by looking at Fig. 2.15: The new state is described by T^1, x_B^α and x_B^β. The reference state is described by T^0, $x_B^{\alpha,e}$ and $x_B^{\beta,e}$. Previously in the derivation of the Gibbs Thomson effect for concentrations we defined the reference state and curved interface system as having the same temperature, thus $T^0 = T^1$. However, we could have equally valid chosen the reference state to be defined by $x_B^\alpha = x_B^{\alpha,e}$ or $x_B^\beta = x_B^{\beta,e}$. In each case we fix one of the three variables and obtain equations for the two others. The result is schematically shown in the right part of Fig. 2.15. This diagram can be understood as follows: For both pure components the Gibbs-Thomson effect that changes the melting temperature is different, since the Gibbs-Thomson coefficient or their surface tensions and melting entropies are different. Thus, the liquidus and solidus lines of a curved system cannot be obtained by a simple shift from the phase diagram as is the case for a dilute system. For arbitrary concentrations $0 < x_B < 1$ the partitition coefficient changes with concentration in a curved interface system in a different manner compared to a system with a planar interface and thus the solidus and liquidus lines in a system with a curved interface have different shapes.

3. Transport of heat and mass

Growth and coarsening of particles or precipitates in a matrix require the transport of heat or mass or both simultaneously. Heat and mass can be transported by diffusion, where heat diffusion in a crystal occurs by the random transport of enthalpy via electrons and phonons, quantized lattice vibrations, with a direction set by the temperature gradient, where heat always flows from high to low temperatures. Mass diffusion occurs in crystals by a random walk of atoms utilizing lattice vacancies. The details on heat and mass transport are discussed in standard textbooks on solid state physics and material sciences. In liquids heat and mass can be transported by diffusion, convection, or both. To describe the diffusion process we derive the conservation equations for heat and mass under diffusive and convective-diffusive conditions. These basic equations are later used for particle growth and coarsening.

3.1 Diffusive heat transport

Let's assume an arbitrarily shaped body with temperature variations inside as shown in figure 3.1. These induce heat fluxes that are proportional to the temperature gradients leading to a flux of heat into or out of the volume which we define here as $\boldsymbol{q}$. If there are no heat sources or sinks inside the body all heat fluxes must go through the surface area A and the change in temperature of the body is due to these heat fluxes only. The total change in heat content per second is then given by integrating the flux going through the surface,

$$Q = \int_A \boldsymbol{q} \cdot \boldsymbol{df}, \tag{3.1}$$

where $\boldsymbol{df}$ denotes an infinitesimal surface element of the body's contour with $\boldsymbol{df} = \boldsymbol{n} dA$ and $\boldsymbol{n}$ being the outward pointing normal from the

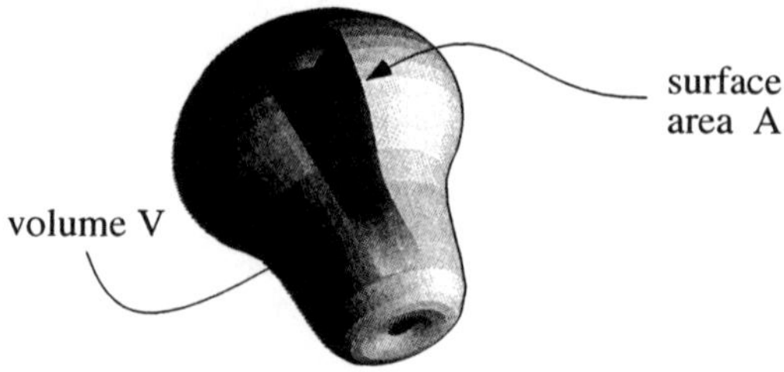

Fig. 3.1. Arbitrarily shaped body with volume V and surface area A. If the surface has a different temperature than the interior a heat flux exists from the interior to the surface. The enthalpy change in the body volume is solely dictated by the flux through the surface.

center of A. The heat or enthalpy content of the body of volume V can be written as

$$H = \int_V H_V dV \tag{3.2}$$

with H_V as the enthalpy per volume as defined in the preceding chapter,

$$H_V = \int_0^T C_P(T)\rho(T)dT. \tag{3.3}$$

The heat flux through the surface changes the enthalpy content of the volume V. Its change with time equals the heat flux through the surface leading to

$$\frac{dH}{dt} + Q = 0 \tag{3.4}$$

or written explicitly,

$$\frac{d}{dt}\left(\int_V \int_0^T C_P(T)\rho(T)dTdV\right) + \int_A \boldsymbol{q} \cdot \boldsymbol{df} = 0. \tag{3.5}$$

Performing the time differentiation we obtain

$$\int_V C_P(T)\rho(T)\frac{\partial T}{\partial t}dV + \int_A \boldsymbol{q} \cdot \boldsymbol{df} = 0. \tag{3.6}$$

Application of Gauss' theorem to the surface integral leads to

$$\int_V \left[C_P(T)\rho(T)\frac{\partial T}{\partial t} + \nabla \cdot \boldsymbol{q}\right] dV = 0. \tag{3.7}$$

This relation can only be true if the integrand itself equals zero and thus we obtain the continuity equation,

$$C_P(T)\rho(T)\frac{\partial T}{\partial t} = -\nabla \cdot \boldsymbol{q}. \tag{3.8}$$

Let us now specify the heat flux more precisely. Heat can diffuse in a body due to the afore mentioned temperature differences or gradients in the body. Fourier's law states that then the heat flux, $\boldsymbol{q}_{\text{diff}}$, is proportional to the temperature,

$$\boldsymbol{q}_{\text{diff}} = -\lambda \nabla T. \tag{3.9}$$

where λ is the *heat conductivity*. This relation is in full accordance with the general concept of irreversible thermodynamics, in which fluxes are related to general forces (here the temperature gradient) in a linear fashion. Inserting this relation into the continuity equation (3.8) and assuming λ is a constant we obtain

$$\rho C_P \frac{\partial T}{\partial t} = \lambda \nabla^2 T, \tag{3.10}$$

or in a more convenient form as

$$\frac{\partial T}{\partial t} = \frac{\lambda}{\rho C_P}\nabla^2 T = \kappa \nabla^2 T \tag{3.11}$$

with

$$\kappa = \frac{\lambda}{\rho C_P} \tag{3.12}$$

being the *heat diffusivity*[1]. Since the temperature of our volume element changes at a constant pressure P this change will result in a density change. Thus, in order to conserve the volume material has to flow into or out of the volume element. This volume flow has been neglected at this point since it is usually small.

3.2 Diffusive mass transport

Mass or species transport in solids or liquids can occur by diffusion and/or bulk fluid flow. In order to clearly formulate the diffusion equation we refer to a volume element small compared to any system dimension and large compared to interatomic distances. In such a volume

[1] In the american literature the heat conductivity often is denoted as k and the heat diffusivity as α.

element to every atom or molecule we can assign a velocity $\boldsymbol{v_{ik}}$, being arbitrary in direction and amount. The local average velocity of one species in such a volume can be calculated from:

$$\langle \boldsymbol{v}_i \rangle = \frac{1}{n_i} \sum_k^{n_i} \boldsymbol{v}_{ik} \ , \tag{3.13}$$

where n_i is the number of molecules or atoms of species i in the sample volume. The $\langle \boldsymbol{v}_i \rangle$ for different species maybe different. The relative motion of (unlike) molecules or atoms is called diffusion. A mathematical way to describe this relative motion centers on the concept of an average flow velocity. A frame of observation is chosen, and the local average velocity $\langle \boldsymbol{v}_i \rangle$ of each species i is measured with respect to this frame. Then, a weighted average velocity is constructed according to the general rule,

$$\langle \boldsymbol{v}^r \rangle = \sum_i w_i \langle \boldsymbol{v}_i \rangle, \tag{3.14}$$

with w_i suitable weighting factors. There are several ways to choose the weighting factors, for example the w_i could be the mass fraction, the molar fraction, the volume fraction etc. (for a detailed discussion see [247]). The diffusion velocity is defined with respect to this average velocity. Throughout this monograph we will use molar concentrations and therefore derive the following equations with respect to the molar average velocity. For simplicity let us consider a binary system consisting of A and B atoms. Then we have

$$\langle \boldsymbol{v}_A \rangle = \frac{1}{n_A} \sum_k^{n_A} \boldsymbol{v}_{Ak} \ , \tag{3.15}$$

$$\langle \boldsymbol{v}_B \rangle = \frac{1}{n_B} \sum_k^{n_B} \boldsymbol{v}_{Bk} \ , \tag{3.16}$$

for the local average species velocities and for the average molar velocity

$$\langle \boldsymbol{v} \rangle = x_A \langle \boldsymbol{v}_A \rangle + x_B \langle \boldsymbol{v}_B \rangle. \tag{3.17}$$

The total flux density $\boldsymbol{q}_A$ of one component, say A, relative to the frame of observation in which $\langle \boldsymbol{v}_A \rangle$ is determined as the sum of a *bulk flow* velocity and an *entrained diffusive* velocity is in our reference frame given by

$$\boldsymbol{q}_A = n_V^A \langle \boldsymbol{v}_A \rangle = n_V^A \langle \boldsymbol{v} \rangle + n_V^A \left(\langle \boldsymbol{v}_A \rangle - \langle \boldsymbol{v} \rangle \right) = n_V^A \langle \boldsymbol{v} \rangle + \boldsymbol{j}_A. \tag{3.18}$$

We call $\boldsymbol{j}_A$ the molar diffusive flux with respect to the reference velocity $\langle \boldsymbol{v} \rangle$. If the bulk flow is zero, i.e. $\langle \boldsymbol{v} \rangle = 0$, then

$$x_A \langle \boldsymbol{v}_A \rangle + x_B \langle \boldsymbol{v}_B \rangle = 0 \tag{3.19}$$

and

$$\begin{aligned} \boldsymbol{j}_A &= n_V^A \langle \boldsymbol{v}_A \rangle, \\ \boldsymbol{j}_B &= n_V^B \langle \boldsymbol{v}_B \rangle. \end{aligned} \tag{3.20}$$

Using eq.(3.19) in eq.(3.20) we get

$$\frac{x_A}{n_V^A} \boldsymbol{j}_A + \frac{x_A}{n_V^B} \boldsymbol{j}_B = 0 \tag{3.21}$$

and now using the definition of the mole fraction we have

$$\frac{1}{n} \boldsymbol{j}_A + \frac{1}{n} \boldsymbol{j}_B = 0 \tag{3.22}$$

or

$$\boldsymbol{j}_A + \boldsymbol{j}_B = 0 \tag{3.23}$$

Thus, the fluxes of both species are not independent from each other. Any flux of one species is counterbalanced by the flux of the other species.

We now have to relate the diffusional flux to the driving forces for diffusion. In solids or liquids diffusion originates from chemical potential differences and not from concentration differences, since in thermodynamic equilibrium the chemical potentials are uniform and, thus, there can be large concentration differences without giving rise to any diffusion of species. Take for example a sample being held isothermally at a temperature below the eutectic temperature. There are two phases of largely different composition in equilibrium and no exchange of matter occurs between them. Therefore, *a difference or gradient in chemical potential is necessary to drive any diffusive mass transport.* For a binary system with components A and B the mass flux densities $\boldsymbol{j}_k$, $k = A, B$ measured as the number of moles passing through a reference plane of unit area is defined by:

$$\begin{aligned} \boldsymbol{j}_A &= -M_{AA} x_A \nabla \mu_A - M_{AB} x_B \nabla \mu_B \\ \boldsymbol{j}_B &= -M_{BA} x_A \nabla \mu_A - M_{BB} x_B \nabla \mu_B \end{aligned} \tag{3.24}$$

where the coefficients M_{kj} are called mobilities. These relations state that the fluxes are first proportional to the thermodynamic force, the gradient of a potential, and second that the flux of one species is generally influenced by the force acting on the other diffusing species. This is expressed via the cross mobilities M_{AB}, M_{BA}. For simplicity neglecting the non-diagonal terms the mobility coefficients are $M_{AA} = M_{BB} \equiv M$. We can arrive at the relation for the diffusion flux of species B (B is the minority component),

$$\boldsymbol{j}_B = -M x_B \nabla \mu_B. \tag{3.25}$$

Using eq.(2.47) for the chemical potential in terms of the molar free energy yields

$$\boldsymbol{j}_B = -M x_B \nabla \left[G_m + (1 - x_B)\frac{\partial G_m}{\partial x_B} \right] \tag{3.26}$$

$$= -M x_B \left[\frac{\partial G_m}{\partial x_B} \nabla x_B + \frac{\partial^2 G_m}{\partial x_B^2} \nabla x_B - \frac{\partial G_m}{\partial x_B} \nabla x_B - x_B \frac{\partial^2 G_m}{\partial x_B^2} \nabla x_B \right] \tag{3.27}$$

$$= -M x_B \left[(1 - x_B) \frac{\partial^2 G_m}{\partial x_B^2} \nabla x_B \right] \tag{3.28}$$

$$= -M x_B \left[(1 - x_B) \frac{\partial^2 G_m}{\partial x_B^2} \nabla x_B \right] \tag{3.29}$$

$$= -M x_B x_A \frac{\partial^2 G_m}{\partial x_B^2} \nabla x_B \tag{3.30}$$

$$\boldsymbol{j}_B = -D^+ \nabla x_B. \tag{3.31}$$

with the diffusion coefficient, D^+, defined as

$$D^+ = \left[M x_A x_B \frac{\partial^2 G_m}{\partial x_B^2} \right]. \tag{3.32}$$

The relationship of eq.(3.31) is called *Fick's first law.* The diffusion coefficient D^+ defined here has somewhat unusual units, not the well known case of m^2/s. This is due to the use of mole fraction as the unit for measuring concentration. We will see below how this definition combines together with a continuity equation for species into the well known diffusion coefficient[2].

To derive a continuity equation for species we follow the same line of arguments as given above for heat transport. Since the diffusion of species into or out of a volume element V changes there its local concentration we have to consider that any time change in local concentration in a volume element is balanced by an appropriate flux of species through the surface of the volume element. The flux through the surface $S(V)$ is given by summing up all diffusion current densities. Let $\boldsymbol{df}$ be a normal surface element of the volume V pointing outward with respect to the center of the volume. Then the total number of moles of B passing that surface per unit of time is given as

$$I_B = \int_{S(V)} \boldsymbol{j}_B \cdot \boldsymbol{df}. \tag{3.33}$$

This leads to a change of number of moles of B in the volume element. Its number is given as[3]

$$n_B = \int_V \frac{1}{V_m} x_B dV. \tag{3.34}$$

The time rate of change of n_B is counterbalanced by the diffusion flux through the surface,

$$\frac{dn_B}{dt} + I_B = \frac{dn_B}{dt} + \int_{S(V)} \boldsymbol{j}_B \cdot \boldsymbol{df} = 0. \tag{3.35}$$

Performing the differentiation on n_B and using Gauss' theorem we obtain

$$\int_V \left[\frac{1}{V_m} \frac{\partial x_B}{\partial t} + \nabla \cdot \boldsymbol{j}_B \right] dV = 0. \tag{3.36}$$

This relation holds if the integrand vanishes itself leading to the mass conservation equation,

$$\frac{\partial x_B}{\partial t} + V_m \nabla \cdot \boldsymbol{j}_B = 0. \tag{3.37}$$

---

[2] Tabulated values of the diffusion coefficient are related to a different reference frame than the one used here. Usually it is based on a volume average velocity. For more details see [247].

[3] Here one should remember that the molar fraction is not a constant but a scalar field depending on space and time, i.e. $x_B = x_B(\boldsymbol{r}, t)$ with $\boldsymbol{r}$ denoting any position in space.

Inserting for the flux density eq.(3.31) we obtain

$$\frac{\partial x_B}{\partial t} = V_m \nabla (D^+ \nabla x_B). \tag{3.38}$$

If the molar volume of B in the alloy is independent of composition, we can write

$$\frac{\partial x_B}{\partial t} = \nabla (V_m D^+ \nabla x_B) = \nabla (D \nabla x_B), \tag{3.39}$$

which is *Fick's second law* with the diffusion coefficient

$$D = D^+ V_m, \tag{3.40}$$

where D has the familiar units of m^2/s. With this definition of the diffusion coefficient we can rewrite *Fick's first law* given in eq.(3.31) as,

$$j_B = -\frac{D}{V_m} \nabla x_B. \tag{3.41}$$

If the diffusion coefficient D is independent of composition x_B equation (3.39) further simplifies to

$$\frac{\partial x_B}{\partial t} = D \nabla^2 x_B. \tag{3.42}$$

The relations (3.39) and (3.42) are most often used in the literature. Throughout this book we will use a concentration independent diffusion coefficient.

It should be noted that the transport equations for heat and mass transport are structurally equal. They are both parabolic partial differential equations of second kind. Both type of equations are treated in standard textbooks like those of Carslaw and Jaeger [55] on heat transport and Crank [72] on diffusion.

3.3 Convective heat and mass transport

In fluids heat and mass can be transported by convection or bulk fluid flow as well as diffusion. Convection can be a much faster way of transport than diffusion. We first will derive the extension of the two continuity equations given above to account for convection and then discuss both especially with respect to the so-called boundary layer approximations used later in this monograph.

Assume that there exists a bulk fluid flow described by a velocity field with vector $\boldsymbol{v}$. This transports heat according to

$$\boldsymbol{q}_{conv} = H_V \boldsymbol{v}. \tag{3.43}$$

Inserting this relation into the general continuity equation, eq.(3.8), we obtain

$$C_P(T)\rho(T)\frac{\partial T}{\partial t} = -\nabla \cdot (H_V \boldsymbol{v}) = -H_V \nabla \cdot \boldsymbol{v} - \boldsymbol{v} \cdot \nabla H_V. \tag{3.44}$$

In an incompressible fluid this can be simplified, since $\nabla \cdot \boldsymbol{v} = 0$. Using eq.(3.3) we then obtain finally the temperature change of the body if heat is only transported by convection,

$$\frac{\partial T}{\partial t} = -\boldsymbol{v} \cdot \nabla T. \tag{3.45}$$

If heat is transported by diffusion and convection both fluxes simply add and we find for the total change in temperature,

$$\frac{\partial T}{\partial t} + \boldsymbol{v} \cdot \nabla T = \kappa \nabla^2 T. \tag{3.46}$$

This is the so-called *transport equation for heat* under convective diffusive conditions. It is valid as long as we have no heat sources which would add an additional term to the RHS.

Similar to the heat transport we can have diffusive and convective mass transport. If we again describe the bulk fluid flow with velocity field denoted as $\boldsymbol{v}$ this will create a molar flux density given by

$$\boldsymbol{j}_{conv} = \frac{x_B}{V_m}\boldsymbol{v}, \tag{3.47}$$

where $\boldsymbol{v} = \langle \boldsymbol{v} \rangle$ of eq.(3.17). Adding this to the diffusion flux density gives a total molar flux density

$$\boldsymbol{j}_{total} = -D^+ \nabla x_B + \frac{x_B}{V_m}\boldsymbol{v}. \tag{3.48}$$

With eq.(3.37) we obtain

$$\frac{\partial x_B}{\partial t} + V_m \nabla \boldsymbol{j}_{total} = 0$$

$$\begin{aligned} \frac{\partial x_B}{\partial t} &= D\nabla^2 x_B - \nabla(x_B \boldsymbol{v}) \\ &= D\nabla^2 x_B - \boldsymbol{v} \cdot \nabla x_B. \end{aligned} \tag{3.49}$$

To obtain the last equation we again assumed incompressibility $\nabla \cdot \boldsymbol{v} = 0$ and also that the diffusion coefficient is independent of concentration. This is the so-called *transport equation for species* under convective diffusive conditions.

Both transport equations are mathematically again similar and we therefore will consider in the following only the equation for mass transport. For a full solution of the continuity equations derived above, we must also know or specify the boundary and the initial conditions to be satisfied by the temperature or concentration field. The equations for diffusion and convective diffusion can greatly be simplified if the partial time derivative vanishes; the fields are then time independent. For a further discussion of the convective diffusion equations, Eqs.(3.46,3.49), we neglect first the time derivative and reduce them to a dimensionless form. This is accomplished by introducing a characteristic length Λ (e.g. for spheres this could be their radius) and a characteristic flow velocity U_0 (e.g. this could be the flow velocity far away from a dispersed body). We scale the spatial coordinate $\boldsymbol{r}$ by the length Λ, to obtain $\hat{\boldsymbol{r}} = \boldsymbol{r}/\Lambda$ and the flow velocity is divided by the characteristic velocity $\hat{\boldsymbol{v}} = \boldsymbol{v}/U_0$. Introducing these relations into eq.(3.49) leads to

$$\hat{\boldsymbol{v}} \cdot \nabla x_B = \frac{D}{U_0 \Lambda} \nabla^2 x_B, \tag{3.50}$$

where the operator ∇ is also written in dimensionless variables. The dimensionless ratio

$$Pe_D = \frac{U_0 \Lambda}{D} \tag{3.51}$$

is known as the *diffusional Peclet number*. With this definition we can rewrite the equation in two forms,

$$Pe_D\,(\hat{\boldsymbol{v}} \cdot \nabla x_B) = \nabla^2 x_B, \tag{3.52}$$

$$\hat{\boldsymbol{v}} \cdot \nabla x_B = \frac{1}{Pe_D} \nabla^2 x_B, \tag{3.53}$$

The left side of the convective diffusion equation represents the convective mass transfer, while the right side that due to molecular or atomic diffusion. The ratio of both is described by a single parameter - the dimensionless Peclet number. When the Peclet number is small the left side of eq.(3.52) is small and thus the right side can be set to zero and we yield the standard diffusion equation without convective influences. If the Peclet number is large the term on the right side (eq.(3.53)) becomes small and thus mass transport occurs essentially by convective transfer and diffusion can be neglected. Thus the value of the Peclet number is essential to differentiate between a situation in which diffusion or convection dominates mass transport[4]. Another important dimensionless number in mass transport is the *Schmidt number* defined as

$$Sc = \frac{\nu}{D}, \tag{3.54}$$

where ν is the kinematic viscosity. For liquid metals the kinematic viscosity is of the order $\nu \approx 10^{-6}\mathrm{m}^2/\mathrm{s}$ and the diffusion coefficient is usually of the order $D \approx 10^{-9}\mathrm{m}^2/\mathrm{s}$ and thus we have for the Schmidt number a value of $Sc \approx 1000$. This means that even at low fluid velocities the transport of matter by convection can be larger than that due to atomic diffusion (in contrast to gases for instance in which $\nu \approx D$).

For a further discussion of the convective diffusion equation we will introduce the so-called boundary layer approximation. This is valid close to a body immersed in a fluid. If the Peclet number is large the diffusional term, eq.(3.49), can be neglected and the convective term reads

$$\boldsymbol{v} \cdot \nabla x_B = v_x \frac{\partial x_B}{\partial x} + v_y \frac{\partial x_B}{\partial y} + v_z \frac{\partial x_B}{\partial z}, \tag{3.55}$$

with $\boldsymbol{v} = (v_x, v_y, v_z)$. Eq.(3.55) is satisfied by the simple solution,

$$x_B = const., \tag{3.56}$$

which obviously fulfills also the condition of constant concentration far from any boundary of the immersed body. The solution does, however, not fulfill any boundary condition at the interface that is different than the far field concentration, e.g. we could require the concentration at the interface of the body to be zero or fix the gradient of the concentration at the interface. Near the interface, therefore, a thin layer exists in which the concentration changes rapidly. The

[4] In heat transport the *thermal Peclet number* is defined as $Pe_T = U_0\Lambda/\kappa$.

derivatives of concentration with respect to distance are in this case very large, with the result that the right side in eq.(3.53) cannot be neglected. Thus, at high Peclet numbers the liquid maybe divided into two regions, the first, a region of nearly constant concentration far away from the immersed body, and the second, a region of rapidly changing concentration in the close vicinity of the body's interface. This layer is called a *diffusion boundary layer*, since in this layer diffusion plays an important role. Its thickness shall be denoted by δ.

Let us evaluate the order of magnitude of the quantities in eq.(3.49) at the border of the diffusion boundary layer. For convenience we use cartesian coordinates. The body shall extend into the $x-z-$plane and the fluid shall flow parallel to the body's surface. The diffusion boundary layer then extends into the y-direction and its extension is $y \approx \delta$. We then have the following set of relations using $\delta \ll \Lambda$:

$$\frac{\partial x_B}{\partial x} \approx \frac{x_B}{\Lambda}, \tag{3.57}$$

$$\frac{\partial x_B}{\partial y} \approx \frac{x_B}{\delta}, \tag{3.58}$$

$$\frac{\partial^2 x_B}{\partial x^2} \approx \frac{x_B}{\Lambda^2}, \tag{3.59}$$

$$\frac{\partial^2 x_B}{\partial y^2} \approx \frac{x_B}{\delta^2}, \tag{3.60}$$

$$\text{hence} \qquad \frac{\partial^2 x_B}{\partial y^2} \gg \frac{\partial^2 x_B}{\partial x^2}, \tag{3.61}$$

and therefore the equation for convective diffusion in a diffusion boundary layer takes a simple form,

$$v_x \frac{\partial x_B}{\partial x} + v_y \frac{\partial x_B}{\partial y} = D \frac{\partial^2 x_B}{\partial y^2} . \tag{3.62}$$

This equation is called the equation of *convective diffusion*, meaning diffusion and convection play both a role in mass transport. Outside the diffusion boundary layer the concentration is nearly constant. Inside the boundary layer the full convective-diffusion equation given in eq.(3.49) is simplified to eq.(3.62) meaning we have to take only diffusion normal to the interface into account. The result obtained here in cartesian coordinates can be extended to other coordinates as well. We will later perform two analyses for solid and liquid spheres moving in a liquid matrix and solve the species transport problem with the approximations given here in order to obtain growth and coarsening laws under diffusive-convective conditions.

4. Growth

In this chapter we will examine the growth of particles in a matrix via a number of mechanisms. First we consider growth or solidification from a supercooled melt, second growth in a supersaturated matrix, and third growth by interface kinetics of first order where mass or heat is transported. We then discuss the growth by interface kinetics of second order and growth of particles or droplets where convective heat or mass fluxes play an important role. The closing section deals with growth under varying, time dependent supersaturation.

4.1 The general growth problem

Let us consider a situation where a single phase material is suddenly placed in a region of the phase diagram in which in equilibrium two phases, α- and β, of different composition coexist, e.g. by quenching from a high to a sufficiently low temperature. Then the material is in a metastable or unstable state (see discussion in section 2.3). We assume as before that we are dealing with a two-component system consisting of A and B atoms. In order to achieve a transformation from a metastable single phase state to a stable two-phase situation second-phase particles must first nucleate and then grow. A description of nucleation is given in Appendix B, and that of a nucleation, growth, and coarsening as an integral process will be given in Chapter 10. Here we study a situation in which the second phase, the β-phase, is already nucleated and is growing in order to achieve thermodynamic equilibrium.

The basic idea for the following is as follows: We have a spherical particle of β-phase growing in a supersaturated matrix (α-phase). We shall assume that at the surface of the particle atoms attach easily resulting in a growth of the particle and a depletion of the matrix. The concentration in the matrix at the interface, thus, is very close to the equilibrium concentration. This is the *local equilibrium approximation.* At distances far from the particle the concentration is still the original value. This results in a concentration gradient and consequently in a mass flow of B atoms towards the growing particle (we assume only diffusion and not convection). The growth of the particle is limited by the flux of B atoms to the particle, which depends on the concentration gradient. The concentration gradient is a function of the difference between the equilibrium concentration at the interface and the far-field concentration as well as the geometry of the particle. The Gibbs-Thomson effect (curvature dependence of the equilibrium concentration) shall be small compared to the concentration differences in the matrix and is neglected. Thus, the entire problem is exactly specified. Since the problem has spherical symmetry, we can simplify the equations by using spherical coordinates.

In any transformation involving long range transport of mass or heat the diffusion equations must be satisfied in all parts of the primary α-phase. We first study growth by mass transport. The transport equation for diffusive mass transport was derived in the preceding chapter,

$$\frac{\partial x_B}{\partial t} = D\nabla^2 x_B. \tag{4.1}$$

The composition x_B is a function of position $\boldsymbol{r}$ and time t. The diffusion equation can be rewritten in spherical coordinates with the origin at the center of the particle. The concentration field has due to the spherical symmetry only a dependence on the radial coordinate leading to

$$\frac{\partial x_B}{\partial t} = D\left(\frac{\partial^2 x_B}{\partial r^2} + \frac{2}{r}\frac{\partial x_B}{\partial r}\right). \tag{4.2}$$

r is the distance from the center of the particle. Let the concentration of solute in the initial α phase be x_B^0. At the interface we assume local thermodynamic equilibrium, i.e.

$$x_B(r = R(t), t) = x_B^\alpha. \tag{4.3}$$

Since the particle grows at the equilibrium concentration, no concentration gradients arise inside the particle and we have $x_B(r < R(t), t) = x_B^\beta$. Far from the interface, located at the particle radius $R(t)$, the matrix concentration remains at its initial value,

$$x_B(\infty, t) = x_B^0. \tag{4.4}$$

The time dependence of the particle radius $R(t)$ is not known a priori, but a solution of the problem itself. The time rate of change of the radius depends on the diffusion of solute. The diffusion flux at the boundary in r direction is given by eq.(3.31), $-D^+(\partial x_B/\partial r)_{r=R(t)}$. The diffusion flux will, as discussed above, balance the amount of B-atoms incorporated into the growing particle. For the particle to grow by an amount δR a certain number of moles of solute atoms is required, given that $x_B^\beta \neq x_B^\alpha$. Since the volume of a spherical shell of thickness δR on a particle of radius R is $4\pi R^2 \delta R$ assuming $\delta R \ll R$, the additional number of moles of solute, δn_B, required to grow a particle of radius R a distance δR is

$$\delta n_B = (x_B^\beta - x_B^\alpha)\frac{4\pi R^2 \delta R}{V_m}, \tag{4.5}$$

where we are assuming that the molar volume V_m is independent of concentration, i.e. $V_m^\beta = V_m^\alpha = V_m$. From eq.(4.5) we see that if $x_B^\beta - x_B^\alpha > 0$ the solute must be added to the particle for it to grow ($\delta R > 0$). This solute must be supplied by diffusion from the matrix. So, in a time increment δt the amount of solute supplied by diffusion is

$$\delta n_B = 4\pi R^2 D^+ \left.\frac{\partial x_B}{\partial r}\right|_{r=R(t)} \delta t. \tag{4.6}$$

The amount is positive since we are interested in the amount entering the particle or the flux in the negative r direction. Thus, at the interface we have the balance

$$(x_B^\beta - x_B^\alpha)\frac{4\pi R^2 \delta R}{V_m} = 4\pi R^2 D^+ \left.\frac{\partial x_B}{\partial r}\right|_{r=R(t)} \delta t. \tag{4.7}$$

Since $V_m D^+ = D$, and taking the limit $\delta t \to 0 \quad \Rightarrow \quad \delta R/\delta t \to dR/dt$, we have the mass balance condition

$$D \left.\frac{\partial x_B}{\partial r}\right|_{r=R(t)} = (x_B^\beta - x_B^\alpha)\frac{dR}{dt}. \tag{4.8}$$

The diffusion equation, eq.(4.2), together with the boundary conditions, eq.(4.3), and eq.(4.4) can be solved using a similarity transformation. The mass balance equation, eq.(4.8) is then used to determine the growth rate of the particle.

When looking at the problem one might notice that if the particle radius is small the concentration gradients around the particle are steep and as the particle becomes larger the gradients become smaller. The reason for this behavior lies in the spherical symmetry of the problem. The consequence will be that as the particle grows and the concentration gradients become smaller, the flux towards the particle decreases and growth slows with time. Thus, our different lengthscales increase with time (R, $(x_B^\beta - x_B^\alpha)/\nabla x_B$, and $\frac{dR}{dt} \times t$). This can be taken as a guess at this point and we will later see that it is indeed correct. Having this guess we can take advantage of it by creating a new variable that captures this scaling behavior,

$$\hat{r} = \frac{r}{\sqrt{t}}. \tag{4.9}$$

We could have tried r/t or other choices as well but such choices are not valid for the diffusion equation with the boundary conditions and the mass balance equation given above. The advantage of our choice is that the partial differential equation for $x_B(r,t)$ will be reduced to an ordinary differential equation (ODE) for $x_B(\hat{r})$ that is much easier to solve.

The differentials in eq.(4.2) can be written in terms of this new variable

$$\frac{\partial x_B}{\partial t} = -\frac{\hat{r}}{t}\frac{dx_B}{d\hat{r}} \tag{4.10}$$

$$\frac{\partial x_B}{\partial r} = \frac{1}{\sqrt{t}}\frac{dx_B}{d\hat{r}} \tag{4.11}$$

$$\frac{\partial^2 x_B}{\partial r^2} = \frac{1}{t}\frac{d^2 x_B}{d\hat{r}} \tag{4.12}$$

We then arrive at an ODE for the composition $x_B(\hat{r})$,

$$x_B'' + (\frac{2}{\hat{r}} + \frac{\hat{r}}{2D})x_B' = 0, \tag{4.13}$$

where the prime indicates differentiation with respect to $\hat{r}$. Denoting $y = x_B'$ the equation becomes an ODE of first order,

$$y' + (\frac{2}{\hat{r}} + \frac{\hat{r}}{2D})y = 0, \tag{4.14}$$

that can be solved by the separation of variables and direct integration,

$$\frac{dy}{y} = -(\frac{2}{\hat{r}} + \frac{\hat{r}}{2D})d\hat{r}, \tag{4.15}$$

leading to

$$y(\hat{r}) = \frac{a_0}{\hat{r}^2} \exp\left(-\frac{\hat{r}^2}{4D}\right). \tag{4.16}$$

a_0 is an arbitrary constant of integration. Then, the concentration field is obtained by a second integration over $\hat{r}$,

$$x_B(\hat{r}) = a_1 + a_0 \int_{\hat{r}}^{\infty} \frac{1}{\hat{r}^2} \exp\left(-\frac{\hat{r}^2}{4D}\right) d\hat{r}, \tag{4.17}$$

with a_1 as another constant of integration. The idea of the $\hat{r}$-variable was to take advantage of a possible scaling behavior of the problem. Consistent with our guess for the nature of this kind of scaling is the guess of

$$R(t) = \lambda_R \sqrt{t}. \tag{4.18}$$

since then our boundary conditions become time independent,

$$x_B(r = R(t), t) = x_B^\alpha \quad \rightarrow \quad x_B(\lambda_R) = x_B^\alpha, \tag{4.19}$$
$$x_B(r \rightarrow \infty, t) = x_B^0 \quad \rightarrow \quad x_B(\infty) = x_B^0. \tag{4.20}$$

Thus, the boundary conditions and mass balance are also consistent with the similarity transformation. The condition in eq.(4.19) leads to $a_1 = x_B^0$ and eq.(4.19) fixes a_0,

$$x_B(\lambda_R) = x_B^\alpha = x_B^0 + a_0 \int_{\lambda_R}^{\infty} \frac{1}{\hat{r}^2} \exp\left(-\frac{\hat{r}^2}{4D}\right) d\hat{r}. \tag{4.21}$$

The integral can be evaluated,

$$\int_{\lambda_R}^{\infty} \frac{1}{\hat{r}^2} e^{-\frac{\hat{r}^2}{4D}} d\hat{r} = -\frac{1}{\hat{r}} e^{-\frac{\hat{r}^2}{4D}} \Bigg|_{\lambda_R}^{\infty} + \int_{\lambda_R}^{\infty} \frac{1}{\hat{r}} \frac{-2\hat{r}}{4D} e^{-\frac{\hat{r}^2}{4D}} d\hat{r} \tag{4.22}$$
$$= \frac{1}{\lambda_R} e^{-\frac{\lambda_R{}^2}{4D}} - \frac{1}{2D} \int_{\lambda_R}^{\infty} e^{-\frac{\hat{r}^2}{4D}} d\hat{r} \tag{4.23}$$
$$= \frac{1}{\lambda_R} e^{-\frac{\lambda_R{}^2}{4D}} - \sqrt{\frac{\pi}{4D}} \operatorname{erfc}\left(\frac{\lambda_R}{2\sqrt{D}}\right), \tag{4.24}$$

and a_0 is equal to

$$a_0 = \frac{x_B^\alpha - x_B^0}{\frac{1}{\lambda_R} e^{-\frac{\lambda_R{}^2}{4D}} - \sqrt{\frac{\pi}{4D}} \operatorname{erfc}\left(\frac{\lambda_R}{2\sqrt{D}}\right)}. \tag{4.25}$$

We can now determine the growth rate of the particle radius using eq.(4.9) and eq.(4.18). Using eq.(4.18) to eliminate dR/dt in eq.(4.8), using eq.(4.11) in eq.(4.8) and taking the derivative of eq.(4.17) leads to an equation for the scaling constant λ_R,

$$\frac{2Da_0}{x_B^\beta - x_B^\alpha} \frac{1}{\lambda_R{}^2} e^{-\frac{\lambda_R{}^2}{4D}} = \lambda_R \tag{4.26}$$

or

$$\left(2D\frac{x_B^0 - x_B^\alpha}{x_B^\beta - x_B^\alpha}\right) \left[\frac{e^{-\frac{\lambda_R{}^2}{4D}}}{\lambda_R e^{-\frac{\lambda_R{}^2}{4D}} - \sqrt{\frac{\pi}{4D}} \lambda_R{}^2 \mathrm{erfc}(\frac{\lambda_R}{2\sqrt{D}})}\right] = \lambda_R. \tag{4.27}$$

If growth is sufficiently slow λ_R is small and $\lambda_R \ll 1$. In this limit the complicated dependence of the term in the square brackets on λ_R can be simplified greatly. We expand the exponentials and the error function in a Taylor series about $\lambda_R = 0$, taking the limit $\lambda_R \ll 1$ and neglect the higher order terms. Then the term in the square bracket can be replaced by $1/\lambda_R$ and we arrive at a simple equation for λ_R,

$$\lambda_R{}^2 = 2D\frac{x_B^0 - x_B^\alpha}{x_B^\beta - x_B^\alpha}. \tag{4.28}$$

We, thus, can write for the time dependence of the particle radius in the limit $\lambda_R \ll 1$

$$R(t) = \left(\frac{2D(x_B^0 - x_B^\alpha)}{x_B^\beta - x_B^\alpha}\right)^{\frac{1}{2}} t^{\frac{1}{2}}. \tag{4.29}$$

If we denote the supersaturation as

$$S_0 = \frac{x_B^0 - x_B^\alpha}{x_B^\beta - x_B^\alpha} \tag{4.30}$$

we can see by eq.(4.28) that the limit $\lambda_R \ll 1$ implies that $S_0 \ll 1$ or small supersaturations. Thus, eq.(4.29) is valid in the limit of small supersaturations or a slow particle growth rate. We can write eq.(4.29) as

$$R(t) = \sqrt{2DS_0 t} \tag{4.31}$$

or for the growth rate of the particle

$$\frac{dR}{dt} = \frac{DS_0}{R}. \tag{4.32}$$

A similar result can be obtained for solidification of a sphere from a pure melt (see [55]) where the supersaturation is replaced by a term that is related to the undercooling or the difference in temperature between that at the interface and the far-field one. We will in the next sections present simpler derivations of eq.(4.32) and its thermal equivalent using the so-called stationary interface approximation for heat and mass transport.

4.2 Growth of a supercooled sphere

In this section we treat the growth of a single sphere of radius R in a supercooled melt as schematically shown in fig. 4.1. Far away from the interface the temperature is kept constant at T_∞ below the equilibrium melting temperature T_m.

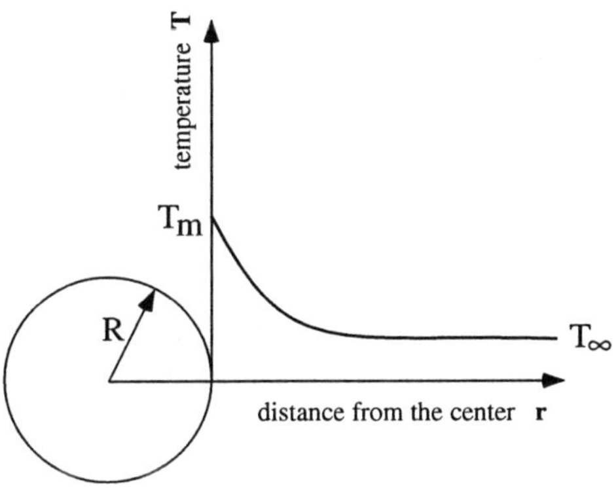

Fig. 4.1. Scheme of the temperature profile ahead of a growing sphere of radius R.

We assume growth to proceed so slow that a stationary interface approximation holds. This means, we separate the calculation of the temperature field from the growth process and do not solve the time dependent heat conduction equation with a moving boundary. We assume that the growth velocity of the particle, $v = dR/dt$, is small enough for the temperature field to be in equilibrium given the particle as a heat source which is the case during solidification. This is always a good approximation if the supercooling, denoted as $\Delta T = T_m - T_0$, is not too large. Specifically, the criterion for the validity of this assumption is that the ratio of the velocity of the interface v to the velocity of the heat flux, given by thermal diffusivity κ divided by the typical length-scale R, is small: $Pe_T = vR/\kappa \ll 1$. This ratio is called *thermal Peclet number*, Pe_T. In cases of rapid solidification with interface velocities in m/s-range this approach is invalid.

For the temperature field outside the particle we solve the Laplaces equation in spherical coordinates,

$$\nabla^2 T = \frac{\partial^2 T}{\partial r^2} + \frac{2}{r}\frac{\partial T}{\partial r} = 0, \tag{4.33}$$

with the boundary conditions

$$T(r \to \infty) = T_0, \tag{4.34}$$

$$T(r = R) = T_m. \tag{4.35}$$

The solution of this equation is

$$T(r) = a_0 + \frac{a_1}{r}. \tag{4.36}$$

The boundary conditions determine the constants: $a_0 = T_0$ and $a_1 = (T_m - T_0)R$. The solution then reads

$$T(r) = T_0 + (T_m - T_0)\frac{R}{r}. \tag{4.37}$$

The latent heat liberated at the interface by the solidification equals the heat transported away due to the temperature gradients. This gives the heat balance at the interface,

$$\rho L \frac{dV}{dt} = -\lambda \left.\frac{\partial T}{\partial r}\right|_{r=R(t)} A, \tag{4.38}$$

with ρ the density, L the latent heat, λ the heat conductivity, A the surface area of the sphere, and V its volume. Taking the derivative of $T(r)$ in eq.(4.37),

$$\left.\frac{\partial T}{\partial r}\right|_{r=R(t)} = -\frac{(T_m - T_0)}{R}, \tag{4.39}$$

and using it in eq.(4.38) gives the growth rate,

$$\frac{dR}{dt} = \frac{\lambda}{\rho L} \frac{T_m - T_0}{R}. \tag{4.40}$$

If the supercooling $\Delta T = T_m - T_0$ is constant during growth, growth proceeds in a parabolic manner,

$$R^2 - R_0^2 = \frac{2\lambda}{\rho L} \Delta T \, (t - t_0). \tag{4.41}$$

Using the thermophysical properties for succinonitrile (SCN), and taking as an undercooling a value of $\Delta T = 0.5$ K, we obtain for the growth rate a value of $(2\lambda\Delta T)/(\rho L) \approx 2 \cdot 10^{-10} \text{m}^2/\text{s}$[1]. This means within 1 second a particle would grow to 10 microns in radius. In 5000 seconds the particle would grow to a size of a millimeter. It is important to notice that, since $dR/dt \approx 1/R$, smaller particles change their radius faster than the bigger ones, see eq.(4.40).

4.3 Growth in a supersaturated matrix

We now reexamine the problem from section 4.1, but treat it in a similar manner to the growth of a particle from a supercooled melt.

A single sphere of radius R shall grow in a supersaturated matrix (which could be either a melt or a solid). This is schematically shown in Fig. 4.2 together with a sketch of a phase diagram. Far from the interface the concentration is kept constant at x_B^0, which is the initial alloy concentration. Such a situation can be realized if for example an alloy is rapidly quenched from a single phase field into the two-phase field. If during quenching no nucleation and phase separation occurs, annealing at a temperature higher then the temperature of the quench leads to the nucleation of particles and their subsequent growth. Age-hardenable alloys are treated in this way to increase their yield and ultimate tensile strengths. If one cools for example an Al-4 wt.% Cu alloy from the single phase field after solution and homogenization treatment at say 480^0C rapidly to room temperature (this usually is done by quenching into water) the slow diffusion in the solid state impedes any nucleation of precipitates. Metastable precipitates are

[1] SCN and alloys like SCN-ethanol, SCN water are the standard material to investigate solidification processes under an optical microscope since the mid 1960's. This is first because SCN is transparent and thus allowing direct optical observation of all the microstructural details during solidification and second it mimics metallic alloys in properties relevant for solidification.

nucleated and grow if the alloy is then heat treated at temperatures ranging from room temperature to approximately 180^{0}C. Normally in age hardenable alloys the stable equilibrium phase does not appear directly during annealing, but metastable phases form which are easier to nucleate. We assume as in the section above, that growth proceeds

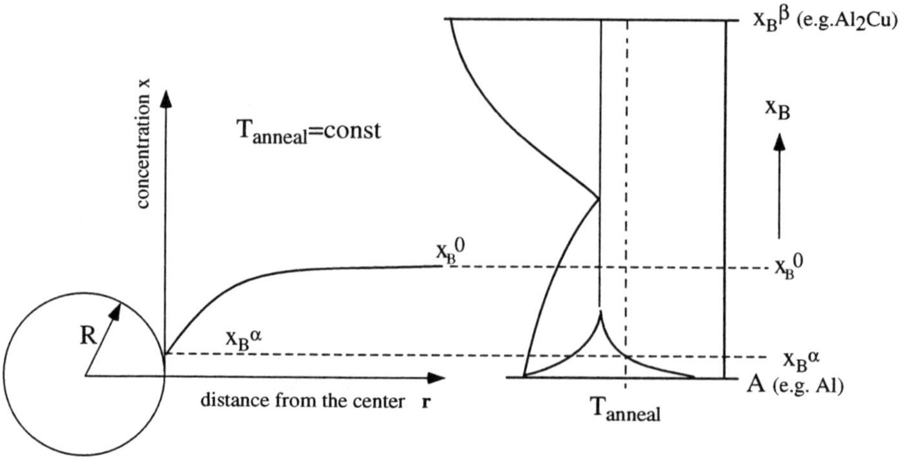

Fig. 4.2. Scheme of the concentration profile ahead of a growing sphere of radius R with a sketch of a phase diagram.

so slow that a stationary interface approximation holds. As before we can formulate a condition for the validity of this assumption: The ratio of the velocity of particle growth to the velocity of mass transport has to be small: $Pe_D = vR/D \ll 1$, where the velocity v of the interface replaces U_0 and R replaces Λ in eq.(3.51).

Again we solve the Laplace equation in spherical coordinates to calculate the concentration field outside the particle,

$$\nabla^2 x_B = \frac{\partial^2 x_B}{\partial r^2} + \frac{2}{r}\frac{\partial x_B}{\partial r} = 0, \tag{4.42}$$

with the boundary conditions

$$x_B(r \to \infty) = x_B^0, \tag{4.43}$$

$$x_B(r = R) = x_B^\alpha. \tag{4.44}$$

As before we assume that at the interface local equilibrium exists, i.e. the concentration of B atoms in the matrix is already down to the value given by the phase diagram. This is justified for several reasons: First, as above, if the Peclet number is small, the diffusion in the matrix is fast enough to equilibrate the interface region; second, the β-phase already has formed and depleted the surrounding matrix from B atoms, whose initial composition was x_B^0, just leaving a region of equilibrium concentration. This approximation breaks down if the Peclet number is high. The solution to equation (4.42) is

$$x_B(r) = a_0 + \frac{a_1}{r}. \tag{4.45}$$

The boundary conditions determine the constants to $a_0 = x_B^0$ and $a_1 = (x_B^\alpha - x_B^0)R$. The solution then reads

$$x_B(r) = x_B^0 + (x_B^\alpha - x_B^0)\frac{R}{r}. \tag{4.46}$$

The mass balance at the interface is given in eq.(4.8). This gives the growth rate, using the same procedure as in the last section,

$$\frac{dR}{dt} = \frac{D}{R}\frac{x_B^0 - x_B^\alpha}{x_B^\beta - x_B^\alpha} = D\frac{S_0}{R}. \tag{4.47}$$

Since the supersaturation $S_0 = (x_B^0 - x_B^\alpha)/(x_B^\beta - x_B^\alpha)$ is constant growth proceeds again in a parabolic manner,

$$R^2 - R_0^2 = 2DS_0 \cdot (t - t_0). \tag{4.48}$$

By choosing $R(t = 0) = 0$ we recover eq.(4.31).

There are indeed experiments performed, which show that such a growth law can be observed. Wiltzius and coworkers used a mixture of polyisoprene - polyethylenepropylene (a system that shows a miscibility gap in the liquid state) and observed after quenching from the single phase melt into the two phase region, that the droplets shortly after nucleation followed a parabolic growth as described here [73]. One result of the experimental observations is shown in Fig. 4.3. Different quench depths into the miscibility gap region induces different supersaturations.

4.4 Interface Kinetics of first order

In deriving the growth laws in the sections above, we implicitly assumed that the speed of growth is solely determined by either heat

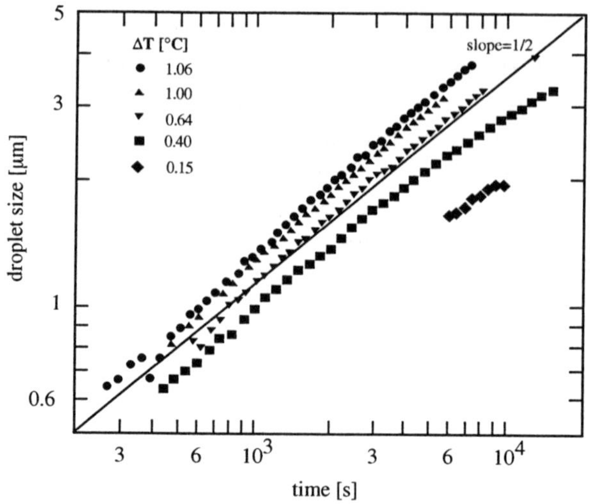

Fig. 4.3. Increase in average size as a function of supercooling obtained by quenching liquids from the single phase field below the line of coexisting in the miscibility gap of the polyisoprene-polyethylenepropylene system. The quench depth below the equilibrium temperature of the binodal line are given in the legend of the figure.

or mass transport in the matrix and also that the incorporation of atoms into the particle does not limit the growth rate. The last assumption means that at the particle interface there are always vacant sites for atoms from the matrix to jump into and that these are easily occupied. We will now treat the opposite case. We assume that mass or heat transport does not determine growth, but instead growth is limited by the atoms incorporating into the particle at the interface. This is modeled by assuming an energy barrier that an atom from the matrix must overcome before it can become part of the particle. Such a model holds for highly anisotropic interfaces with terraces, ledges, kinks, i.e. smooth interfaces instead of diffuse ones. The energetic picture of this situation is shown in Fig. 4.4. A movement of an atom from the α-phase to the β-phase leads to a gain in Gibbs free energy denoted by $\Delta G_{atom} = \Delta G/N_A$. There is, however, an energy barrier to overcome, which is denoted by G_A. Let us call this energy an activation energy. Let n_a^α be the number of atoms per area at the interface in the matrix and let ν be the number of jumps an atom performs per second (Imagine this as thermally activated jumps, which are for example due to lattice vibrations. Then ν would be of the order of the Debye frequency). The thermally activated flux from the α-phase to the β-phase is

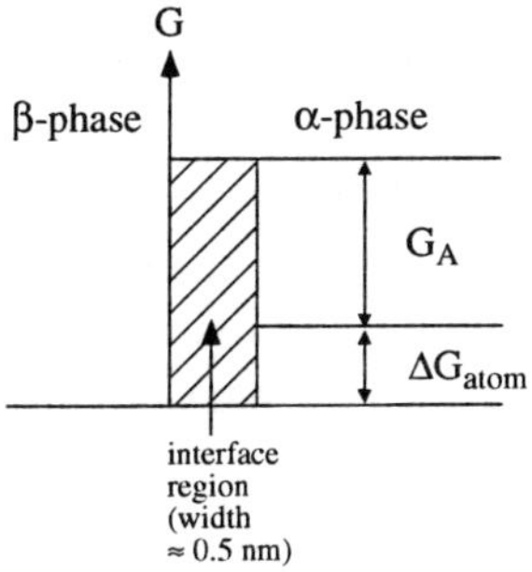

Fig. 4.4. Scheme of the energetic situation at a smooth interface where attachment to it is impeded by an energy barrier.

$$I_{\alpha\to\beta} = n_a^\alpha \nu \exp\left(-\frac{G_A}{k_B T}\right). \tag{4.49}$$

Here we introduced the exponential factor (Boltzmann factor) that takes into account how many atoms have an energy greater than G_A. There is, however, also a back-flux from the β-phase to the α-phase,

$$I_{\beta\to\alpha} = n_a^\beta \nu \exp\left(-\frac{G_A + \Delta G_{atom}}{k_B T}\right). \tag{4.50}$$

Note, that the energy barrier is increased by ΔG_{atom}. The difference in these fluxes multiplied by the atomic volume Ω gives just the velocity of the interface. The number of atoms or molecules per area generally depends on the crystallographic orientation. Low index planes should have a higher number of available lattice sites then high index planes. We assume for simplicity that $n_a^\alpha = n_a^\beta = n_a$. With this assumption we obtain the local interface velocity.

$$\mathrm{v} = \Omega(I_{\alpha\to\beta} - I_{\beta\to\alpha}) = \Omega n_a \nu \exp\left(-\frac{G_A}{k_B T}\right)\left[1 - \exp\left(-\frac{\Delta G_{atom}}{k_B T}\right)\right]. \tag{4.51}$$

If $\Delta G_{atom}/k_B T \ll 1$ we can expand the expression in the square bracket to first order and obtain

$$\mathrm{v} = \Omega n_a \nu \exp\left(-\frac{G_A}{k_B T}\right)\frac{\Delta G_{atom}}{k_B T} \tag{4.52}$$

or

$$\mathrm{v} = \frac{dR}{dt} = M \Delta G_{atom} \tag{4.53}$$

with the interface mobility

$$M = \Omega n_a \nu \exp\left(-\frac{G_A}{k_B T}\right)\frac{1}{k_B T} \tag{4.54}$$

Thus, the growth rate depends on the gain in energy obtained from transferring an atom from the matrix to the particle. If this energy gain does not depend on the particle radius the growth law of eq.(4.52) or eq.(4.53) already implies that the particle radius will grow linear in time.

4.4.1 Interface kinetics in a supersaturated matrix

The growth rate of a precipitate is directly related to the total flux of atoms across the interface, eq.(4.52). This flux, in turn, is related to Gibbs free energy difference between the α (matrix) and β (precipitate) phases at the interface, ΔG_{atom}. This Gibbs free energy difference is the energy change on moving an infinitesimal amount of material of interface composition x_B^i in the α phase to the β phase of composition $x_B^{\beta,e}$. Here we are assuming for simplicity that the composition of the β phase is the equilibrium composition and neglecting the possibility of solute drag. In general this is not the case as the composition in the β phase is a function of the velocity of the interface and the process of transferring atoms from the α to β phases dissipates energy, or, induces a drag. Nevertheless, the change in Gibbs free energy on transferring atoms is given by eq.(2.64) since it gives the energy change on removing a small amount of material of a certain composition, which in this case is the material removed from the α phase of composition given by the interface composition, and replacing it with a small amount of material of a different composition, which in this case is a small amount of β of its equilibrium composition. Since two different phases are present in this case, the Gibbs free energy functions appearing in eq.(2.64) will be evaluated in two different phases. Specifically, the composition of the material removed is replaced by x_B^i, the composition of the material formed is $x_B^{\beta,e}$, $G_m\left(x_B^i\right)$ is replaced by $G_m^\beta(x_B^{\beta,e})$ and $G_m\left(x'_B\right)$ is replaced by $G_m^\alpha\left(x_B^i\right)$. Thus, the Gibbs free energy change or driving force for interface motion is,

$$\Delta G_m = G_m^\alpha\left(x_B^i\right) - G_m^\beta\left(x_B^{\beta,e}\right) - \left.\frac{\partial G_m^\alpha}{\partial x_B}\right|_{x_B = x_B^i}\left(x_B^i - x_B^{\beta,e}\right) \tag{4.55}$$

Substituting eq.(4.55) into eq.(4.51) gives a relationship between the interface velocity and the compositions at the interface in the α- and

β-phases[2]. This, however, gives a nonlinear relationship between v and $x_B^i - x_B^{\beta,e}$. To simplify the nonlinear relationship we shall assume that the composition of the α-phase at the interface is close to the equilibrium composition of the α-phase. Thus we shall determine ΔG_m to first order in the difference $x_B^i - x_B^{\alpha,e}$ by expanding the Gibbs free energy of the α-phase and its first derivative with respect to composition in a Taylor's series,

$$G_m^\alpha(x_B) = G_m^\alpha(x_B^{\alpha,e}) + \left.\frac{\partial G_m^\alpha}{\partial x_B}\right|_{x_B=x_B^{\alpha,e}} (x_B - x_B^{\alpha,e}) + \frac{1}{2}\left.\frac{\partial^2 G_m^\alpha}{\partial x_B^2}\right|_{x_B=x_B^{\alpha,e}} (x_B - x_B^{\alpha,e})^2 \tag{4.56}$$

Taking the derivative of eq.(4.56) yields an approximation to the first derivative of first order in $x_B^i - x_B^{\alpha,e}$

$$\frac{\partial G_m}{\partial x_B} = \left.\frac{\partial G_m^\alpha}{\partial x_B}\right|_{x_B=x_B^{\alpha,e}} + \left.\frac{\partial^2 G_m^\alpha}{\partial x_B^2}\right|_{x_B=x_B^{\alpha,e}} (x_B - x_B^{\alpha,e}) \tag{4.57}$$

Evaluating eq.(4.56) and eq.(4.57) at $x_B = x_B^i$, keeping terms to first order in $x_B^i - x_B^{\alpha,e}$, and substituting the result in eq.(4.55), gives

$$\Delta G_m = G_m^\alpha(x_B^{\alpha,e}) - G_m^\beta(x_B^{\beta,e}) + \left.\frac{\partial G_m^\alpha}{\partial x_B}\right|_{x_B^{\alpha,e}} (x_B^i - x_B^{\alpha,e}) - \left[\left.\frac{\partial G_m^\alpha}{\partial x_B}\right|_{x_B^{\alpha,e}} + \left.\frac{\partial^2 G_m^\alpha}{\partial x_B^2}\right|_{x_B^{\alpha,e}} (x_B^i - x_B^{\alpha,e})\right] (x_B^i - x_B^{\beta,e}) \tag{4.58}$$

Thus,

$$\Delta G_m = G_m^\alpha(x_B^{\alpha,e}) - G_m^\beta\left(x_B^{\beta,e}\right) + \left.\frac{\partial G_m^\alpha}{\partial x_B}\right|_{x_B=x_B^{\alpha,e}} \left(x_B^{\beta,e} - x_B^{\alpha,e}\right) - \left.\frac{\partial^2 G_m^\alpha}{\partial x_B^2}\right|_{x_B=x_B^{\alpha,e}} \left(x_B^i - x_B^{\alpha,e}\right)\left(x_B^i - x_B^{\beta,e}\right) \tag{4.59}$$

The first three terms sum to zero, since by the common tangent construction,

[2] The derivation made in the preceding section was based on the energy per atom. Here we use energy per mol. Both can be converted into each other using instead of Boltzmann's constant the universal gas constant and instead of the atomic volume the molar volume.

$$\frac{G_m^\alpha\left(x_B^{\alpha,e}\right)-G_m^\beta\left(x_B^{\beta,e}\right)}{x_B^{\alpha,e}-x_B^{\beta,e}}=\left.\frac{\partial G_m^\alpha}{\partial x_B}\right|_{x_B=x_B^{\alpha,e}} \tag{4.60}$$

Equation 4.60 can be derived by drawing a common tangent line between two Gibbs free energy curves. Now, considering just the last term,

$$(x_B^i-x_B^{\alpha,e})(x_B^i-x_B^{\beta,e})=(x_B^i-x_B^{\alpha,e})(x_B^i-x_B^{\alpha,e}+x_B^{\alpha,e}-x_B^{\beta,e}) \tag{4.61}$$

Thus, to first order in $(x_B^i-x_B^{\alpha,e})$ eq.(4.61) becomes,

$$\left(x_B^i-x_B^{\alpha,e}\right)\left(x_B^i-x_B^{\beta,e}\right)=\left(x_B^i-x_B^{\alpha,e}\right)\left(x_B^{\alpha,e}-x_B^{\beta,e}\right) \tag{4.62}$$

Using eq.(4.60) and eq.(4.62) in eq.(4.59) yields the Gibbs free energy change driving interface migration to first order in $(x_B^i-x_B^{\alpha,e})$,

$$\Delta G_m=-\left.\frac{\partial^2 G_m^\alpha}{\partial x_B^2}\right|_{x_B=x_B^{\alpha,e}}\left(x_B^i-x_B^{\alpha,e}\right)\left(x_B^{\alpha,e}-x_B^{\beta,e}\right) \tag{4.63}$$

Thus, in the small supersaturation limit the change in Gibbs free energy is linearly proportional to the difference in composition at the interface from the equilibrium composition. The sign of the change in Gibbs free energy is clearly a function of the difference in composition between the precipitate and matrix phases as well as the value of the interfacial composition relative to the equilibrium matrix phase composition.

If the interface is moving very slowly on the time scale associated with diffusion then there is no composition gradient in the α phase, and thus the composition at the interface must be equal to that at infinity, and thus eq.(4.63) can be written as,

$$\Delta G_m=\zeta\left(x_B^0-x_B^{\alpha,e}\right) \tag{4.64}$$

where

$$\zeta=-\left.\frac{\partial^2 G_m^\alpha}{\partial x_B^2}\right|_{x_B=x_B^{\alpha,e}}\left(x_B^{\alpha,e}-x_B^{\beta,e}\right) \tag{4.65}$$

With this expression we obtain for the growth rate of a particle,

$$\frac{dR}{dt}=\mathrm{v}=k_s(T)(x_B^0-x_B^{\alpha,e}), \tag{4.66}$$

where,

$$k_s(T)=\Omega n_a \nu\exp\left(-\frac{G_A}{k_B T}\right)\frac{\zeta}{k_B T} \tag{4.67}$$

4.4.2 Interface kinetics in a supercooled melt

A similar situation as the one just treated for a supersaturated matrix can be found in growth from a supercooled melt, especially if facetted crystalline phases grow. Let us calculate the Gibbs free energy gain if the temperature of the melt is close to the melting point. The Gibbs free energy change can be expressed to first order around the melting point as (see section 2.2)

$$G_{sol.}(T) - G_{liq.}(T) = (H_{sol.}(T) - H_{liq.}(T)) - T(S_s(T) - S_l(T)) \approx L_m - T\frac{L_m}{T_m} \quad (4.68)$$

or

$$G_{sol.}(T) - G_{liq.}(T) \approx \frac{L_m}{T_m}(T_m - T). \quad (4.69)$$

We can now insert this into the general expression of eq.(4.52). Since the phase transformation is driven by a temperature difference rather than a concentration difference we just replace ΔG_{atom} with $(G_{solid}(T) - G_{liquid}(T))/N_A$ and obtain

$$\frac{dR}{dt} = \Omega n_a \nu \exp\left(-\frac{G_A}{k_B T}\right)\frac{1}{\mathcal{R}T}\frac{L_m}{T_m}(T_m - T) \quad (4.70)$$

where we have used the identity $\mathcal{R} = k_B N_A$. Identifying now T with the interface temperature T_I we obtain

$$\frac{dR}{dt} = \Omega n_a \nu \exp\left(-\frac{G_A}{k_B T}\right)\frac{1}{\mathcal{R}T}\frac{L}{T_m}(T_m - T_I) = k_T(T)(T_m - T_I) \quad (4.71)$$

with the abbreviation

$$k_T(T) = \Omega n_a \nu \exp\left(-\frac{G_A}{k_B T}\right)\frac{1}{\mathcal{R}T}\frac{L}{T_m} \quad (4.72)$$

Thus, also in solidification the particle would grow linearly with the interfacial supercooling. This result holds if, first, the supercooling is not too large and, second, if there is a kinetic impediment at the interface (leading to a deviation of the interface temperature from the equilibrium melting temperature) that can be taken into account by the energy barrier model described above.

4.5 Interface kinetics of second order

Interface kinetics of first order are a special example of attachment kinetics controlled growth of particles or crystals in a solution or melt or from the vapor phase. There are other possible modes of growth which are described by different growth laws. Especially for crystals with directed bonding like semiconductors or ionic crystal, but also for some metals like Cu, a non-linear relation between growth rate and supersaturation or supercooling is observed. This was discovered and described theoretically 50 years ago by Burton, Cabrera and Frank [50]. Their essential assumption was that no crystal grows with a perfect crystal lattice. All crystals have imperfections such as dislocations. A crystal plane growing is not atomically flat but has steps or ledges and distributed along these steps are kinks. The situation of a screw dislocation from which a spiral develops is shown schematically in Fig. 4.5. If the growing crystal interface would be atomically flat growth in a direction perpendicular to the interface would be hindered. Once a crystal surface has no edges left a new plane has to be started for instance by the formation of a small pill box. This is called a two dimensional nucleation and has some similarities to the nucleation of a particle in a supercooled melt – including an energy barrier. Growth would stop once the plane is completed. A new nucleation event would be necessary. It can be shown that this type of growth would need large supercoolings or supersaturation levels usually not observed in crystal growth. The reason why such crystals grow at smaller undercoolings or supersaturations is the existence of crystal defects like terraces, ledges, and kinks (see Fig. 4.5) and, especially, dislocations.

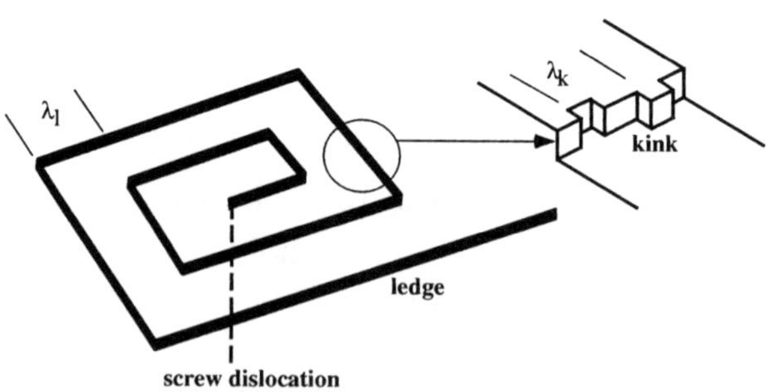

Fig. 4.5. Facet of a crystal with ledges of separation λ_l and kinks distributed at the ledges with a separation λ_k.

An example of screw dislocations intersecting a surface and inducing so-called growth spirals is shown in Fig. 4.6. A description of growth when crystal defects are present is rather complicated and must account for a series of possible mechanisms that result from different crystal imperfections. We will here only discuss an interface kinetics of second order due to a screw dislocation intersecting a crystal surface. Other growth laws are discussed by Tiller [269]. Let us consider

Fig. 4.6. Growth spirals on the surface of a SiC crystal.

a growing or dissolving crystal dispersed in a solid or liquid matrix as depicted in Fig. 4.5. The interface of the crystal is assumed to be intersected by screw dislocations. At each screw dislocation a growth spiral evolves which consists of ledges with a separation λ_l and along the ledges kinks are distributed with a separation λ_k. The expression for interface kinetics limited growth derived in Section 4.4 gives the growth velocity v at small driving forces ΔG,

$$\mathrm{v} = \Omega n_a \nu \exp\left(-\frac{G_A}{k_B T}\right) \frac{\Delta G_{atom}}{k_B T} \tag{4.73}$$

Now the number of atoms per interface area n_a has to be replaced by the density of ledges and kinks available for atoms to attach at the crystal surface and leading, thus, to growth. We generalize eq.(4.73) and write

$$\mathrm{v} = \Omega n_a^D \nu \exp\left(-\frac{G_A}{k_B T}\right) \frac{\Delta G_{atom}}{k_B T} \tag{4.74}$$

with n_a^D the number of defect sites available for attachment per interface area. This is not a constant and it can not be derived from crystallographic considerations but depends on the driving force itself. For a screw dislocation intersecting the surface the number of defects is given by

$$n_a^D = \frac{1}{\lambda_l \lambda_k}. \tag{4.75}$$

Whereas the density or separation of kinks depends on temperature (kinks are defects occurring along a ledge or step at an interface in thermal equilibrium. They are in a certain sense comparable to vacancies in the bulk crystal) the ledge distance depends on the supercooling or supersaturation expressed in the change in Gibbs free energy ΔG_{atom}. The theory of two-dimensional nucleation (see [269]) shows that the nose of the screw dislocation has a radius of curvature equal to the critical sized pill-box, i.e. $r_c = \sigma \Delta G_{atom}$, with σ the solid-liquid interfacial energy. The spacing between the layers far from the nose – assuming it to be represented by an Archimedean spiral – is given by

$$\lambda_l = 4\pi r_c = \frac{4\pi\sigma}{\Delta G_{atom}}. \tag{4.76}$$

Inserting this relation into eqs.(4.74,4.75) and collecting terms we obtain

$$\mathrm{v} = \frac{\Omega \nu \Delta G_{atom}}{4\pi\sigma\lambda_k} \exp\left(-\frac{G_A}{k_B T}\right) \frac{\Delta G_{atom}}{k_B T} = k^*(T)|\Delta G_{atom}|\Delta G_{atom}. \tag{4.77}$$

where we have introduced an absolute value in one of the ΔG_{atom} to illustrate that if the driving force is negative the crystal dissolves. Although this result was derived for the screw dislocation mechanism of particle growth, we will use this equation as a general form for the growth of a particle by a second-order reaction. Then $k^*(T)$ is an arbitrary constant. Consistent with this general form for the growth of a particle by a second-order reaction, we shall assume for simplicity that the particles are spherical. This approximation is appropriate, if the number of facets or screw dislocations intersecting the surface is large.

4.5.1 Growth rates in second order kinetics

The growth rate of a particle can now be written down using the results of the section on interface kinetics of first order. There we had derived for the driving force the relation

$$\Delta G_{atom} = \zeta(x_B^0 - x_B^I). \tag{4.78}$$

Inserting this into eq.(4.77) yields

$$\frac{dR}{dt} = k^S(T)|x_B^0 - x_B^I|(x_B^0 - x_B^I). \tag{4.79}$$

k^S is the rate constant for concentration driven growth. Thus, we have a simple result: Although the growth rate depends on the square of the supersaturation it is linear in time (this statement is valid as long as the concentration at the interface is constant). A very similar result can easily be obtained if the solution or melt is supercooled and the crystal grows by a second order reaction. Inserting the relation for the driving force given above leads to

$$\frac{dR}{dt} = k^T(T)|T_m - T_I|(T_m - T_I). \tag{4.80}$$

k^T is the rate constant for temperature driven growth We will later see that in thc case of coarsening the quadratic dependence on supersaturation or supercooling leads to a different time behavior than the linear dependence in the case of interface kinetics of first order.

4.6 Convective contributions to heat and mass transfer

In this section we will deal with the situation of both, convective and diffusive transport. This is e.g. the case if particles or droplets are dispersed in a liquid matrix. There are several origins of convective contributions to mass or heat transport. A dispersed phase with a different density than the matrix moves in the gravity field with Stokes settling speed. Temperature gradients in the liquid matrix inclined to the gravity vector will produce natural convection taking the dispersed phase with it. Fluid flow may also be induced artificially by stirring mechanically or electromagnetically. Convection may affect heat and mass transport and as a result the growth or dissolution of particles. A well known example of this is the dissolution of sugar grains in tea or coffee. If the tea is stirred with a spoon the sugar grains dissolve much faster than without stirring. The next section gives a general overview without deriving all the details. For those familiar with convective contributions to heat and mass transport this is a sufficient review within the context of this monograph. Thereafter we derive in detail the growth laws for the mass transport to moving spheres.

4.6.1 General growth laws at large Peclet numbers

The growth laws derived in the sections above for heat and mass transport could generally be written as

$$\frac{dR}{dt} \propto \left.\frac{\partial(x_B, T)}{\partial r}\right|_R \propto \frac{\Delta(x_B, T)}{R}. \tag{4.81}$$

where the notation (x_B, T) means, that either the temperature or the concentration field shall be taken and $\Delta(x_B, T)$ means either constant supercooling or constant supersaturation. Fluid flow enhances mass and heat transport. The general effect of fluid flow towards a spherical particle is shown schematically in Fig. 4.7. Considering the thermal or

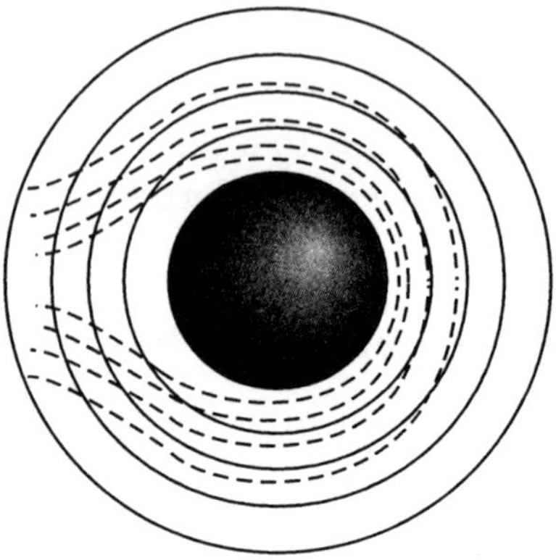

Fig. 4.7. Fluid flow changes and squeezes the isotherms and the isoconcentration surfaces. The flow is directed from the right side to the left side of the figure.

the (mass) diffusional Peclet number, Pe_T or Pe_D respectively, reveals much of the physics underlying the transport processes in the liquid: The Peclet numbers are the ratio of convective to diffusive heat and mass transport,

$$Pe_T = \frac{v_c R}{\kappa} \qquad Pe_D = \frac{v_c R}{D} \tag{4.82}$$

with v_c a characteristic velocity of the flow field. We have treated the case of low Peclet numbers (diffusional heat and mass transport) before. For large Peclet numbers convective heat and mass transport dominate (see chapter 3 and [22, 182, 193]) and one can show that the concentration or temperature gradient at the interface can be approximated by

$$\left.\frac{\partial(x_B, T)}{\partial r}\right|_R \propto \frac{\Delta(x_B, T)}{\delta}, \tag{4.83}$$

where δ is the so-called *boundary layer thickness.* Inside a sphere of extension δ mass and heat transport are governed by diffusion and convection. Outside this sphere convective flow dominates the transport and the temperature or concentration is a constant. Thus, the effect of convection is to squeeze the isoconcentration lines or isotherms closer to the interface and thus increase the gradient. In fluid dynamics it is shown that the boundary layer varies with Peclet number in the following way (we ignore the difference between heat and mass transport):

$$\delta_{s/l}/R \propto Pe^{-1/3} \quad \text{valid for solid/liquid interfaces} \tag{4.84}$$

$$\delta_{l/l}/R \propto Pe^{-1/2} \quad \text{valid for liquid/liquid interfaces} \tag{4.85}$$

The different dependence of the boundary layer thickness near solid/fluid and fluid/fluid interfaces is due to the difference in the fluid velocities at a solid/fluid interface. At a solid/fluid interface there is no slip at the interface, i.e. the normal and tangential fluid flow velocity components are zero, whereas at fluid-fluid interfaces the flow velocity vectors in both liquid phases are equal. Thus, heat and mass is transported more effectively to liquid/liquid interfaces than to solid/liquid interfaces.

Particles dispersed in a liquid matrix can move via a number of mechanisms. If the particle has a density different from the matrix it will move in a gravity field according to Stokes law,

$$v_{Stokes} \propto R^2, \tag{4.86}$$

where the proportionality factor includes the density difference, viscosities, and gravity acceleration (for details see appendix A). The relation holds for solid and liquid particles. A liquid droplet surrounded by a matrix in a temperature or concentration gradient will undergo Marangoni motion. The Marangoni velocity is proportional to the radius,

$$v_{Marangoni} \propto R, \tag{4.87}$$

The proportionality factor contains the viscosities, the temperature gradient, and the dependence of interfacial tension on temperature (for details see appendix A). Using the different Peclet numbers for Stokes- and Marangoni-motion we obtain the boundary layer thickness for the difference cases,

$$\delta^{sol.}_{Stokes} \propto R, \qquad \text{Stokes, solid particles} \tag{4.88}$$

$$\delta^{liq.}_{Stokes} \propto R^{-1/2}, \qquad \text{Stokes, liquid particles} \tag{4.89}$$

$$\delta_{Marangoni} \propto R \qquad \text{Marangoni, liquid particles.} \tag{4.90}$$

Inserting these relations into the expression of eq.(4.81) and using eq.(4.83) we obtain the growth laws,

$$\frac{dR}{dt} \propto \Delta(c,T), \qquad \text{Stokes, solid particles} \tag{4.91}$$

$$\frac{dR}{dt} \propto \Delta(c,T)R^{1/2}, \qquad \text{Stokes, liquid particles} \tag{4.92}$$

$$\frac{dR}{dt} \propto \Delta(c,T). \qquad \text{Marangoni, liquid particles} \tag{4.93}$$

Integrating these relations gives either linear growth of radius with time or quadratic growth (not parabolic as in the diffusion case). In each case the convective transport of heat and mass has changed growth from parabolic to linear or even quadratic. The proportionality constants are not written down here, but for two special cases this will be done in the next subsections.

4.6.2 Growth of a falling solid sphere

Let us consider in this section the mass transport to a solid particle moving with a constant velocity in a liquid. We can equally think of the particle as being at rest and the fluid moving around it. The fluid flow velocity at infinity shall have the finite value v_0. For large Peclet numbers the change in concentration occurs only in a thin boundary layer despite the fact that the fluid flow velocity gradually varies (see chapter 3). The convective diffusion eq.(3.62) for the boundary layer in spherical coordinates r, θ, ϕ becomes[3]

[3] We use here polar coordinates r, θ, ϕ. The flow problem is in principle a 3D problem, but due to the rotational symmetry around the polar axis its reduced to a problem in r, θ alone.

$$v_r \frac{\partial x_B}{\partial r} + \frac{v_\theta}{r}\frac{\partial x_B}{\partial \theta} = D\left(\frac{\partial^2 x_B}{\partial r^2} - \frac{2}{r}\frac{\partial x_B}{\partial r}\right). \tag{4.94}$$

In the Laplacian on the right term we have omitted all terms in the azimuthal angle ϕ because the problem is symmetrical around the polar axis. We have further omitted the term with a derivative with respect to θ, since the derivative along the sphere surface is small compared to the radial derivative (as was shown in chapter 3). The stream function for Stokes flow (see Appendix A) is given by

$$\Psi = -\frac{v_0}{2}\sin^2\theta\left(r^2 - \frac{3}{2}Rr + \frac{1}{2}\frac{R^3}{r}\right). \tag{4.95}$$

Since we are interested only in the flow field close to the particles surface we define a new coordinate $\tilde{r} = r - R$ measuring distance from the particles surface. Close to the surface, i.e. at small values of $\tilde{r}$ the stream function can be approximated as

$$\Psi \approx -\frac{3}{4}v_0\tilde{r}^2\sin^2\theta. \tag{4.96}$$

In order to get this relation, one inserts the expression $r = R + \tilde{r}$ into eq.(4.95) and expands the expression $R^3/2(R+\tilde{r})$ up to second order in $\tilde{r}/R$. This shows that expression (4.96) is correct to second order in $\tilde{r}$. As shown in the appendix A, the fluid flow velocity components can be expressed via the stream function as

$$v_r = -\frac{1}{r^2\sin\theta}\frac{\partial\Psi}{\partial\theta} \quad \text{and} \quad v_\theta = -\frac{1}{r\sin\theta}\frac{\partial\Psi}{\partial r} \tag{4.97}$$

Using the approximated stream function (4.96) the tangential velocity component is

$$v_\theta = -\frac{1}{r\sin\theta}\frac{\partial\Psi}{\partial\tilde{r}} \approx -\frac{1}{R\sin\theta}\frac{\partial\Psi}{\partial\tilde{r}} = \frac{3}{2}v_0\frac{\tilde{r}}{R}\sin\theta. \tag{4.98}$$

and the radial velocity component is:

$$v_r = -\frac{1}{r^2\sin\theta}\frac{\partial\Psi}{\partial\theta} \approx -\frac{1}{R^2\sin\theta}\frac{\partial\Psi}{\partial\theta} = \frac{3}{2}v_0\frac{\tilde{r}^2}{R^2}\cos\theta. \tag{4.99}$$

In contrast to the θ-component of the velocity the radial one depends quadratically on $\tilde{r}/R$ and since $\tilde{r} \ll R$ we neglect it compared to the term linear in $\tilde{r}/R$. Close to the surface of the sphere, we have for the derivatives of the concentration the relation

$$\frac{\partial^2 x_B}{\partial \tilde{r}^2} \gg \frac{2}{R}\frac{\partial x_B}{\partial \tilde{r}}, \tag{4.100}$$

which can be understood in the same way as done for cartesian coordinates in chapter 3, namely, approximate $\partial^2 \Delta x_B/\partial \tilde{r}^2 \approx \Delta x_B/\delta^2$ and the first derivative with $\Delta x_B/\delta$, with Δx_B some arbitrary concentration difference between for example the particles surface and the far field concentration value. Since the boundary layer thickness δ shall be a small quantity, the second order derivative is by an order of $1/\delta$ larger than the first order one. We now can proceed to write eq.(4.94) in new coordinates, so-called stream function coordinates, meaning we will transform from (r, θ) to (Ψ, θ) coordinates. Let us first collect the results obtained so far with the estimates and approximations used. Eq.(4.94) reads now in the vicinity of the particles surface:

$$\frac{v_\theta}{R}\frac{\partial x_B}{\partial \theta} = D\frac{\partial^2 x_B}{\partial \tilde{r}^2}. \tag{4.101}$$

Transforming from $\tilde{r}$ to Ψ means replacing the partial derivatives by:

$$\frac{\partial}{\partial \tilde{r}} = \frac{\partial}{\partial \Psi}\frac{\partial \Psi}{\partial \tilde{r}} = R\sin\theta v_\theta \frac{\partial}{\partial \Psi} \tag{4.102}$$

and

$$\frac{\partial^2}{\partial \tilde{r}^2} = \frac{\partial \Psi}{\partial \tilde{r}}\frac{\partial}{\partial \Psi}\left(\frac{\partial \Psi}{\partial \tilde{r}}\frac{\partial}{\partial \Psi}\right) = R\sin\theta v_\theta \frac{\partial}{\partial \Psi}\left(R\sin\theta v_\theta \frac{\partial}{\partial \Psi}\right) \tag{4.103}$$

Inserting these transformations into eq.(4.101) leads to

$$\frac{1}{R}\frac{\partial x_B}{\partial \theta} = DR\sin^2\theta \frac{\partial}{\partial \Psi}\left(Rv_\theta \frac{\partial x_B}{\partial \Psi}\right). \tag{4.104}$$

Eq.(4.98) can be rewritten using eq.(4.96):

$$Rv_\theta = \sqrt{-3v_0\Psi} \tag{4.105}$$

leading to a new equation for the concentration field close to the particles interface:

$$\frac{\partial x_B}{\partial \theta} = DR^2\sin^2\theta\sqrt{3v_0}\frac{\partial}{\partial \Psi}\left(\sqrt{-\Psi}\frac{\partial x_B}{\partial \Psi}\right) \tag{4.106}$$

The boundary conditions are

$$x_B = 0 \;\text{ at } \Psi = 0 \quad \text{on the surface of the sphere}(\tilde{r} = 0), \quad (4.107)$$
$$x_B = x_B^0 \text{ at } \Psi \to -\infty \qquad \text{far from the sphere}, \quad (4.108)$$
$$x_B = x_B^0 \text{ at } \theta = 0, \quad \Psi = 0. \quad (4.109)$$

The last condition means that the point $\theta = 0, \Psi = 0$ represents the point of incidence of the fluid flow on the sphere. At this point the concentration is equal to that in the bulk liquid. In order to integrate eq.(4.106) we introduce a new variable,

$$\tilde{\theta} = DR^2\sqrt{3v_0}\int \sin^2\theta d\theta = \frac{DR^2\sqrt{3v_0}}{2}\left(\theta - \frac{\sin 2\theta}{2}\right) + a_0. \quad (4.110)$$

This coordinate transformation $\theta \to \tilde{\theta}$ changes eq.(4.106) to

$$\frac{\partial x_B}{\partial \tilde{\theta}} = \frac{\partial}{\partial \Psi}\left(\sqrt{-\Psi}\frac{\partial x_B}{\partial \Psi}\right). \quad (4.111)$$

This equation looks very similar to a diffusion equation with a variable diffusion coefficient. In most cases diffusion scales with distance over square root of time. We try here another scaling: $\tilde{\Psi} = -\Psi/\tilde{\theta}^{2/3}$ and assume a solution of the type $x_B = x_B(\tilde{\Psi})$. Insertion this into eq.(4.111) and evaluating all derivatives leads to an ordinary differential equation for x_B dependent on the single variable $\tilde{\Psi}$,

$$\frac{2}{3}\tilde{\Psi}\frac{dx_B}{d\tilde{\Psi}} = \frac{d}{d\tilde{\Psi}}\left(\sqrt{\tilde{\Psi}}\frac{dx_B}{d\tilde{\Psi}}\right). \quad (4.112)$$

Introducing again a new variable $\breve{\Psi} = \sqrt{\tilde{\Psi}}$ we obtain a simple ordinary differential equation whose solution can be found in standard textbooks [322],

$$\frac{d^2x_B}{d\breve{\Psi}^2} + \frac{4}{3}\breve{\Psi}^2\frac{dx_B}{d\breve{\Psi}} = 0. \quad (4.113)$$

The solution reads

$$x_B(\breve{\Psi}) = a_1\int_0^{\breve{\Psi}} \exp(-4/9\breve{\Psi}^3)d\breve{\Psi} + a_2. \quad (4.114)$$

If we insert all transformation used so far, we can express $\breve{\Psi}$ in the original coordinates as

$$\breve{\Psi} = \sqrt{\frac{3v_0}{4}} \frac{\tilde{r}\sin\theta}{DR^2\sqrt{\frac{3v_0}{4}\left(\theta - \frac{\sin 2\theta}{2}\right) - a_0}}. \qquad (4.115)$$

The constants are determined from the boundary conditions. We will not go through all the calculations for the integration constants but instead give the final result,

$$\breve{\Psi} = \sqrt[3]{\frac{3v_0}{4DR^2}} \frac{\tilde{r}\sin\theta}{\left(\theta - \frac{\sin 2\theta}{2}\right)^{1/3}}, \qquad (4.116)$$

$$x_B = \frac{x_B^0 - x_B^I}{(9/4)^{1/3} 1/3\Gamma(1/3)} \int_0^{\breve{\Psi}} \exp(-4/9\breve{\Psi}^3) d\breve{\Psi}. \qquad (4.117)$$

x_B^I is the concentration at the interface. Γ is the so-called Gamma-function and can be found in mathematical reference books [4]. The flux density at the surface can be calculated from

$$j(\theta) = D \left.\frac{\partial x_B}{\partial \tilde{r}}\right|_{\tilde{r}=0}. \qquad (4.118)$$

The total mass flux to the sphere is obtained from

$$I = \int_{surface} j da = 2\pi R^2 \int_0^{\pi} j \sin\theta d\theta \qquad (4.119)$$

Performing the integration using eqs.(4.116, 4.117, 4.118) and approximating the Γ-function with a two digit numerical value in eq.(4.117) we obtain the final result,

$$I = 7.98(x_B^0 - x_B^I) D^{2/3} v_0^{1/3} R^{4/3} \qquad (4.120)$$

Here x_B^0 is the concentration far away from the interface and x_B^I the concentration at the interface. We do not write this in terms of the diffusional Peclet number, since the drop velocity could depend on radius too and that could change the dependence on radius and thus Peclet num ber quite significantly. It is interesting to note that the flux is not proportional to the diffusion coefficient but to a power of D. Also the velocity enters with a power of 1/3 and the particle radius enters also to a fractional power. This is in contrast to the equations derived for pure diffusion or interface kinetics.

The concentration gradient, however, is a function of θ, see eq.(4.118). Thus, the particle would not remain spherically symmetric during growth. We shall assume that this distortion is small and thus employ eq.(4.119) at all time.

Inserting into eq.(4.120) the expression for the settling speed of a solid sphere in a gravity field eq.(A.58) and using the relation for the growth rate,

$$(x_B^\beta - x_B^\alpha)\frac{dV}{dt} = I, \tag{4.121}$$

we obtain for the growth rate the expression,

$$\frac{dR}{dt} = 0.385\frac{x_B^0 - x_B^I}{x_B^\beta - x_B^\alpha} D^{2/3}\frac{\rho_p - \rho}{\eta} g \tag{4.122}$$

The growth rate does not depend on the radius and thus is formally identical to an interface reaction of first order. If the supersaturation would be time independent the particle size would change linearly with time.

4.6.3 Growth of a falling liquid sphere

We will solve the stationary convective diffusion equation eq.(3.49) for a liquid particle moving with constant velocity v_0 in a liquid matrix (for example with the Stokes settling speed) in the boundary layer approximation given in eq.(3.62). The fluid flow field $\boldsymbol{v}(r,\theta)$ can also be described by a stream function (for the derivation of the stream function see appendix A and also notice the footnote 3 on page 80),

$$\Psi(r,\theta) = -\frac{1}{2}v_0 r^2 \sin^2\theta \left[1 - \frac{1}{2}\frac{2+3k_\eta}{1+k_\eta}\frac{R}{r} + \frac{1}{2}\frac{k_\eta}{1+k_\eta}\left(\frac{R}{r}\right)^3\right], \tag{4.123}$$

where k_η is the ratio of the viscosities $k_\eta = \eta_{particle}/\eta_{matrix}$. The radial and azimuthal components of the velocity field are defined by

$$v_r = -\frac{1}{r^2\sin\theta}\frac{\partial\Psi}{\partial\theta} \quad \text{and} \quad v_\theta = -\frac{1}{r\sin\theta}\frac{\partial\Psi}{\partial r}. \tag{4.124}$$

In order to solve the convective diffusion equation we change the coordinate system from (r, θ, x_B) to a stream function system $(r \to \Psi(r,\theta)$, θ, $x_B)$. Performing the transformation in eq.(3.49) leads after some lengthy manipulations to

$$D\left(\frac{\partial^2 x_B}{\partial\Psi^2}\left(\frac{\partial\Psi}{\partial r}\right)^2 + \frac{\partial x_B}{\partial\Psi}\frac{\partial^2\Psi}{\partial r^2} + \frac{2}{r}\frac{\partial x_B}{\partial\Psi}\frac{\partial\Psi}{\partial r}\right) \tag{4.125}$$

$$-\frac{\partial x_B}{\partial\Psi}\frac{\partial\Psi}{\partial r}v_r + \frac{v_\theta}{r}\frac{\partial x_B}{\partial\theta} + \frac{v_\theta}{r}\frac{\partial x_B}{\partial\Psi}\frac{\partial\Psi}{\partial\theta} = 0. \tag{4.126}$$

Although this equation looks more frightening than the original one it is easier to solve if we make some suitable approximations. We assume that the concentration field far away from the interface is constant (large Peclet number, i.e. $\nabla x_B = 0$, therefore $x_B(r \gg R) = \text{const.}$, see chapter 3). Therefore, we expand the stream function in terms of distance from the particle interface introducing the new coordinate $\tilde{r} = r - R$. Let us rewrite the stream function in a slightly different way as

$$\Psi(r,\theta) = -\frac{1}{2} v_0 R^2 \sin^2\theta \left[\frac{r^2}{R^2} - \frac{1}{2} a \frac{r}{R} + \frac{1}{2} b \frac{R}{r} \right] \tag{4.127}$$

with $a = (2 + 3k_\eta)/(1 + k_\eta)$ and $b = k_\eta/(1 + k_\eta)$. We evaluate the expression in the square brackets using $r = R + \tilde{r}$

$$[\ldots] = (\frac{\tilde{r}}{R} + 1)^2 + \frac{1}{2} a (\frac{\tilde{r}}{R} + 1) + \frac{1}{2} b \frac{R}{\tilde{r} + R}. \tag{4.128}$$

Expanding the last expression up to second order and collecting terms of equal order in $\tilde{r}/R$ leads to:

$$[\ldots] = \frac{1}{1 + k_\eta} \frac{\tilde{r}}{R} + \frac{1 + 3/2k_\eta}{1 + k_{eta}} \left(\frac{\tilde{r}}{R} \right)^2 \tag{4.129}$$

If the droplet would have an extremely high viscosity, let us say it would become solid and we could approximate a solid as a liquid with infinite viscosity, then the first term would vanish, since $k_\eta \to \infty$. The second term is exactly the expression we used in the preceding section for a falling solid sphere. In the case under consideration here we can, however, neglect the second order expansion since the first order is non-zero. Thus, the stream function close to the interface can be approximated by

$$\Psi \approx -\frac{1}{2(1 + k_\eta)} v_0 R \tilde{r} \sin^2\theta. \tag{4.130}$$

Calculating the derivatives of the stream function with respect to $\tilde{r}$ and θ[4] eq.(4.126) reduces to

$$D \left(\frac{\partial^2 x_B}{\partial \Psi^2} \left(\frac{\partial \Psi}{\partial \tilde{r}} \right)^2 + \frac{2}{R} \frac{\partial x_B}{\partial \Psi} \frac{\partial \Psi}{\partial \tilde{r}} \right) = \frac{v_\theta}{r} \frac{\partial x_B}{\partial \theta}. \tag{4.131}$$

[4] Notice: $\partial/\partial r = \partial/\partial \tilde{r}$.

This can further be simplified. Close to the interface we neglect the second term, i.e. we neglect the first derivative of x_B with respect to Ψ as compared to the term with the second derivative using the same arguments as in the preceding section. We then approximate the term on the right side as

$$\frac{v_\theta}{r}\frac{\partial x_B}{\partial \theta} \approx \frac{v_\theta}{R}\frac{\partial x_B}{\partial \theta} \tag{4.132}$$

and further write the velocity as

$$v_\theta \approx \frac{1}{R\sin\theta}\frac{\partial \Psi}{\partial \tilde{r}} = \frac{v_0 \sin\theta}{2(1+k_\eta)} \tag{4.133}$$

leading to the continuity equation

$$\frac{\partial x_B}{\partial \theta} = \frac{D v_0 R^3 \sin^3\theta}{2(1+k_\eta)}\frac{\partial^2 x_B}{\partial \Psi^2}. \tag{4.134}$$

This looks formally like a classical diffusion equation. θ plays the role of a time and Ψ would be a spatial variable. The prefactor on the right side would be a θ dependent diffusion constant. To remove this dependence we write the equation in a slightly different way,

$$\frac{1}{\sin^3\theta}\frac{\partial x_B}{\partial \theta} = \frac{D v_0 R^3}{2(1+k_\eta)}\frac{\partial^2 x_B}{\partial \Psi^2} = D^*\frac{\partial^2 x_B}{\partial \Psi^2}. \tag{4.135}$$

This leads us to introduce a new variable $\breve{\theta}$ such that we have

$$\frac{1}{\sin^3\theta}\frac{\partial x_B}{\partial \breve{\theta}}\frac{\partial \breve{\theta}}{\partial \theta} = \frac{\partial x_B}{\partial \breve{\theta}}. \tag{4.136}$$

This relation can be fulfilled if we introduce $\breve{\theta}$ by

$$\frac{d\breve{\theta}}{d\theta} = \sin^3\theta \tag{4.137}$$

or

$$\breve{\theta} = \int \sin^3\theta d\theta = \frac{\cos^3\theta}{3} - \cos\theta. \tag{4.138}$$

In the new variable the continuity equation reads now

$$\frac{\partial x_B}{\partial \breve{\theta}} = D^*\frac{\partial^2 x_B}{\partial \Psi^2} \tag{4.139}$$

with the boundary conditions

$$x_B = x_B^0 \quad \text{for} \quad -\Psi \to \infty \qquad \text{and} \qquad x_B = x_B^\alpha \quad \text{for} \quad \Psi \to 0. \tag{4.140}$$

Here x_B^0 is the concentration far away from the particle and x_B^α is the surface concentration assuming local equilibrium. The initial condition is

$$x_B(\theta = 0, \Psi) = x_B^0 \tag{4.141}$$

The solution of this equation can be found in standard textbooks on heat transfer like Carlsaw and Jaeger ([55], page 29),

$$x_B(\Psi, \breve{\theta}) = (x_B^0 - x_B^\alpha)\mathrm{erf}\left(\frac{-\Psi}{2\sqrt{D^*\breve{\theta}}}\right) + x_B^\alpha. \tag{4.142}$$

Inserting for Ψ the approximate expression of the stream function (eq.(4.130)) and transforming back to $\tilde{r}$ and θ using the expression of eq.(4.138) yields the final solution of the problem,

$$x_B(\tilde{r}, \theta) = x_B^\alpha + (x_B^0 - x_B^\alpha)\mathrm{erf}\left(\frac{-v_0 R \sin^2\theta \tilde{r}}{(1+k_\eta} \sqrt{D^* \frac{\cos^3\theta}{3} - \cos\theta}\right) \tag{4.143}$$

The solute current density at the particle interface, $\tilde{r} = 0$, is obtained by taking the derivative and multiplication with the diffusion coefficient,

$$j = -D \frac{\partial x_B}{\partial r}\bigg|_{\tilde{r}=0} = (x_B^0 - x_B^\alpha)\frac{2}{\pi} \frac{D v_0 R \sin^2\theta}{(1+k_\eta)\sqrt{\frac{D v_0 R^3}{2} \frac{\cos^3\theta}{3} - \cos\theta}} \tag{4.144}$$

Integrating this expression over θ from 0 to π along the sphere surface gives the total flux of solute through the surface

$$J = 4\pi R^2 \int_0^\pi j \sin\theta d\theta. \tag{4.145}$$

The result[5] is

$$J = 4\pi(x_B^0 - x_B^\alpha)\frac{8}{\sqrt{3\pi}(1+k_\eta)} D^{1/2} v_0^{1/2} R^{3/2}. \tag{4.146}$$

Using the mass balance at the interface given above written here in a slightly different manner,

[5] In order to perform the integration over the angle θ use the transformation given above to convert θ into $\breve{\theta}$. Then, a very simple integral is obtained.

$$(x_B^\beta - x_B^\alpha)\frac{dV}{dt} = J, \tag{4.147}$$

we can calculate the growth rate as

$$\frac{dR}{dt} = \frac{x_B^0 - x_B^\alpha}{x_B^\beta - x_B^\alpha}\frac{8}{\sqrt{3\pi}} D^{1/2} v_0^{1/2} R^{-1/2}. \tag{4.148}$$

It is interesting to note several features of this equation. The dependence on the diffusion coefficient is again non-linear, but here with a different power. The dependence on the radius would lead on integration at constant supersaturation to a growth of $R^{3/2} \propto t$, i.e. strong than linear. Both effects are due to the different boundary condition of the fluid flow field at the liquid-liquid interface. Note also that this solution does not converge to the solution given above for a solid particle, ($k_\eta \to \infty$), since then we must use a higher order approximation of the stream function inside the diffusion boundary layer.

Inserting into eq.(4.148) the expression for the settling speed of a droplet in the gravity field eq.(A.90) leads to the expression,

$$\frac{dR}{dt} = \frac{8}{3}\sqrt{\frac{2}{\pi}} \frac{x_B^0 - x_B^I}{x_B^\beta - x_B^\alpha} D^{1/2} \sqrt{\frac{(\eta + \eta_p)(\rho_p - \rho)g}{\eta(2\eta + 3\eta_p)}} R^{1/2} \tag{4.149}$$

The growth rate depends on the square root of the particle radius. If the supersaturation would be time independent the particle size would grow with time like $R^{3/2} \propto t$.

Inserting into eq.(4.148) the expression for the Marangoni velocity of a drop in a temperature field with a constant temperature gradient dT/dx, eq.(A.94), leads to the expression,

$$\frac{dR}{dt} = \frac{8}{3}\sqrt{\frac{2}{\pi}} \frac{x_B^0 - x_B^I}{x_B^\beta - x_B^\alpha} D^{1/2} \sqrt{\frac{dT/dx(d\sigma/dT)(\eta + \eta_p)}{\eta(2\eta + 3\eta_p)(1 + \lambda_p/\lambda)}} \tag{4.150}$$

The growth rate does not depend on the particle radius. If the supersaturation would be time independent the particle size would grow linearly with time similar to a first order interface reaction.

4.6.4 Growth at arbitrary Peclet numbers

A complete calculation of the problem of convective and diffusive transport of mass to particles valid for arbitrary Peclet numbers does not need to introduce the concept of a boundary layer but rather solves the continuity equation for solute and the Navier-Stokes equation simultaneously. This has been done by Brian and Hales for solid particles moving in a fluid with an arbitrary but constant velocity at low Reynolds numbers. The result of their numerical integration of the coupled problem of convective growth (or dissolution) and the Navier Stokes equation can be expressed by simple fit functions and is shown in Fig. 4.8.

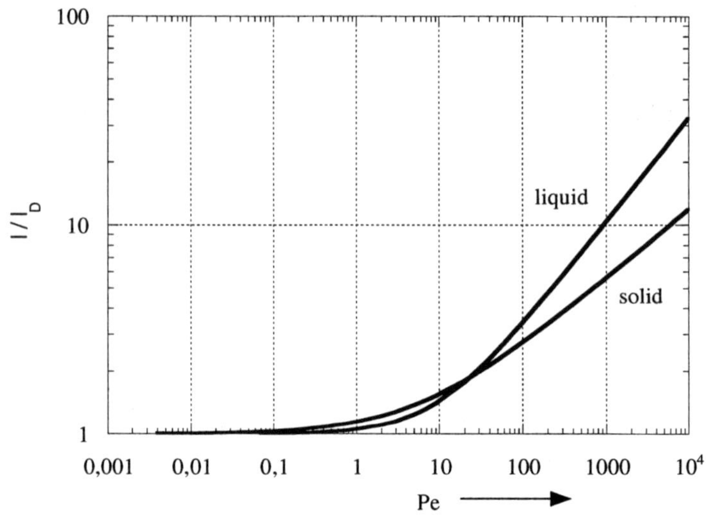

Fig. 4.8. Solute current I= 'number of atoms entering the particle surface per unit of time' as a function of Peclet number for liquid and solid particles. For liquid particles the viscosity in the particle $\eta_{particle}$ is assumed to be equal to that of the matrix η_{matrix}.

The same method can be applied for liquid particles and yields also a simple fit function, shown in Fig. 4.8 [241]. The transport currents can be cast into the expressions

$$I_{solid} = I_D \left(1 + \frac{1.21}{4} Pe^{\frac{2}{3}}\right)^{\frac{1}{2}}, \tag{4.151}$$

$$I_{liquid} = I_D \left(1 + \frac{2}{3\pi} \frac{\eta_{matrix}}{\eta_{particle} + \eta_{matrix}} Pe\right)^{\frac{1}{2}}, \tag{4.152}$$

where I_D is the transport current due to diffusion. The differential growth laws for sedimentation due to gravity can then be written as

$$\frac{d\tilde{R}}{d\tau} = \frac{1}{\tilde{R}}(1 + a_0\tilde{R}^2)^{\frac{1}{2}}, \qquad \text{solid particles} \qquad (4.153)$$

$$\frac{d\tilde{R}}{d\tau} = \frac{1}{\tilde{R}}(1 + a_1\tilde{R}^3)^{\frac{1}{2}}. \qquad \text{droplets} \qquad (4.154)$$

with $\tilde{R} = R/R_0$ a dimensionless radius, $\tau = DS_0/R_0^2 t$ a dimensionless time, and the constants a_0 and a_1 are related to the mean initial Peclet number in a dispersion $a_0 \sim Pe_0^{2/3}$ and $a_1 \sim Pe_0$. For liquid

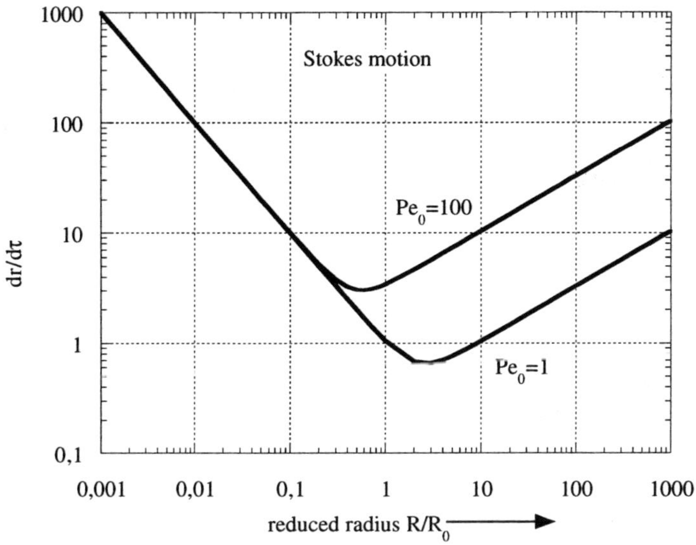

Fig. 4.9. Rate of droplet growth in convective diffusion as a function of dimensionless radius for different initial Peclet numbers assuming that the drops move with Stokes motion.

particles moving in a temperature gradient, as described by Young *et al.* [314], the radius dependence of the growth law is similar to that for solid particles; the constant, however, changes containing then the viscosities (for details of the derivation of (4.151),(4.152) see [241]). The growth law for liquid particles is shown in Fig. 4.9. The important result sketched here without any detailed derivation is simply that once the growth law under both, nonconvective (diffusive) conditions and that for very large Peclet numbers, is known, then the intermediate range of Peclet numbers can be obtained by simply adding both solute currents in a Pythagorean way. Thus, a combination of the two limiting solutions is sufficient to describe the whole range of Peclet numbers.

4.7 Growth with varying supersaturation

In the preceding sections we have always set the supersaturation or supercooling as a constant, time independent value and, therefore, could solve the differential equations by simple integration. In metallic alloys it is nearly impossible to keep such a condition for an extended time. In solution growth one can easily install such conditions if the solvent (matrix) has a sufficient vapor pressure like water. One would simply evaporate continuously solvent liquid at a proper rate and could keep the supersaturation level at a constant value. There are, however, numerous situations in which this is not possible. In those cases the supersaturation changes with time until it finally is zero and growth is finished (usually coarsening becomes important at low levels of supersaturation and supercooling). One can solve the problem of time dependent supersaturation and, thus, extend the treatment presented in section 4.3.

Consider a sample with an alloy of composition x_B^0 that is suddenly moved from a single phase field region. The equilibrium concentrations of both phases in which such a material would decompose shall be x_B^α (as the matrix phase) and x_B^β for the particle phase. The volume fraction Φ_{final} of β-phase after complete decomposition would then be

$$\Phi_{final} = \frac{x_B^0 - x_B^\alpha}{x_B^\beta - x_B^\alpha}. \qquad (4.155)$$

The instantaneous change from the single into the two phase field will lead to a material having the initial composition x_B^0 everywhere. After completion of decomposition into two phases, α and β, this composition has decreased to the equilibrium composition of the matrix x_B^α. The average matrix composition thus changes with time from x_B^0 to x_B^α, i.e. $< x_B(t) >= x_B^0$ at $t = 0$ and $< x_B(t) >= x_B^\alpha$ at $t \to \infty$. The time dependent volume fraction can be written as

$$\Phi(t) = \frac{< x_B(t) > - x_B^\alpha}{x_B^\beta - x_B^\alpha}. \qquad (4.156)$$

If the particles have all the same size the volume fraction can be expressed, however, by the radius of the particles and their number density n_V,

$$\Phi(t) = \frac{4\pi}{3} R^3(t) n_V. \qquad (4.157)$$

Here one should note that during growth the number density of particles is constant (see chapter 5). Eqs.(4.156) and (4.157) allow to express the average matrix concentration through the particle radius, leading to

$$< x_B(t) >= x_B^0 - x_B^\alpha - \frac{4\pi}{3} R(t)^3 n_V (x_B^\beta - x_B^\alpha). \tag{4.158}$$

The equation for the growth rate given above (eq.(4.47)) can be rewritten as

$$\frac{dR}{dt} = D \frac{< x_B(t) > - x_B^\alpha}{x_B^\beta - x_B^\alpha} \frac{1}{R} \tag{4.159}$$

where x_B^0 has been replaced by $< x_B(t) >$. Then,

$$\begin{aligned} \frac{dR}{dt} &= \frac{D}{R} \frac{x_B^0 - x_B^\alpha - \frac{4\pi}{3} R(t)^3 n_V (x_B^\beta - x_B^\alpha) - x_B^\alpha}{x_B^\beta - x_B^\alpha} \\ &= \frac{D}{R} \frac{x_B^0 - x_B^\alpha}{x_B^\beta - x_B^\alpha} - \frac{4\pi}{3} R(t)^2 n_V D. \end{aligned} \tag{4.160}$$

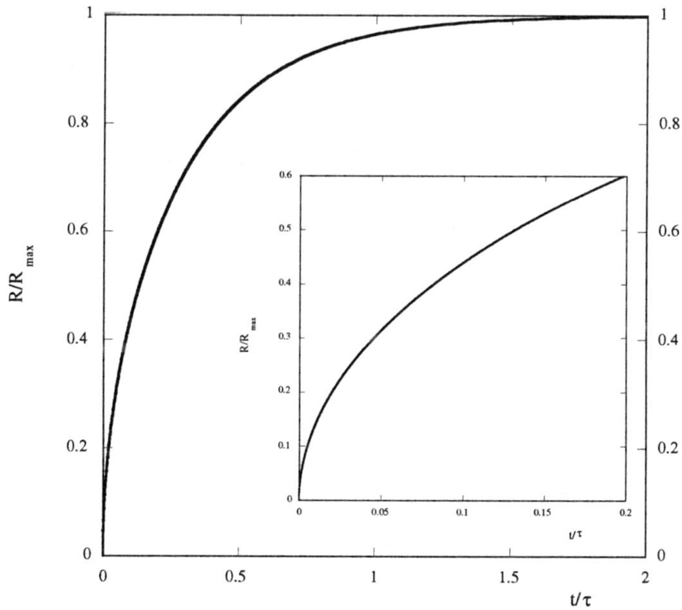

Fig. 4.10. Growth of a sphere according to a numerical solution of eq.(4.169).

Denoting

$$S_0 = \frac{x_B^0 - x_B^\alpha}{x_B^\beta - x_B^\alpha} \tag{4.161}$$

and

$$\tilde{D} = \frac{4\pi}{3} n_V D \tag{4.162}$$

we can write this equation as

$$\frac{dR}{dt} = \frac{DS_0}{R} - \tilde{D}R^2. \tag{4.163}$$

This ordinary differential equation for the radius can be solved by separation of variables,

$$\frac{RdR}{DS_0 - \tilde{D}R^3} = dt, \tag{4.164}$$

and integrating from zero to R,

$$\int_0^R \frac{RdR}{DS_0 - \tilde{D}R^3} = t. \tag{4.165}$$

Instead of performing the integration directly let us do some manipulations and then introduce a new variable. The integral can be written as

$$\int_0^R \frac{RdR}{DS_0 - \tilde{D}R^3} = \frac{1}{DS_0} \int_0^R \frac{RdR}{1 - \frac{\tilde{D}}{DS_0} R^3} \tag{4.166}$$

Using $\tilde{D}/(DS_0) = 4\pi n_V/(3S_0)$ and

$$S_0 = \frac{4\pi}{3} n_V R_{max}^3 \tag{4.167}$$

we obtain

$$\int_0^R \frac{RdR}{DS_0 - \tilde{D}R^3} = \frac{R_{max}^2}{DS_0} \int_0^R \frac{RdR}{1 - \frac{R^3}{R_{max}^3}}. \tag{4.168}$$

Defining $\hat{r} = R/R_{max}$ and $\tau = R_{max}^2/(DS_0)$ and using eq.(4.168) we can write eq.(4.165) as

$$\int_0^{\hat{r}} \frac{\hat{r}d\hat{r}}{1 - \hat{r}^3} = \frac{t}{\tau}. \tag{4.169}$$

The integration can be performed easily, yielding

$$\frac{1}{6}\ln\left(\frac{1+\hat{r}+\hat{r}^2}{(1+\hat{r})^2}\right)-\frac{1}{\sqrt{3}}\left(\arctan\frac{2\hat{r}+1}{\sqrt{3}}+\arctan\frac{1}{\sqrt{3}}\right)=\frac{t}{\tau}. \quad (4.170)$$

The result is shown graphically in figure 4.10. The central result of this diagram is that the radius saturates to a maximum level. The radius increases with time parabolically for $\hat{r} \ll 1$ and then increases at a slower rate to reach, in the end, the maximum radius which is set by the initial supersaturation and the number density of particles.

5. Statistics of growth

In this chapter we will study the evolution of a number of spherical particles described by a particle size density distribution. We assume that the particles do not interact, meaning for example if they grow by heat or mass diffusion at constant supercooling or supersaturation the diffusion fields do never overlap. The particles never touch each other and grow without any limit. Thus, the description of a particle ensemble does not take into account for example mass conservation applied to the entire sample. This simplification allows us to treat the statistics in a simple and analytical way. However, the situation described is not purely theoretical: E.g. when cooling a metal one can have for a limited period of time following nucleation just growth of these particles - very much as described here. As the particles grow the matrix concentration decreases and competitive growth – coarsening – sets in. This situation will be treated in chapters 7,8. The general statistical concepts presented below are useful in these chapters.

5.1 Continuity equation and moments

In the following sections we investigate the time evolution of a dispersion of spheres that change their size due to the different growth laws derived in the previous chapter. The particle size density distribution, $f(R,t)dR$, obeys a continuity equation in the size space,

$$\frac{\partial f}{\partial t} + \frac{\partial}{\partial R}\left(\frac{dR}{dt}f\right) = 0. \tag{5.1}$$

This equation can be understood as follows: dR/dt could be viewed as the velocity with which a particle moves in radius space. $f(R,t)$ is a number density per radius interval. Therefore, the flux density in radius space is just the product of both: *flux density* $= dR/dt \cdot f(R,t)$. If no sources for particles exist, then there is only a motion from one radius class $R, R + dR$ to another. Therefore, a continuity equation holds, i.e. the divergence of the flux density equals the time derivative of the density in the interval under consideration. The interval here is the radius class, and the divergence the partial derivative of the flux density with respect to radius. This consideration leads to eq.(5.1).

We introduce dimensionless coordinates,

$$\rho = \frac{R}{r_c}, \tag{5.2}$$

$$\tau = \frac{K_j}{r_c^m} \cdot t, \tag{5.3}$$

in which r_c is a suitable reference length (e.g. the initial mean radius or the capillary length) and K_j is one of the rate constants used in growth laws derived in the preceding sections and m is the exponent in the differential growth law (e.g. if $dR/dt \propto 1/R$, then $m = 1$). Then, all the growth laws can be written as

$$\frac{d\rho}{d\tau} = \frac{1}{\rho^m}. \tag{5.4}$$

with m either -1/2, 0 or 1. As a reminder $m = -1/2$ is the exponent of the growth law, when droplets are moving in a fluid with Stokes settling velocity. $m = 0$ is the case of either an interface limited growth situation (first or second order) or a solid particle moving with Stokes velocity in a matrix or a droplet moving with Marangoni velocity. The $m = 1$ case is achieved when the growth is controlled by either diffusion of heat or mass. Transforming the continuity equation to these dimensionless coordinates leads to

$$\frac{\partial f}{\partial \tau} + \frac{\partial}{\partial \rho}\left(\frac{d\rho}{d\tau} f\right) = 0. \tag{5.5}$$

The moments are defined as usual, but some have special meanings,

$$\frac{1}{\rho^*} = \frac{1}{n_V}\int_0^\infty \frac{1}{\rho} f(\rho,\tau)d\rho \qquad \text{harmonic mean,} \tag{5.6}$$

$$n_V(\tau) = \int_0^\infty f(\rho,\tau)d\rho \qquad \text{number density,} \tag{5.7}$$

$$\langle\rho\rangle = \frac{1}{n_V}\int_0^\infty \rho f(\rho,\tau)d\rho \qquad \text{average dimensionless radius,} \tag{5.8}$$

$$\Phi(\tau) = \frac{4\pi}{3} r_c^3 \int_0^\infty \rho^3 f(\rho,\tau)d\rho \qquad \text{particle volume fraction.} \tag{5.9}$$

Depending on the functional characteristics of $dR/dt = \psi(R,t)$ the fraction solid or the number density can be time independent. Generally both are not conserved quantities. This will be discussed later in context with special growth laws.

5.2 Statistics at constant growth rate

The continuity eq.(5.1) has a simple solution if the growth rate is constant,

$$dR/dt = V_0 = const. \tag{5.10}$$

The solution is simply given by a shift in radius space that is linear in time,

$$f(R,t) = f_0(R - V_0 t), \tag{5.11}$$

where f_0 is an arbitrary function with the property $f(R, t = 0) = f_0(R)$. All growth laws derived above for interface kinetics without curvature and for convective heat or mass transport of particles moving with Stokes velocity or drops moving with Marangoni velocity fall into this class. This means a given size distribution shifts in radius space simply to larger values (see Fig. 5.1). Its spread, its standard deviation, remains unchanged only the average radius increases linearly with time and the volume or solid fraction increases proportional to the 3rd power of time.

5.3 Superposition of growth modes

If a growth behavior treated in the preceding section would be superimposed by any other growth law of the type $dR_{add.}/dt = \psi(R,t)$, i.e.

$$\frac{dR}{dt} = V_0 + \psi(R,t). \tag{5.12}$$

The continuity eq.(5.1) can always be transformed to a moving frame. Define

$$\hat{R}(t) = R(t) - V_0 t. \tag{5.13}$$

We then have

$$d\hat{R}/dt = dR/dt - V_0 = \psi(R,t) \tag{5.14}$$

and, thus, the continuity equation looks like that given in eq.(5.5) with $d\hat{R}/dt = \psi(\hat{R},t)$. This leaves us with the problem to solve the continuity equation for cases in which $m \neq 0$ in eq.(5.4).

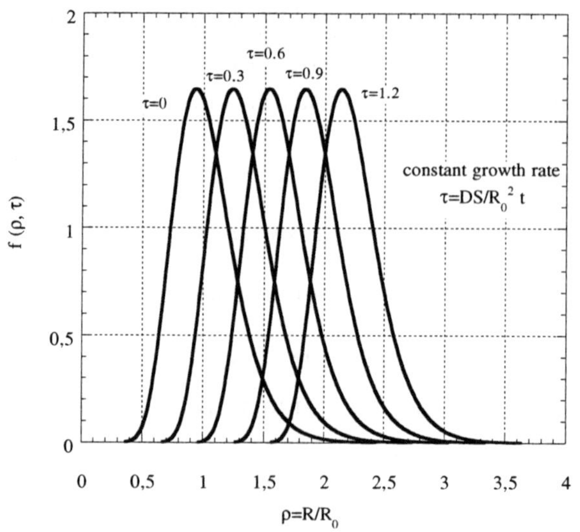

Fig. 5.1. Size distribution as a function of time eq.(5.11). The simple shift shown is observed for growth due to interface reactions of first or second order, Marangoni motion of drops in a constant temperature gradient or convective diffusion to solid particle moving with Stokes settling velocity. R_0 denotes an arbitrary radius to normalize the radius coordinate, e.g. it could be the initial mean radius.

5.4 Statistic of growth at constant supercooling or supersaturation

Let us first discuss the evolution of the size distribution if the growth is fixed by a constant supercooling or supersaturation. Then the growth rate is given by eq.(5.4) with $m = 1$,

$$\frac{d\rho}{d\tau} = \frac{1}{\rho} = \psi(\rho). \tag{5.15}$$

Generally partial differential equations of first order like eq.(5.5) are solved by the so-called method of characteristics.

Before we proceed some comments shall be made. The continuity equation has different kinds of solutions. One type of solution can sometimes be obtained by scaling the radius properly with time, i.e. introducing a scaling variable instead of the two variables ρ and τ. This will be done in section 5.4.2 below. We also solve the continuity equation using a separation ansatz, i.e. splitting the size distribution $f(\rho, \tau)$ into a product like $f(\rho, \tau) = g(\tau)h(\rho)$. Inserting this into eq.(5.5) would yield a solution. It will be treated in the section 5.5. Both techniques can be found in the literature for various physical problems, especially in aerosol mechanics and cloud physics [83, 246].

5.4.1 Initial value problem - Cauchy problem

In the following we solve instead the so-called Cauchy problem, i.e. we look for a solution of the continuity equation that coincides at an arbitrary time τ_0 with a given or preset initial function. Thus, the Cauchy problem is similar to the initial value problem of ordinary first order differential equations. It was proven by Cauchy and Kowalewski that partial differential equations like the continuity equation possess a unique solution provided that the functions and their derivatives are analytical.

Let us rewrite the continuity equation as

$$0 = \frac{\partial f}{\partial \tau} + \frac{\partial}{\partial \rho}\left(\frac{d\rho}{d\tau} f\right) = \frac{\partial f}{\partial \tau} + \frac{\partial \psi}{\partial \rho} f + \psi \frac{\partial f}{\partial \rho}. \qquad (5.16)$$

With $d\rho/d\tau = \psi(\rho, \tau) = 1/\rho$ this yields

$$\frac{\partial f}{\partial \tau} + \frac{1}{\rho}\frac{\partial f}{\partial \rho} = \frac{1}{\rho^2} f. \qquad (5.17)$$

The characteristic equations are given by

$$d\tau = \frac{d\rho}{\frac{1}{\rho}} = \frac{df}{\frac{1}{\rho^2} f}. \qquad (5.18)$$

The solution of the first equation reads

$$d\tau = \rho d\rho, \qquad (5.19)$$

$$\tau - \tau_0 = \frac{1}{2}\rho^2 - \frac{1}{2}\rho_0^2, \qquad (5.20)$$

$$\tau_0 = \tau + \frac{1}{2}\rho_0^2 - \frac{1}{2}\rho^2, \qquad (5.21)$$

$$\rho_0^2 = \rho^2 + 2\tau_0 - 2\tau, \qquad (5.22)$$

$$\rho^2 = \rho_0^2 + 2\tau - 2\tau_0. \qquad (5.23)$$

The second equation has the solution

$$\frac{df}{d\tau} = \frac{1}{\rho^2} f, \tag{5.24}$$

$$= \frac{f}{\rho_0^2 + 2(\tau - \tau_0)}, \tag{5.25}$$

$$\frac{df}{f} = \frac{d\tau}{\rho_0^2 + 2(\tau - \tau_0)} \tag{5.26}$$

with $\rho_0 = \rho(\tau_0)$. This equation can be integrated readily to give

$$\ln \frac{f}{f_0} = \int_{\tau_0}^{\tau} \frac{d'\tau}{\rho_0^2 + 2('\tau - \tau_0)} \tag{5.27}$$

$$= \frac{1}{2} \ln \left(\rho_0^2 + 2(\tau - \tau_0) \right) \Big|_{\tau_0}^{\tau} \tag{5.28}$$

$$= \frac{1}{2} \ln \frac{\rho_0^2 + 2(\tau - \tau_0)}{\rho_0^2}, \tag{5.29}$$

$$\rightarrow \quad f = f_0 \sqrt{\frac{\rho_0^2 + 2(\tau - \tau_0)}{\rho_0^2}}. \tag{5.30}$$

Inserting the relations we derived above from the solution of the first characteristic equation yields

$$f(\rho, \tau) = f(\rho_0, \tau_0) \frac{\rho}{\sqrt{\rho^2 + 2(\tau_0 - \tau)}}. \tag{5.31}$$

Setting $\tau_0 = 0$ we obtain

$$f(\rho, \tau) = f(\rho_0) \frac{\rho}{\sqrt{\rho^2 - 2\tau}} \tag{5.32}$$

and replacing ρ_0 with the result from eq.(5.22) we obtain the final result,

$$f(\rho, \tau) = \frac{\rho}{\sqrt{\rho^2 - 2\tau}} f_0(\sqrt{\rho^2 - 2\tau}) \tag{5.33}$$

Another way to solve the continuity equation is to define a new function as

$$\zeta(\rho) = - \int_0^{\rho} \frac{d\rho'}{\psi(\rho')}. \tag{5.34}$$

and define its inverse via the relation

$$\zeta^{-1}(\zeta(\rho)) = \rho. \tag{5.35}$$

Evaluating eqs.(5.34,5.35) we obtain

$$\zeta(\rho) = -\frac{\rho^2}{2} \qquad \text{and} \qquad \zeta^{-1}(x) = \sqrt{-2x}. \tag{5.36}$$

Then, the solution of the Cauchy problem of eq.(5.5) reads

$$f(\rho, \tau) = \frac{\psi(\zeta^{-1}(\zeta(\rho) + \tau))}{\psi(\rho)} f_0(\zeta^{-1}(\zeta(\rho) + \tau)) \tag{5.37}$$

with $f_0(\rho, 0)$ denotes an arbitrary initial size distribution. After inserting eqs.(5.36) we have

$$f(\rho, \tau) = \frac{\rho}{\sqrt{\rho^2 - 2\tau}} f_0(\sqrt{\rho^2 - 2\tau}) \tag{5.38}$$

Both results state that after a certain time τ the minimum particle size is given by $\rho_{min.} = \sqrt{2\tau}$. By evaluating eq.(5.7)) it can be checked that this size distribution conserves the number of particles. If one normalizes the initial size density distribution, the evolving distributions remain normalized. Let us check the other moments. We generally obtain the result

$$M_m(\tau) = \int_{\rho_{min.}}^{\infty} \rho^m f(\rho, \tau) d\rho = \int_0^{\infty} (\chi^2 + 2\tau)^{m/2} f_0(\chi) d\chi. \tag{5.39}$$

with $\chi = \sqrt{\rho^2 - 2\tau}$. Unfortunately the odd moments cannot be evaluated further without specifying the initial size distribution. The even moments can be calculated. For example the second moment reads

$$M_2 = M_2^0 + 2\tau \cdot n_V. \tag{5.40}$$

where the index "0" denotes the respective moment of the initial size distribution and n_V is the number density of particles (being a constant). Thus, the second moment increases linearly with time. For the odd moments the relation

$$M_m = \frac{1}{(m+2)} \frac{dM_{m+2}}{d\tau} \tag{5.41}$$

holds. It can be used to relate the average radius with the solid fraction,

$$\langle \rho \rangle = \frac{M_1}{n_V} = \frac{1}{3n_V} \frac{d\Phi}{d\tau}. \tag{5.42}$$

To illustrate the results we choose as a special example the initial distribution,

$$f_0(\rho) = \rho^2 \exp\left(-\frac{(\rho - \langle\rho\rangle_0)^2}{2\sigma^2}\right). \tag{5.43}$$

We choose $\langle\rho\rangle_0 = 0.5$ initially, and for the standard deviation of the size distribution a value of $\sigma = 0.25$. Then, the evolution of the size distribution is given by eq.(5.38),

$$f(\rho,\tau) = \rho\sqrt{\rho^2 - 2\tau}\exp\left(-\frac{(\sqrt{\rho^2 - 2\tau} - \langle\rho\rangle_0)^2}{2\sigma^2}\right). \tag{5.44}$$

The time evolution is shown in fig.5.2 and the evolution of the average radius in fig.5.3. Note that although the individual particle grows ac-

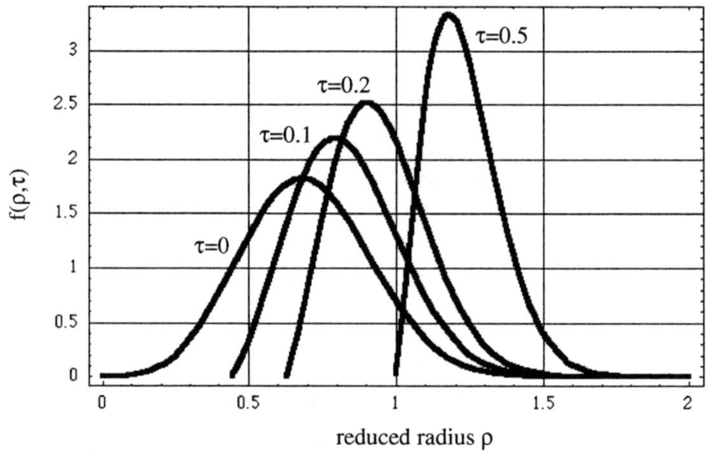

Fig. 5.2. Size distribution as a function of time eq.(5.44).

cording to a simple power law, the average radius initially grows only slowly with time and only later follows the power law. This initial transient depends much on the shape of the initial size distribution.

A third way to solve the continuity equation is simply to use a symbolic computations program like *Mathematica*™, *Maple*™or *Macsyma*™. These programs easily solve the continuity equation in a few seconds and less. Here is a small program for *Mathematica*™:

$$\frac{\partial f(t,r)}{\partial t} = f^{(1,0)}[t,r] \tag{5.45}$$

Here $f^{(1,0)}[t,r]$ is the Mathematica notation for the partial derivative

$$\frac{\partial f(t,r)}{\partial r} = f^{(0,1)}[t,r] \tag{5.46}$$

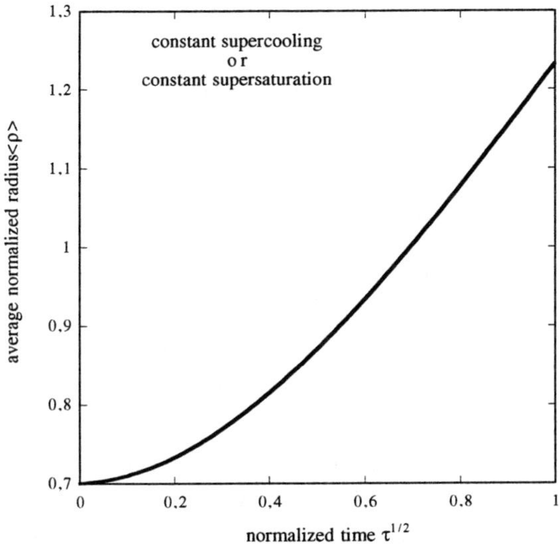

Fig. 5.3. The average radius grows proportional to the square root of time after an initial transient time.

$$In[1] := \text{DSolve}[f^{(1,0)}[t,r] == \frac{f[t,r]}{r^2} - \frac{f^{(0,)]}[t,r]}{r}, f[t,r], \{t,r\}] \quad (5.47)$$

$$Out[1] = \{\{f[t,r] \rightarrow \frac{\sqrt{r^2} C[1][-\frac{r^2}{2}+t]}{\sqrt{2}}\}\} \quad (5.48)$$

Thus, the general solution of the continuity equation is obtained writing a program with one line. Some comments might be necessary: First, $C[1][-\frac{r^2}{2}+t]$ denotes an arbitrary function of the argument given in the square bracket. Second, instead of writing ρ or τ we simply wrote r, t. Third, This solution is the general solution and not the solution of the initial value problem. That can be obtained specifying the function $C[1]$ using

$$f(r,0) = f_0(r) = \frac{r}{\sqrt{2}} C[1][-\frac{r^2}{2}]. \quad (5.49)$$

We obtain

$$C[1][-\frac{r^2}{2}] = \frac{\sqrt{2}}{r} f_0(r). \quad (5.50)$$

Defining a temporary variable $\delta = -\frac{r^2}{2}$ or $r = \sqrt{-2\delta}$ we obtain

$$C[1][\delta] = \frac{\sqrt{2}}{\sqrt{-2\delta}} f_0(\sqrt{-2\delta}) \quad (5.51)$$

and thus

$$C[1][-\frac{r^2}{2} + t] = \frac{\sqrt{2}}{\sqrt{r^2 - 2t}} f_0(\sqrt{r^2 - 2t}). \tag{5.52}$$

This leads to the result we already derived above.

5.4.2 Scaling solutions

The solution of the continuity equation can be a difficult task especially if the growth law of an individual particle is not a simple power law. It might also be possible that the Cauchy problem has no simple analytical solution (which often is due to the fact that the inverse function defined in eq.(5.35) is not unique). Then, so-called scaling solutions are often considered, since they allow insight into the behavior of the solution to be obtained under special conditions, for instance at short or long time scales. Interestingly enough, the continuity equation, eq.(5.5), also has scaling or similarity solutions in the case for power law growth. These can generally be obtained from the rules of the so-called one parameter group scalings. For this problem two different scaling solutions exist. We write them down here in order to illustrate the manner in which the technique can be applied in such cases.

Scaling with square root of time. A first scaling solution can be obtained defining

$$\xi = \frac{\rho}{\sqrt{\tau}} \tag{5.53}$$

and assuming that the size distribution can be written as

$$f = \bar{f}(\xi). \tag{5.54}$$

insertion of these expressions into the continuity equation, eq.(5.5), yields after some manipulations, similar to those used to arrive at eq.(4.13), an ordinary differential equation for $\bar{f}$,

$$\frac{d\bar{f}}{d\xi}\left(\frac{1}{\xi} - \frac{\xi}{2}\right) = \frac{1}{\xi^2}\bar{f} \tag{5.55}$$

This equation has two solutions,

$$\bar{f}(\xi) = \bar{f}_N \frac{\xi}{\sqrt{2 - \xi^2}} \qquad \text{for} \quad \xi < \sqrt{2}, \tag{5.56}$$

$$\bar{f}(\xi) = \bar{f}_N \frac{\xi}{\sqrt{\xi^2 - 2}} \qquad \text{for} \quad \xi > \sqrt{2}. \tag{5.57}$$

with f_N a normalization constant[1]. The first solution is shown in fig.5.4. and the second solution is shown in fig.5.5.

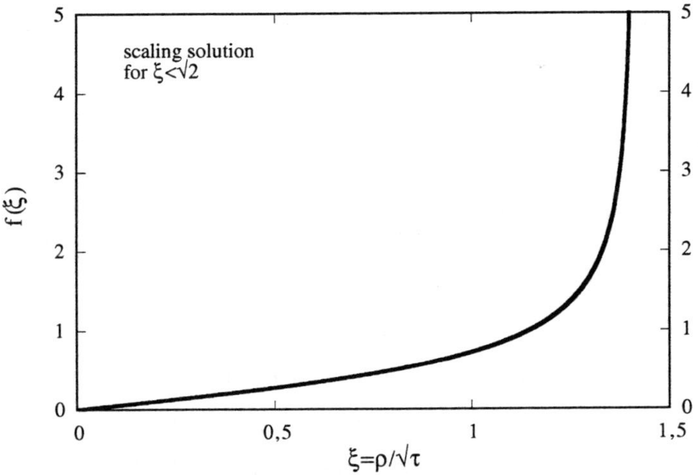

Fig. 5.4. Normalized scaled size distribution of eq.(5.56).

Discussion of the first scaling solution. Let us discuss the first scaling solution. We first determine the normalization constant by simply integrating the size distribution. The constant is determined to be $f_N = \sqrt{2}$. All other moments can also be calculated directly,

$$M_m = \frac{1}{\sqrt{2}} \int_0^{\sqrt{2}} \frac{\xi\, \xi^m}{\sqrt{2-\xi^2}} d\xi = 2^{m/2} \frac{\sqrt{\pi}}{2} \frac{n}{n+1} \frac{\Gamma(m/2)}{\Gamma((m+1)/2)}, \quad (5.58)$$

with an arbitrary number m and $\Gamma(z)$ the Gamma-function. The relation is valid for all $m \geq 1$. We obtain for instance

$$\langle \rho \rangle = \frac{\pi}{4}\sqrt{2\tau}, \quad (5.59)$$

$$\langle \rho^2 \rangle \approx \frac{4}{3}\tau, \quad (5.60)$$

$$\langle \rho^3 \rangle \approx \frac{3\pi}{16}(2\tau)^{3/2}. \quad (5.61)$$

Thus, with this scaling the time dependence of the moments is set immediately and in full agreement with the expectation from the growth

[1] The second solution can be obtained by rewriting eq.(5.55) as $-\xi/2(\xi - \sqrt{2})(\xi + \sqrt{2})df/d\xi = \bar{f}$ and solving this differential equation.

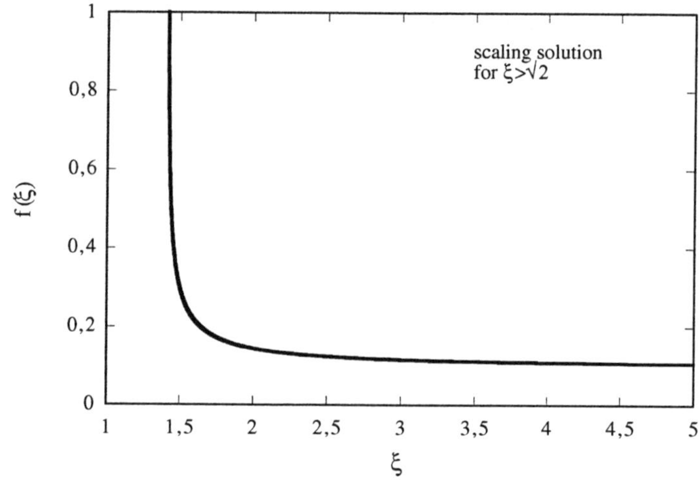

Fig. 5.5. Normalized scaled size distribution of eq.(5.57).

law itself. Although everything seems to be in agreement with the expectations, this scaling solution predicts that there are no particles larger than $\xi = \sqrt{2}$ and there always exist particles having smaller radii. This is in contrast to the differential growth law, that states that the small particles grow faster than the bigger ones and, thus, after long times their size classes are empty. It also is in contrast to the full solution of the Cauchy problem given above, which shows that after a certain time τ there are no particles left in the system having a small radius. Even normalizing the radii with their time dependent average the distribution always is skewed to the right, but never to the left. The scaling with the average size is, besides a constant factor, identical to the scaling used here. We, therefore, are lead to the conclusion that the scaling solution of eq.(5.56) though mathematically correct, is not physically meaningful.

Discussion of the second scaling solution. The second scaling solution has first the remarkable property: it looks like the prefactor of the full Cauchy solution (compare eq.(5.38) the factor in front of the initial size distribution and use the transformation $\xi = \rho/\sqrt{\tau}$). Thus, the second scaling solution has the appeal as an asymptotic distribution. It poses, however, a serious problem:

$$n_{V,0} = \int_{\sqrt{2}}^{\infty} \frac{\xi d\xi}{\sqrt{\xi^2 - 2}} = \frac{1}{2} \int_0^{\infty} \frac{d\chi}{\sqrt{\chi}} = \frac{1}{2} \sqrt{\chi}|_0^{\infty} \tag{5.62}$$

diverges at infinity (here we introduced $\chi = \xi^2 - 2$) and, thus, the number density is not constant and finite. This problem can only be avoided if there is a maximum reduced particle size, which shall be denoted as ξ_{max}. Note that the real maximum increases with time as $\rho_{max} \propto \sqrt{\tau}$. We then have

$$n_{V,0} = \frac{1}{2}(\xi_{max}^2 - 2)^{1/2}. \tag{5.63}$$

Thus, the number density in the system defines the maximum particle size uniquely. With this definition the scaled size distribution is

$$\bar{f}(\xi) = \frac{2}{\sqrt{\xi_{max}^2 - 2}} \frac{\xi}{\sqrt{\xi^2 - 2}} \qquad \text{for} \quad \xi > \sqrt{2}. \tag{5.64}$$

This is the distribution already plotted in Fig. 5.5. Using such a cut-off radius all higher moments of the size distribution are also finite. Therefore, the average radius grows again proportional to time and the volume fraction proportional to a power of 3/2 of time. This size distribution seems physically correct since it reflects somehow the long time behavior of the initial value problem. It only remains unsatisfactory that the size distribution does not tend to zero at large scaled radii.

5.4.3 Scaling with the inverse square root of time

In the second scaling we define a scaled coordinate as: $\xi = \sqrt{\tau}/\rho$ and assume that the size distribution has the form

$$f(\rho, \tau) = \tilde{f}(\xi) \tag{5.65}$$

Inserting this expression for f into the continuity equation and performing the derivatives leads to the ordinary differential equation for $\tilde{f}$,

$$\frac{d\tilde{f}}{d\xi}\left(\frac{1}{2} - \xi^2\right) = \xi \tilde{f}. \tag{5.66}$$

This equation has the general solution

$$\tilde{f}(\xi) = \tilde{f}_0 \frac{1}{\sqrt{|1 - 2\xi^2|}} \tag{5.67}$$

with $\tilde{f}_0$ being a normalization constant. Whereas the integral converges for $\xi < 1/\sqrt{2}$ to

$$\tilde{f}_0 = \int_0^{1/\sqrt{2}} \sqrt{|1-2\xi^2|} d\xi = \frac{\pi}{2\sqrt{2}} \tag{5.68}$$

it diverges for $\xi \to \infty$,

$$\tilde{f}_0 = \int_{1/\sqrt{2}}^{\xi_m} \sqrt{|1-2\xi^2|} d\xi = \frac{\pi}{2\sqrt{2}} + \ln\left(4\xi_m^2 + 2\xi_m\sqrt{4\xi_m^2 - 2}\right) \tag{5.69}$$

Transforming this solution into ρ, τ coordinates, we obtain

$$f(\rho,\tau) = \frac{\rho}{\sqrt{|\rho^2 - 2\tau|}}. \tag{5.70}$$

This result looks exactly like the prefactor obtained solving the Cauchy problem and is identical to the second scaling solution discussed in the preceding paragraph. Therefore, all comments to that solution apply also to this, meaning, we have to introduce a final cut-off radius ρ_{max} etc. The behavior of $f(\xi)/\tilde{f}_0$ is shown in Fig. 5.6

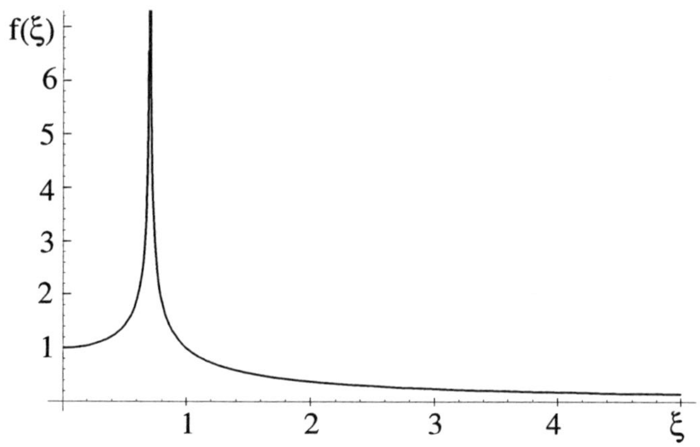

Fig. 5.6. Normalized scaled size distribution of eq.(5.70). The sharp peak is at $\xi = 1/\sqrt{2}$.

5.4.4 Comments on scaling

The essential point of these subsections on scaling was to show that different type of solutions exist and that especially with scaling solutions care must be taken in using these. It has to be mentioned that the validity of a scaling solution depends on a fact not mentioned so far. Although scaling may be done easily and the resulting equation solved as done above, the solution obtained is only a solution to a given physical problem if the boundary or initial conditions can also be scaled properly. In the discussions above we neglected this point, simply, because the initial value problem stated in section 5.4.1 cannot be scaled with the scalings used here. It can be shown, using the techniques of the one parameter scaling group theory, that always a relation like $f(\rho,0) = \tilde{f}(\xi \to \infty)$ or $f(\rho,0) = \tilde{f}(\xi = 0)$ is valid leading to the problem that the scaling solution cannot be adopted for an arbitrary initial function. The initial value problem can, however, be solved with these scaling techniques, if a proper initial function is chosen. The results presented suggest that this function has the properties: $f_{initial}(\rho) = 1 \quad \text{for} \quad \rho < \rho_{max}$. Since the function is unity, it is obviously scalable.

5.5 Separation ansatz solution for growth

The simplest way to solve the continuity equation would be to use a so-called separation ansatz as often done with linear partial differential equations. This technique is commonly used in coarsening studies (see chapter 7). We will therefore see if such an approach leads to valid results in the case of power law growth. Let us rewrite the continuity eq.(5.5),

$$\frac{\partial f}{\partial \tau} + \frac{\partial}{\partial \rho}\left(\frac{1}{\rho} f\right) = 0, \tag{5.71}$$

where we have simply used for the growth rate that for diffusional heat or mass transport in reduced coordinates, i.e. $d\rho/d\tau = 1/\rho$. We apply the separation ansatz to this equation, meaning we write a solution as

$$f(\rho,\tau) = g(\tau)h(\rho). \tag{5.72}$$

We then obtain

$$h\frac{dg}{d\tau} + g\frac{d}{d\rho}\left(\frac{1}{\rho}h\right) = 0, \tag{5.73}$$

$$-\frac{1}{g}\frac{dg}{d\tau} = \frac{1}{h}\frac{d}{d\rho}\left(\frac{1}{\rho}h\right). \tag{5.74}$$

Setting both side equal to a separation constant, denoted as $-\lambda$, we obtain solutions for both functions,

$$g(\tau) = \exp(\lambda\tau), \tag{5.75}$$
$$h(\rho) = \rho\exp(-\lambda\rho^2/2). \tag{5.76}$$

Thus, the distribution function is then

$$f(\rho,\tau;\lambda) = f_0\exp(\lambda\tau)\rho\exp(-\lambda\rho^2/2) \tag{5.77}$$

with f_0 an arbitrary constant of integration. The function depends on the arbitrary separation constant λ. This can be dealt with using the linearity of the continuity equation, which means, if f_1 and f_2 are solutions of eq.(5.71), then their sum is also a solution. We, therefore, integrate $f_\lambda(\rho,\tau)$ over λ,

$$f(\rho,\tau) = \int_0^\infty f(\rho,\tau;\lambda)d\lambda = f_0\frac{2\rho}{\rho^2-2\tau}. \tag{5.78}$$

The integration requires that $\rho^2 - 2\tau > 0$ otherwise the integral diverges. The solution fulfills the continuity equation exactly and, thus, is a solution. The solution has, however, a disadvantage: it does not conserve the number density in the system or in other words, the constant f_0 cannot be fixed as a normalization constant. This follows from

$$n_V = 2f_0\int_{\rho^2=2\tau}^\infty \frac{\rho d\rho}{\rho^2-2\tau} \to \infty. \tag{5.79}$$

The integral only exists if the upper limit is not ∞, but set to a finite maximum radius ρ_m, a cut-off radius. We then obtain a time dependent number density,

$$n_V(\tau) = 2f_0\int_{\rho^2=2\tau}^{\rho_m} \frac{\rho d\rho}{\rho^2-2\tau} = \ln\left(\frac{\rho_m^2-2\tau}{2\tau(2\tau-1)}\right). \tag{5.80}$$

Thus the normalization constant f_0 would be time dependent, in contradiction to its meaning and in contradiction to the physics of growth in which the number density of particles is fixed and independent of time. Thus, a separation ansatz does not lead to a valid solution.

5.6 General solution for power law growth

Equations (5.4,5.5) can be solved generally. Inserting eq.(5.4) into eq.(5.34) yields

$$\zeta(\rho) = -\frac{1}{m+1}\rho^{m+1} \tag{5.81}$$

and for the inverse we obtain

$$\zeta^{-1}(x) = [-(m+1)x]^{1/(m+1)}. \tag{5.82}$$

Inserting these relations and eq.(5.4) into eq.(5.37) we obtain for the evolution of the size distribution,

$$f(\rho,\tau) = \frac{\rho^m}{(\rho^{m+1} - (m+1)\tau)^{1/(m+1)}} f_0((\rho^{m+1} - (m+1)\tau)^{1/(m+1)}). \tag{5.83}$$

Thus, the initial distribution $f_0(x)$ and the growth exponent m in eq.(5.4) define completely its evolution. The general solution has some interesting properties. First, it states that after a time $\tau > 0$ there will be a minimum particle size $\rho_{min} = ((m+1)\tau)^{1/(m+1)}$. Secondly, one could speculate about physically meaningful size distributions. A suitable condition for a meaningful physical distribution would be that $f(\rho_{min}, \tau) = 0$. One could even make a stronger statement that $\partial f/\partial \rho = 0$ at $\rho = \rho_{min}$. The first condition means that a size distribution does not have a jump discontinuity from zero to nonzero values just at $\rho = \rho_{min}$. Such a discontinuity might be questioned physically. The second condition would furthermore require that the distribution function goes smoothly to zero as it approaches ρ_{min} from values larger than ρ_{min}. Interestingly enough, a Gaussian distribution does not fulfill any of these conditions since it is defined on the interval $]-\infty, \infty[$ and thus allows for negative radii, which are physically meaningless. There are other standard distributions fulfilling both conditions, a log-normal distribution and Maxwellian distributions.

Let us finally prove that all types of power law growth keep the particle number or the zeroth moment of the distribution constant. To prove this we integrate the continuity eq.(5.5),

$$\int_{\rho_{min}}^{\infty} \frac{\partial f}{\partial \tau} d\rho + \int_{\rho_{min}}^{\infty} \frac{\partial}{\partial \rho}\left(\frac{1}{\rho^m} f\right) d\rho =$$
$$\frac{dn_V}{d\tau} + \frac{d\rho_{min}}{dt} f(\rho_{min}, \tau) + \frac{1}{\rho^m} f \Big|_{\rho_{min}}^{\infty} = 0. \tag{5.84}$$

Evaluating the terms yields

$$\frac{dn_V}{d\tau} = \lim_{\rho \to \rho_{min}} \frac{f(\rho,\tau)}{\rho^m} + \frac{d\rho_{min}}{dt} f(\rho_{min},\tau) \tag{5.85}$$

Using the general result of eq.(5.83) and the expression for the minimum particle size $\rho_{min} = ((m+1)\tau)^{1/(m+1)}$ we obtain

$$\begin{aligned} \frac{dn_V}{d\tau} &= 2 \lim_{\rho \to \rho_{min}} \frac{1}{\sqrt[m+1]{\rho^{m+1} - (m+1)\tau}} f_0\left(\sqrt[m+1]{\rho^{m+1} - (m+1)\cdot\tau}\right) \\ &= 0 \end{aligned} \tag{5.86}$$

The last equality holds if we require that the size distribution tends to zero faster than linearly in ρ. Then number density always is a conserved quantity for any type of power law growth independent of the value of the growth exponent m.

5.7 Convective diffusion to dispersed drops

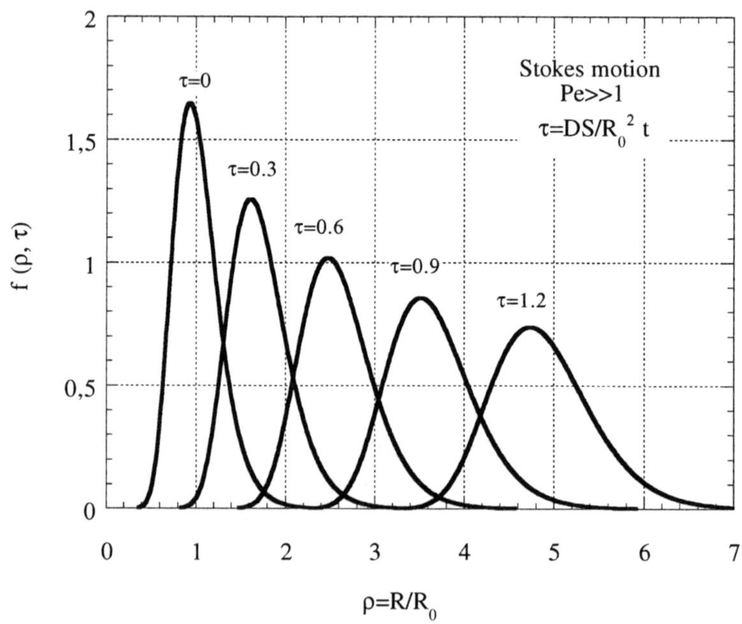

Fig. 5.7. Evolution of an initial distribution with time if dispersed droplets move in the gravity field with Stokes settling velocity. R_0 is again an arbitrary normalization constant, e.g. the initial mean radius.

We derived in chapter 4.6.1 and 4.6.3 the growth velocity of liquid particles moving in a liquid matrix with Stokes settling velocity (i.e. dispersed drops settling in the gravity field),

$$\frac{dR}{dt} = K_{Stokes} R^{1/2}, \tag{5.87}$$

where K_{Stokes} is the growth constant. Using the normalized coordinates defined in this chapter we can write the equation for the growth rate as

$$\frac{d\rho}{d\tau} = \rho^{1/2}. \tag{5.88}$$

Note that this growth is different from all others since it states that bigger particles grow faster than smaller ones. In all other cases treated so far we had the opposite result. This difference is due to the convective mass transport. The larger the liquid particle the larger the mass transport due to the increasing Peclet number (it grows proportional to the cube of the radius). Using the general result derived in the preceding section 5.6 and noting that now $m = -1/2$ the size distribution develops as

$$f(\rho, \tau) = \frac{1}{\rho^{1/2}(\rho^{1/2} - 1/2 \cdot \tau)^2} f_0((\rho^{1/2} - 1/2 \cdot \tau)^2). \tag{5.89}$$

This result is illustrated in figure 5.7. The evolution of the size distribution with time demonstrates that bigger liquid particles grow faster than smaller ones. Therefore, the standard deviation of the distribution increases with time.

6. Coarsening - Basics and Growth Laws

6.1 General considerations on coarsening

Imagine a system consisting of a huge number of particles of a certain phase with different radii embedded in a matrix. Some examples are shown in the introduction in figs.1.3 and 1.6. Assume further that the alloy is held at a constant annealing temperature. The concentration of solute in the matrix shall be at a level given approximately by the phase diagram and the annealing temperature. What will happen to the dispersion of particles? Generally one will observe in experiments that the particle radii change with annealing time, the average radius will grow, the number of particles decreases, but the overall volume fraction remains nearly a constant. An example of such a behavior is shown in Fig. 6.1. Why do these changes

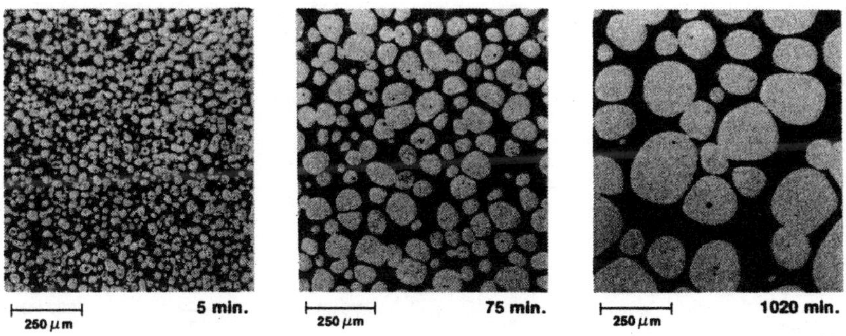

Fig. 6.1. Micrographs of solid tin particles embedded in a eutectic Pb-Sn matrix. The alloy was annealed just above the eutectic temperature for different times as shown in the textlines below the figures.

in particle size occur? Generally speaking, a dispersion of solid or liquid particle in a matrix increases the overall energy due to the existence of interfacial energy associated with the interfacial area. Since the system tries to minimize the energy, the amount of interfacial area decreases with time. A large number of small particles has, however, more interfacial area than one large particle with the same volume. Thus, the energy contained in the interfacial area of a dispersion increases with the number of particles in the dispersion. Reducing the number of particles decreases the excess energy associated with the interfaces. Therefore, a dispersion coarsens or ripens.

Another important aspect of competitive growth or Ostwald ripening (as explained in more detail in the next section) is its self-similar nature. This *self-similarity*, which occurs after sufficiently long coarsening time, can be understood as the following: take a section through a dispersion (as in figs. 6.1 and 6.2) at two different times and change the magnification by the ratio of the average particle size at these times. Then, the microstructures look similar, see Fig. 6.2. Only the *lengthscale* of the system changes. The lengthscale just mentioned is the average particle size, but other lengthscales like e.g. the average particle distance or even the inverse of the specific surface area per volume $1/S_V$ could be used as well. Scaling the radii with the lengthscale, one measures the same scaled size distribution. In the following we will set up the growth laws for coarsening. We will use these laws to derive te kinetics and particle size distributions in later chapters.

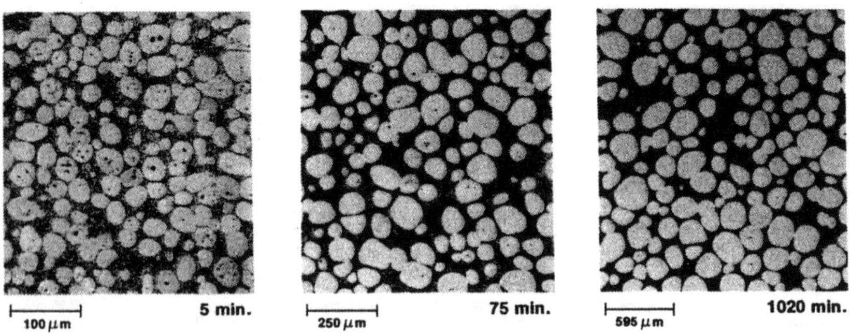

Fig. 6.2. Micrographs of solid tin particles embedded in a eutectic Pb-Sn matrix as already shown in Fig. 6.1 but here we scaled the magnification of each picture by a factor related to the average particle size. All microstructures then look similar.

6.2 Ostwald ripening due to curvature supercooling

In the preceding chapters on growth we have discussed the situation of a single sphere growing in a matrix at constant supercooling. Far from the solid particle interface the matrix had a fixed temperature, denoted by T_0 and the interface temperature was given by the equilibrium melting temperature T_m. We found that the radius of the solid sphere would change proportional to the square root of time. We now examine a different situation. Let us assume that we have a dispersion of spheres with a spectrum of radii embedded in a matrix. The size spectrum is described by a size density distribution $f(R,t)dR$ (note that fdR has the units of numbers of particles per volume) at a given time t. In this case the interface temperature of all particle is no longer equal to the equilibrium melting temperature, but depends on the radius according to the Gibbs-Thomson relation. This has as a consequence that in a closed, adiabatic system the matrix temperature far away from the particle interface is also no longer fixed but determined by a mean field value set by the whole ensemble of spheres. We denote this temperature by T_∞. For simplicity we assume here that all solid spheres are so far away from each other that a single sphere approximation can be used with a different boundary condition at infinity, i.e. we assume that the dispersion is very diluted - like the universe is a diluted dispersion of stars. Then, the temperature field looks as depicted in Fig. 6.3. All interfaces of the particles in the dispersion are

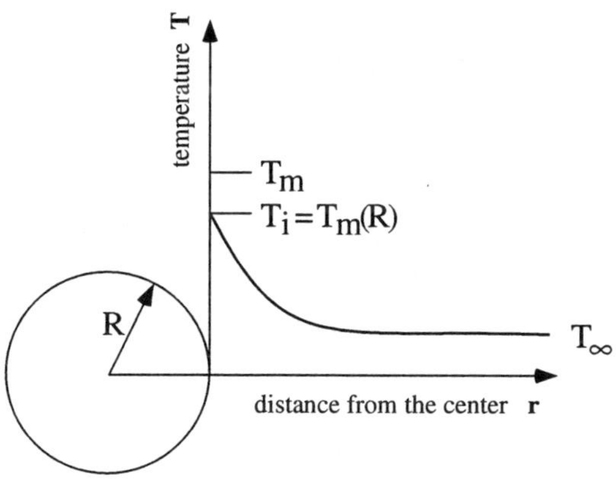

Fig. 6.3. Temperature field around a sphere undergoing radius change due to the combined effect of all other spheres and the Gibbs-Thomson effect.

at a different temperature since the radii are different. This induces locally varying temperature gradients in the matrix. Heat is exchanged within the system between the spheres of different radii. We then have for the growth rate (see eq.(4.40))

$$\frac{dR}{dt} = \frac{\lambda}{\rho L} \frac{T_I(R) - T_\infty}{R}, \tag{6.1}$$

with $T_I(R)$ as the interface temperature depending on the particle radius according to eq.(2.101)

$$T_I(R) = T_m - \frac{2\Gamma}{R}. \tag{6.2}$$

Inserting this expression into eq.(6.1) leads to

$$\frac{dR}{dt} = \frac{\lambda}{\rho L R} T_m - T_\infty - \frac{2\Gamma}{R}, \tag{6.3}$$

This equation can be brought into simpler form if we introduce artificially the following expression for the far field (mean value) temperature,

$$T_\infty = T_m - \frac{2\Gamma}{R_c}, \tag{6.4}$$

with an as yet unknown radius denoted as R_c.[1] We will call this radius the critical radius, a notation becoming clearer soon. Let us insert this expression into eq.(6.3). We obtain

$$\frac{dR}{dt} = \frac{2\Gamma\lambda}{\rho L R} \left(\frac{1}{R_c} - \frac{1}{R} \right) \tag{6.5}$$

The meaning of the critical radius becomes clear from this expression. Spheres having that radius have exactly a zero growth rate. The temperature field in a dispersion with many spheres and different radii, and thus interface temperatures according to the Gibbs-Thomson effect, is shown schematically in Fig. 6.4. Assume first that we have the matrix set at a given temperature, let us say the equilibrium melting temperature, and assume that there are no heat losses at the boundary of the matrix, i.e. we isolate it from the environment.

[1] We will show in the next chapter 7 dealing with the statistical analysis of coarsening that the far field temperature is besides a numerical constant equal to the average interface temperature.

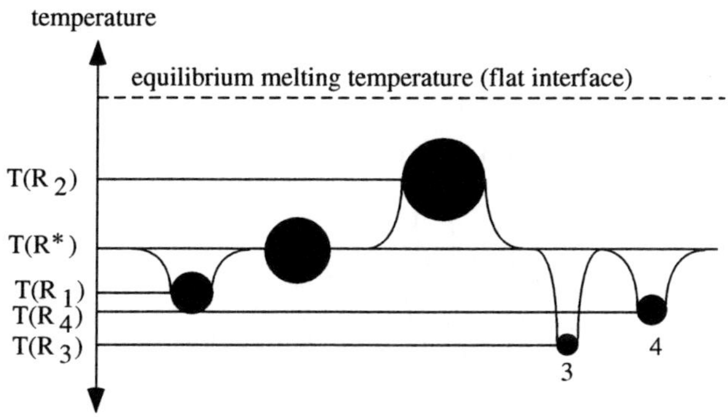

Fig. 6.4. Scheme of the temperature distribution in a dispersion, in which the temperature of the system is defined by the sensible heat exchange between spheres of different melting temperatures according to the Gibbs-Thomson effect.

We then put into the matrix one sphere of radius R. This sphere has, according to the Gibbs-Thomson relation, a melting temperature slightly lower than the equilibrium melting temperature. Therefore, the sphere will melt thereby consuming, i.e. extracting, heat from the matrix and, thus, if the sphere is sufficiently large lowering the temperature of the matrix until an equilibrium is reached, i.e. the matrix temperature is equal to the radius dependent melting temperature of the sphere. We now proceed and put a huge number of spheres into the matrix. The matrix temperature cannot adjust itself to be in equilibrium with all spheres (and their radius dependent melting temperature) at the same time. Eventually it is able to adjust its temperature to just a value as given in eq.(6.4). A few particles will then be in equilibrium, but others have higher or lower melting temperatures. Thus, the temperature gradients existing in the matrix lead to melting of the smaller spheres and growth of the bigger ones, as shown in Fig. 6.4. Since the smaller ones (smaller than the critical radius) will completely melt, the critical radius has to be redefined (the size distribution changes continuously), leading to a new situation: deciding which particles will now dissolve and which will grow at the expense of the smaller ones. This picture also tells that the problem of ripening cannot be solved without knowledge of the evolution of the size distribution since it defines the critical sphere size. It is also obvious that the mean-field temperature increases continuously towards the equilibrium melting temperature.

As just explained, R_c is not a constant but changes with time, i.e. $R_c = R_c(t)$. Its exact dependence on time will be derived in the following chapters, since it depends on the evolution of the size distribution. Eq.(6.5) can be rearranged to yield

$$\begin{aligned}\frac{dR}{dt} &= \frac{2\lambda\Gamma}{\rho L R}\left(\frac{1}{R_c} - \frac{1}{R}\right)\\ &= \frac{2\lambda\Gamma}{\rho L R^2}\left(\frac{R}{R_c} - 1\right) = \frac{K^*_{sl}}{R^2}\left(\frac{R}{R_c} - 1\right),\end{aligned} \tag{6.6}$$

where

$$K^*_{sl} = \frac{2\lambda\Gamma}{\rho L}. \tag{6.7}$$

In contrast to the case with is a constant supercooling, this equation cannot be integrated without knowledge of the time dependence of the critical radius.

6.3 Ostwald ripening due to solubility changes

The equations governing the coarsening of a two phase mixture consisting of a dispersion of particles in a matrix, in which solute transport governs the change in radius, are quite similar to those for a pure solid-liquid mixture given in the preceeding section. The concentration field is sketched in Fig. 6.5. As in the thermally controlled ripening pro-

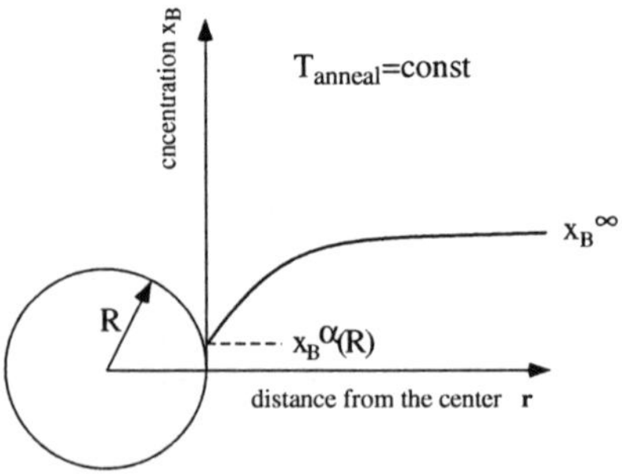

Fig. 6.5. Concentration field around a sphere undergoing radius change due to the combined effect of all other spheres.

cess the concentration at infinity is related to a mean value set by the other particles in the system. We assume again that the whole system is closed: Mass or species are only exchanged within the system, i.e. between the spheres of different radii. Fig. 6.6 shows how the system establishes a mean field concentration if particles are present with different equilibrium solubilities. The growth rate can be written as (see eq.(4.47))

$$\frac{dR}{dt} = \frac{D}{R}\frac{x_B^\infty - x_B^I(R)}{x_B^\beta - x_B^\alpha}. \tag{6.8}$$

In principle, the difference $x_B^\beta - x_B^\alpha$ depends on the radius. Generally, however, the radius dependence of this difference is small compared to the magnitude of the difference itself and thus can be neglected. We again introduce a critical radius R_c via

$$x_B^\infty = x_B^{\alpha,e}(1 + \frac{\ell^\alpha}{R_c}). \tag{6.9}$$

Inserting this expression into the growth rate eq.(6.8) and using eq.(2.124) to express $x_B^I(R)$ leads to

$$\frac{dR}{dt} = \frac{D\ell^\alpha}{(x_B^\beta - x_B^\alpha)R}\left(\frac{1}{R_c} - \frac{1}{R}\right) = \frac{K_{ss}^*}{R^2}(\frac{R}{R_c} - 1), \tag{6.10}$$

where

$$K_{ss}^* = \frac{D\ell^\alpha}{x_B^\beta - x_B^\alpha}. \tag{6.11}$$

Again we find: In contrast to the situation of constant supersaturation, this equation cannot be integrated without knowledge of the time dependence of the critical radius. Looking at the eqs.(6.6,6.10) we notice that both have the same functional dependence on the radius and the critical radius. The general form for diffusional transport of matter or heat is therefore

$$\frac{dR}{dt} = \frac{K_{diff.}^*}{R^2}(\frac{R}{R_c} - 1). \tag{6.12}$$

The difference lies only in the physical meaning of the rate constant, $K_{diff.}^* = K_{sl}^*$ for ripening in a pure solid-liquid system and $K_{diff.}^* = K_{ss}^*$ for ripening in an isothermal two-phase alloy.

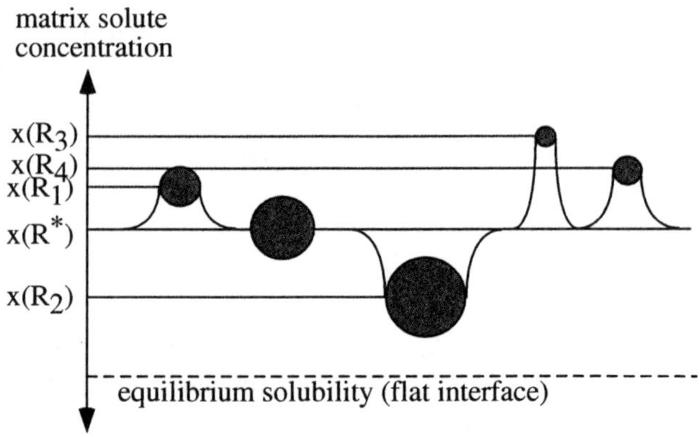

Fig. 6.6. Scheme of the concentration field in a dispersion at constant temperature. The only source of supersaturation shall be the excess solubility of the spheres due to the Gibbs-Thomson effect.

6.4 Ostwald ripening in interface kinetics or general first order reactions

The growth laws derived for interface kinetics of first order for a sphere growing due to concentration and temperature differences were (see eqs.(4.66,4.71))

$$\frac{dR}{dt} = k_s(x_B^0 - x_B^I) \tag{6.13}$$

and

$$\frac{dR}{dt} = k_T(T_I - T_0) \tag{6.14}$$

respectively. Identifying x_B^0 with the far field concentration, eq.(6.9), and T_0 with the far field temperature, eq.(6.4), and using for the interface concentration respectively temperature eqs.(2.124,2.100), we obtain

$$\frac{dR}{dt} = k_s \ell^\alpha (\frac{1}{R_c} - \frac{1}{R}) \tag{6.15}$$

and

$$\frac{dR}{dt} = 2k_T \Gamma (\frac{1}{R_c} - \frac{1}{R}). \tag{6.16}$$

In both cases the equation has the general form

$$\frac{dR}{dt} = K_{int.} \frac{1}{R} \left(\frac{R}{R_c} - 1 \right). \tag{6.17}$$

The difference lies only in the physical meaning of the rate constant $K_{int.}$,

$$K_{int.} = \begin{cases} k_s \ell^\alpha & \text{isothermal two-phase alloy} \\ 2k_T \Gamma & \text{pure solid-liquid mixture} \end{cases} . \tag{6.18}$$

6.5 Coarsening due to a second-order reaction

Now let's assume that the dissolution or growth of a particle is completely controlled by a second-order interface reaction without further qualifying the special microscopic nature of this reaction. In chapter 4 we derived the following relations for the growth rate at constant supersaturation or supercooling:

$$\frac{dR}{dt} = k_s^{2nd} |x_B^0 - x_B^I| (x_B^0 - x_B^I) \tag{6.19}$$

$$\frac{dR}{dt} = k_T^{2nd} |T_m - T_I| (T_m - T_I) \tag{6.20}$$

with the nomenclature described there. Identifying again x_B^0 with the far field concentration eq.(6.9) and T_m with the far field temperature, eq.(6.4), and using for the interface concentration respectively temperature the eqs.(2.124,2.101), we obtain

$$\frac{dR}{dt} = \frac{k_s^{2nd}(\ell^\alpha)^2}{x_B^{\beta,e} - x_B^{\alpha,e}} \frac{1}{R^2} (\rho - 1)^2 \mathrm{sgn}(\rho - 1) \tag{6.21}$$

and thus

$$\frac{dR}{dt} = \frac{K_s^{2nd}}{R^2} (\rho - 1)^2 \mathrm{sgn}(\rho - 1) \tag{6.22}$$

with

$$K_s^{2nd} = \frac{k_s^{2nd}(\ell^\alpha)^2}{x_B^{\beta,e} - x_B^{\alpha,e}} \tag{6.23}$$

and $\rho = R/R_c$. This was simplified using the relations $x_B^I - x_B^\alpha(R) = \ell^\alpha(1/R_c - 1/R) = \ell^\alpha R(\rho - 1)$. The sgn-function switches the sign when both concentrations are equal, this occurs when $\rho = 1$, and thus $\mathrm{sgn}(x_B^I - x_B^\alpha(R)) = \mathrm{sgn}(\rho - 1)$. For growth in an undercooled melt we obtain

$$\frac{dR}{dt} = 2k_T^{2nd} \Gamma (\frac{1}{R_c} - \frac{1}{R}) 2\Gamma \left| \frac{1}{R_c} - \frac{1}{R} \right| \tag{6.24}$$

or

$$\frac{dR}{dt} = \frac{K_T^{2nd} 1}{R^2} (\rho - 1)^2 \mathrm{sgn}(\rho - 1) \tag{6.25}$$

with

$$K_T^{2nd} = 2k_T^{2nd} \Gamma^2 . \tag{6.26}$$

6.6 Coarsening under convective transport

In section 4.6 we derived growth laws of the following kind for cases in which convection outside a diffusion boundary layer dominates mass or heat transport. The growth laws had the general form

$$\frac{dR}{dt} \propto \Delta(x_B, T) \tag{6.27}$$

for Stokes motion of solid particles or Marangoni motion of droplets and

$$\frac{dR}{dt} \propto \Delta(x_B, T) R^{1/2} \tag{6.28}$$

for Stokes motion of droplets. In these equations $\Delta(x_B, T)$ denoted the difference in concentration or temperature between the particle interface and the far field values, i.e. for supersaturation it means $\Delta(x_B) = (x_B^\infty - x_B^\alpha)/(x_B^\beta - x_B^\alpha)$. Using the same procedures as before, i.e. identifying x_B^∞ with the far field concentration in the matrix and using the Gibbs-Thomson relation for x_B^α we obtain relations of the form

$$\frac{dR}{dt} = \frac{K_1}{R}\left(\frac{R}{R_c} - 1\right), \tag{6.29}$$

$$\frac{dR}{dt} = \frac{K_2}{R^{1/2}}\left(\frac{R}{R_c} - 1\right). \tag{6.30}$$

Here the constants K_1, K_2 contain all the physics of the problem. Looking at both equations from a mathematical point of view we see that the first equation, valid for solid particles moving with Stokes velocity or liquid particles moving with Marangoni velocity, is formally identical to that for a first order reaction. Only the physical meaning of the rate constants in all three cases is different. Eq.(6.30), however, gives a new coarsening law and has to be treated separately. We will specify the rate constants after we have solved the complete coarsening problem (see chapter 7).

7. Ostwald ripening - Wagner analysis

In this chapter we treat coarsening under different conditions of heat and mass transport using mainly the Wagner technique to analyse the problem [295], which is somewhat different from the Lifshitz-Slyozov analysis [194]. We will not present the LS analysis in detail since the Marqusee and Ross analysis presented in the next chapter covers the LS analysis and also is much more complete and easier to understand.

A description of the coarsening or Ostwald ripening process could in principle be done as follows: Take the coarsening laws derived in the preceeding chapter 6, the continuity equation together with a conservation equation (for heat or mass) and try to solve the set of equations as we did in the on the statistics of growth chapter 5. If we would go along that line we would realize that an analytical solution of the fully time dependent Ostwald ripening problem, i.e. the time evolution of the particle size distribution starting from an arbitrary initial distribution, is an impossible task or at least it looks like that. Therefore we seek suitable approximate solutions. A classical way to analyse such complex problems is to look for a scaling and to investigate the conditions for which a scaling solution applies [21]. Such a scaling solution was discovered by Lifshitz, Slyozov and Wagner in 1961.

7.1 Diffusion driven coarsening

We already mentioned above (see chapter 6) that for diffusive exchange of heat or mass in a dispersion a particle coarsens according to:

$$\frac{dR}{dt} = \frac{K^*_{diff}}{R^2}\left(\frac{R}{R_c} - 1\right) \tag{7.1}$$

This equation has to be solved together with the continuity equation derived in chapter 5, which we write here again for convenience:

$$\frac{\partial f}{\partial t} + \frac{\partial}{\partial R}\left(\frac{dR}{dt} f\right) = 0 \tag{7.2}$$

Lifshitz, Slyozov and Wagner (LSW) perform the following analysis of this problem. Whereas Lifshitz and Slyozov analyse the problem self-consistently using mass conservation and then solve the problem for the case of very small supersaturations and extended annealing times (LS analysis). Wagner simplifies the analysis by assuming that the volume fraction of particles is a conserved quantity. Lifshitz, Slyozov and Wagner both come to the equations given above and use the fact that the coarsening law in eq.(7.1) suggests a scaling of the particle size R by R_c. We therefore introduce dimensionless coordinates

$$\rho = \frac{R}{R_c} \tag{7.3}$$

$$\tau = \ln R_c / R_c(0) \tag{7.4}$$

with $R_c(0)$ the critical radius at time zero and then transform the equations into these coordinates. We differentiate ρ with respect to time and perform some substitutions:

$$\begin{aligned} \frac{d\rho}{dt} &= \frac{1}{R_c}\frac{dR}{dt} - \frac{R}{(R_c)^2}\frac{dR_c}{dt} \\ &= \frac{1}{R_c} K^*_{diff} \frac{1}{R^2}(\rho - 1) - \frac{\rho}{R_c}\frac{dR_c}{dt} \\ &= \frac{K^*_{diff}}{R_c^3}\frac{1}{\rho^2}(\rho - 1) - \frac{\rho}{R_c}\frac{dR_c}{dt} \end{aligned} \tag{7.5}$$

Using

$$\begin{aligned} \frac{d\rho}{dt} &= \frac{d\rho}{d\tau}\frac{d\tau}{dt} \\ \frac{d\tau}{dt} &= \frac{1}{R_c}\frac{dR_c}{dt} \end{aligned}$$

eq.(7.5) takes the form

$$\begin{aligned} \frac{d\rho}{d\tau} &= \frac{K^*_{diff}}{R_c^2 \frac{dR_c}{dt}}\frac{1}{\rho^2}(\rho - 1) - \rho \\ &= \nu\frac{\rho - 1}{\rho^2} - \rho \end{aligned} \tag{7.6}$$

Here we introduced:

$$\nu = \frac{K^*_{diff}}{R_c^2 dR_c/dt} \tag{7.7}$$

The essential point of the Wagner and the LS analysis is to show, that ν is a constant and takes a unique value at long times, i.e. for τ large. This is done by a suitable analysis of eq.(7.6) and the continuity equation. Before we can proceed to show that ν is a constant and determine it, we solve the size distribution to obtain some general information about the distribution (before we later will write down the solution exactly).

Wagner solved eq.(7.2) rewritten in the new coordinates defined in eqs.(7.3,7.4)

$$\frac{\partial f}{\partial \tau} + \frac{\partial}{\partial \rho}\left(\frac{d\rho}{d\tau} f\right) = 0 \tag{7.8}$$

with a separation ansatz:

$$f(\rho, \tau) = g(\tau) h(\rho) \tag{7.9}$$

to obtain:

$$h\frac{dg}{d\tau} + g\frac{d}{d\rho}\left(\frac{d\rho}{d\tau}h\right) = 0 \tag{7.10}$$

$$-\frac{1}{g}\frac{dg}{d\tau} = \frac{1}{h}\frac{d}{d\rho}\left(\frac{d\rho}{d\tau}h\right) \tag{7.11}$$

The left side depends only on τ, the right only on ρ using eq.(7.6). Therefore they are equal to a separation constant, β. We solve the equation for g:

$$\begin{aligned} \frac{dg}{g} &= -\beta d\tau \\ \rightarrow\ g(\tau) &\propto \exp(-\beta\tau) \\ &\propto R_c^{-\beta} \end{aligned} \tag{7.12}$$

The value of β is determined in the Wagner analysis from the following consideration. As explained above, Ostwald ripening or coarsening means that the big particles grow on the expense of the small ones, i.e. all heat and mass is transferred or exchanged between the particles and therefore the volume fraction or fraction solid is essentially constant. There is a small change in volume fraction, since either the average temperature tends towards the equilibrium melting temperature or the supersaturation decreases to zero. Since the absolute value of these changes is small (due to the smallness of the Gibbs-Thomson effect), the volume fraction can be assumed to be constant (we will see in the next chapter that an extended analysis shows how the volume fraction changes with time). The volume fraction is given as:

$$\Phi = \frac{4\pi}{3}\int_0^\infty R^3 f(R,t)dR \simeq \frac{4\pi}{3}R_c^3 g(\tau)\int_0^\infty \rho^3 h(\rho)d\rho \qquad (7.13)$$

We use the approximate equality sign $\simeq$ to clearly denote that the volume fraction is only approximately constant if the reduced time is large. Assuming Φ to be constant means that the following relation must hold:

$$R_c^3 g(\tau) = R_c^3 R_c^{-\beta} = R_c^{3-\beta} = 1 \qquad (7.14)$$

This determines the separation constant β to $\beta = 3$. We now can proceed to the equation for $h(\rho)$:

$$\frac{d}{d\rho}\left(\frac{d\rho}{d\tau}h\right) = 3h$$

$$\rightarrow \qquad h\frac{d^2\rho}{d\rho d\tau} + \frac{d\rho}{d\tau}\frac{dh}{d\rho} = 3h$$

Inserting the expression for the coarsening rate in reduced coordinates given in eq.(7.6) leads to:

$$h\left[\frac{\nu}{\rho^2} - \frac{2\nu(\rho-1)}{\rho^3} - 1\right] = 3h - \left(\frac{\nu(\rho-1)}{\rho^2} - \rho\right)\frac{dh}{d\rho}$$

If $\rho \to \infty$ we obtain:

$$-h - 3h - \rho\frac{dh}{d\rho} = 0$$

$$\rightarrow \qquad -\rho\frac{dh}{d\rho} = 4h$$

$$\rightarrow \qquad h \propto \rho^{-4} \quad \text{for large } \rho \qquad (7.15)$$

This longtime behavior of the tail of the size distribution is essential. With the result of eq.(7.15) we can show that the size distribution must have a finite cut-off, i.e. the radius distribution cannot extend to infinity. From eq.(7.13) we have:

$$\Phi \simeq \frac{4\pi}{3} \int_0^\infty \rho^3 h(\rho) d\rho \tag{7.16}$$

but from the behavior at large ρ using eq.(7.15):

$$\Phi \propto \int_0^\infty \rho^3 \frac{1}{\rho^4} d\rho \propto \lim_{\rho \to \infty} \ln \rho \tag{7.17}$$

Thus the volume fraction would diverge logarithmically, which is in contradiction with its required constant, and finite, value. Therefore the size distribution must have an upper cut-off size, which we denote as ρ_m, meaning $h(\rho > \rho_m) = 0$. Notice that this conclusion is valid only under the conditions when scaling exists, i.e. for extended annealing times, large τ. In addition we assume that ρ_m is independent of time. The size distribution can be fixed further. We evaluate the differential equation for $h(\rho)$ at the cut-off value ρ_m:

$$h(\rho_m) \left. \frac{d}{d\rho} \frac{d\rho}{d\tau} \right|_{\rho_m} + \left. \frac{d\rho}{d\tau} \right|_{\rho_m} \left. \frac{dh}{d\rho} \right|_{\rho_m} = 3h(\rho_m) \tag{7.18}$$

Mathematically we can draw two alternative conclusions from this equality. First, we could require that the size distribution takes a zero value at ρ_m, i.e. $h(\rho_m) = 0$. Then it follows from eq.(7.18) that

$$\left. \frac{dh}{d\rho} \right|_{\rho_m} = 0 \quad \text{or} \quad \left. \frac{d\rho}{d\tau} \right|_{\rho_m} = 0 \tag{7.19}$$

The first conclusion would mean that the size distribution has a zero slope at the cut-off, thus it smoothly goes to zero there. Second, we could also satisfy eq.(7.18) by requiring that

$$\left. \frac{d}{d\rho} \frac{d\rho}{d\tau} \right|_{\rho_m} = 0$$
$$\left. \frac{d\rho}{d\tau} \right|_{\rho_m} = 0$$

and then automatically we would have $h(\rho_m) = 0$. In both cases the size distribution is equal to zero at the cut-off value. The value of ρ_m and ν defined in equations (7.6) and (7.7) still is undetermined. A unique solution requires two independent equations for both. The determination of these factors turns out to be essential for both the Wagner and the Lifshitz Slyozov analysis. The validity of the manner in which LSW fix both values uniquely is still controversial [43].

To obtain suitable constraints on ν and ρ_m we integrate the continuity equation eq.(7.8) after multiplication with ρ^3 and obtain:

$$\int_0^{\rho_m} \rho^3 \frac{\partial f}{\partial \tau} d\rho = -\int_0^{\rho_m} \rho^3 \frac{\partial}{\partial \rho}\left(\frac{d\rho}{d\tau} f\right) d\rho \tag{7.20}$$

The left side is proportional to the volume fraction

$$\int_0^{\rho_m} \rho^3 \frac{\partial f}{\partial \tau} d\rho = \int_0^{\rho_m} \frac{\partial}{\partial \tau}(\rho^3 f) d\rho = \frac{d}{d\tau}\int_0^{\rho_m} \rho^3 f d\rho \simeq \frac{d\Phi}{d\tau}$$

since ρ is not explicitly dependent on τ. We therefore obtain using $f(\rho, \tau) = g(\tau)h(\rho)$ and integrating the right hand side of eq.(7.20) by parts:

$$\frac{1}{g(\tau)}\frac{d\Phi}{d\tau} \approx \left[\rho^3 \frac{d\rho}{d\tau} h\right]_0^{\rho_m} - 3\int_0^{\rho_m} \rho^2 \frac{d\rho}{d\tau} h d\rho \tag{7.21}$$

The term in the square bracket is zero, since $\rho^3 d\rho/d\tau = 0$ at $\rho = 0$ and as derived above $h(\rho) = 0$ at ρ_m. We assumed earlier that the volume fraction shall be constant and therefore $d\Phi/d\tau = 0$. Then the relation

$$\int_0^{\rho_m} \rho^2 \frac{d\rho}{d\tau} h d\rho = 0 \tag{7.22}$$

must be valid. For a further discussion we have to examine how the function $d\rho/d\tau$ varies with ρ and ν. Fig.7.1 shows four different cases. For negative ν-values the relative growth rate is positive for small relative particle radii ρ and negative for bigger ones, meaning, the small ones would grow and the large ones dissolve. Such a behavior definitively is opposite to the real physical process. Therefore negative values of ν are excluded. This case can also be shown on a pure mathematical basis. Let us insert into eq.(7.22)the expression for the coarsening rate in eq.(7.6). We then yield:

$$\int_0^{\rho_m} [\nu(\rho - 1) - \rho^3] h d\rho = 0 \tag{7.23}$$

or

$$\nu \int_0^{\rho_m} (\rho - 1) h d\rho = \int_0^{\rho_m} \rho^3 h d\rho \tag{7.24}$$

Using the fact that $\rho - 1$ is a monotonic increasing function of ρ we can apply the second mean value theorem of integral calculus on the left side of eq.(7.24) and we have

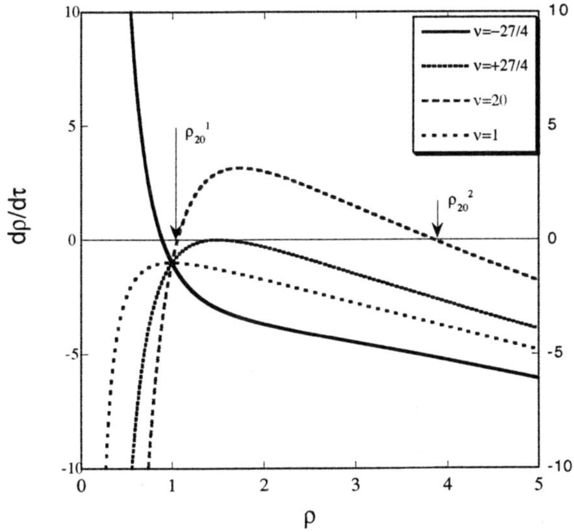

Fig. 7.1. Dependence of $d\rho/d\tau$ on ρ and ν.

$$\nu \int_0^{\rho_m} (\rho - 1) h d\rho = \nu(\rho_m - 1) \cdot \int_x^{\rho_m} h d\rho \quad \text{for} \quad 0 < x < \rho_m \qquad (7.25)$$

The right hand side of eq.(7.24), however, always is positive (the third moment of the distribution). Thus provided $(\rho_m - 1) > 0$ it follows that ν must be positive. Taking a positive value of ν as a given fact we could conclude from this consideration that $\rho_m > 1$.

Having now established that ν must be positive, we see from the examples shown in fig.7.1 that $d\rho/d\tau$ generally has a maximum and can have zeros with the axis $d\rho/d\tau = 0$. Assume that ν is that large that two zeros occur denoted in with ρ_{20}^1 and ρ_{20}^2. Then particles with reduced radii in between these zeros would have a positive growth rate and particles with $\rho > \rho_{20}^2$ and $\rho < \rho_{20}^1$ would dissolve. A particle with a value $\rho > \rho_{20}^1$ would grow until it reaches a size of ρ_{20}^2 and then stop changing its size, since a further change in size always yields $d\rho/d\tau < 0$. At the end we would have a dispersion of spheres having all the same size ρ_{20}^2 that would be stable with respect to coarsening. This is in contradiction to thermodynamics, since equilibrium, which shall be achieved after infinite times, requires a continuous dissolution of the small ones on the expense of the bigger ones. Therefore any ν-value leading to a functional behavior of $d\rho/d\tau$ with two zeros is not possible.

Small positive ν-values, like $\nu = 1$ (see fig. 7.1) do not have any intercept with the axis where $d\rho/d\tau = 0$ meaning that all particles have negative growth rates and thus all particles would dissolve. There is only one solution at which the particles with the biggest relative size, namely $\rho = \rho_m$ have a zero growth rate. This occurs if just the maximum touches the axis $d\rho/d\tau = 0$. Expressing these considerations mathematically leads to two conditions for the function $d\rho/d\tau$. Intercepting the axis $d\rho/d\tau = 0$ at ρ_m means

$$\left.\frac{d\rho}{d\tau}\right|_{\rho_m} = 0\,, \tag{7.26}$$

and having a maximum at that point means

$$\left.\frac{d}{d\rho}\frac{d\rho}{d\tau}\right|_{\rho_m} = 0 \tag{7.27}$$

These conditions fix the value of both ν and ρ_m. Its determination is as follows. Using eq.(7.6) in eq.(7.26) gives

$$\nu\frac{\rho_m - 1}{\rho_m^2} - \rho_m = 0 \tag{7.28}$$

and in eq.(7.27),

$$\frac{\nu}{\rho_m^2} - 2\frac{\nu(\rho_m - 1)}{\rho_m^3} - 1 = 0 \tag{7.29}$$

These are two equations for two unknowns. Solving them leads to:

$$\rho_m = \frac{3}{2} \qquad \text{and} \qquad \nu = \frac{27}{4} \tag{7.30}$$

Having now assembled all the bits and pieces of the analysis, we only have to calculate the size distribution $h(\rho)$. The solution of eq.(7.11) generally is:

$$h(\rho) = \frac{1}{h_0}\frac{d\rho}{d\tau}\exp\left(3\int(\frac{d\rho}{d\tau})^{-1}d\rho\right) \tag{7.31}$$

$$= h_0\frac{\rho^2}{\frac{27}{4}(\rho-1)-\rho^3}\exp\left(-3\int\frac{\rho^2 d\rho}{\rho^3-\frac{27}{4}(\rho-1)}\right) \tag{7.32}$$

with h_0 a normalization constant. Let us abbreviate the denominator in the integral as:

$$P(\rho) = \rho^3 - \frac{27}{4}(\rho - 1) \tag{7.33}$$

The evaluation of the integral can be made easier, if we write $P(\rho)$ as a product of its zeros. Remembering that we already know a zero, namely the cut-off radius 3/2, the polynomial P can be written as a product of an unknown full polynomial of second degree multiplied by $\rho - 3/2$. Determination of the coefficients of this quadratic polynomial and its zeros leads to:

$$P(\rho) = (\rho - 3/2)^2(\rho + 3) \tag{7.34}$$

Then the integral in eq.(7.32) can be solved simply by looking into a table of integrals. Collecting finally all terms together the result is

$$h(\rho) = \frac{1}{h_0} \frac{\rho^2}{(\rho + 3)^{7/3}(3/2 - \rho)^{11/3}} \exp\left(\frac{3}{2\rho - 3}\right) \tag{7.35}$$

where the normalization constant is determined numerically by the condition

$$\int_0^{\rho_m} h(\rho) d\rho = 1 \qquad \rightarrow \qquad h_0 = 0.014419 \tag{7.36}$$

The normalized function $h(\rho)$ is plotted in figure 7.2. We will now

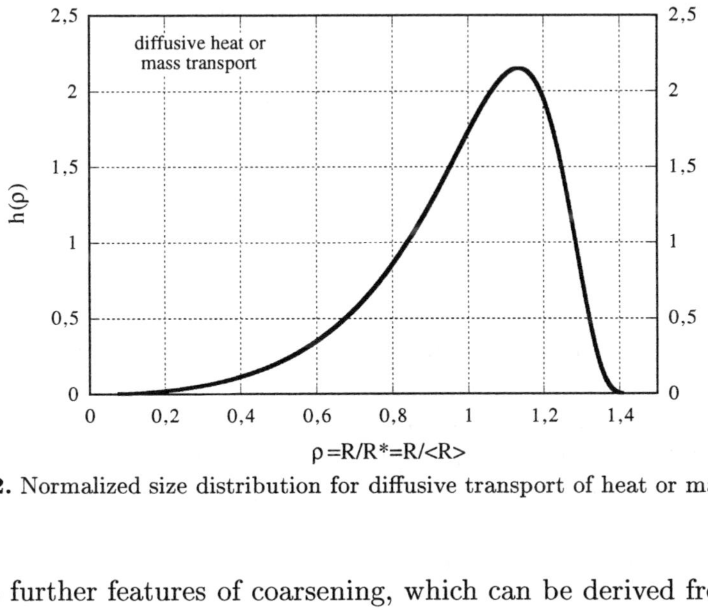

Fig. 7.2. Normalized size distribution for diffusive transport of heat or mass after LSW.

discuss further features of coarsening, which can be derived from the solution obtained. In eq.(7.7) we defined ν by:

$$\nu = \frac{K^*_{diff}}{R_c^2 dR_c/dt} \tag{7.37}$$

Since we have derived the value of ν and showed, that is indeed a constant, we can integrate this equation with respect to time and obtain one of the important results of coarsening theories:

$$R_c^3 = \frac{12}{27} K^*_{diff} t \tag{7.38}$$

Recalling the physical meaning of the constant K^*_{diff} yields for heat transport:

$$R_c^3 = \frac{8\lambda\Gamma}{9\rho L} t \tag{7.39}$$

where we note that in this equation ρ is the density. In a supersaturated matrix we obtain:

$$R_c^3 = \frac{4\ell^\alpha}{9(x_B^{\beta,e} - x_B^{\alpha,e})} t \tag{7.40}$$

Unfortunately the critical radius is not a quantity that can be measured experimentally. There is, however, a unique relation between the average radius that can be measured and the critical one. The average radius can be calculated from:

$$\begin{aligned} \langle R \rangle &= \frac{1}{n_V} \int_0^\infty R f(R,t) dR \\ &= \frac{g(\tau)}{n_V} R_c \int_0^{\rho_m} \rho h(\rho) d\rho \\ &= R_c \int_0^{\rho_m} \rho h(\rho) d\rho \\ &= R_c m_1^h \end{aligned} \tag{7.41}$$

where we have used that the number density is given by:

$$n_V(t) = \int_0^\infty f(R,t) dR = g(\tau) \int_0^{\rho_m} h(\rho) d\rho = g(\tau) \tag{7.42}$$

We used for probability distributions the transformation rule $f(R,t)dR = g(\tau)h(\rho)d\rho$. In eq.(7.41) m_1^h is the first moment of the distribution $h(\rho)$. Evaluating numerically the integral (with for example Mathematica^{TM}) in this relation yields a surprising and important result for diffusive coarsening:

$$\langle R \rangle = R_c \tag{7.43}$$

Therefore we obtain the most famous result for diffusive coarsening:

$$\langle R\rangle^3 = \frac{12}{27} K^*_{diff} t \tag{7.44}$$

Let us evaluate further moments of the size distribution. The standard deviation of a distribution is generally defined as:

$$\sigma^2 = \langle R^2\rangle - \langle R\rangle^2 \tag{7.45}$$

The second moment of the size distribution is calculated from:

$$\begin{aligned}\langle R^2\rangle &= \frac{1}{n_V}\int_0^\infty R^2 f(R,t)dR \\ &= R_c^2 \int_0^{\rho_m} \rho^2 h\rho)d\rho \\ &= R_c^2 m_2^h \end{aligned} \tag{7.46}$$

This yields for the standard deviation the expression:

$$\sigma^2 = R_c^2(m_2^h - 1) = 0.1296 \cdot R_c^2 \tag{7.47}$$

The average interface temperature can also be calculated exactly. It is related to the harmonic mean of the size distribution. The average interface temperature is defined by the relation:

$$\begin{aligned}\langle T_I\rangle &= \frac{1}{n_V}\int_0^\infty T_I(R) f(R,t)dR \\ &= \frac{1}{n_V}\int_0^\infty (T_m - \frac{2\Gamma}{R}) f(R,t)dR \\ &= T_m - \frac{2\Gamma}{R_c}\int_0^{\rho_m} \frac{1}{\rho} h(\rho)d\rho \\ &= T_m - \frac{2\Gamma}{R_c} m^h_{-1} \\ &= T_m - \frac{2\Gamma}{R_c} 1.0775 \\ &= T_m - \frac{2\Gamma 1.0775}{\langle R\rangle} \end{aligned} \tag{7.48}$$

Therefore we draw the conclusion that the average temperature in the system is not given by the Gibbs-Thomson equation averaged with the particle size distribution. The average interface temperature is strictly related by a numerical constant to the far field temperature as defined in eq.(6.4):

$$\frac{\langle T_I \rangle - T_\infty}{T_m - T_\infty} = 1 - m^h_{-1} = -0.0775 \tag{7.49}$$

The average interface temperature is thus around 7% smaller than the far field temperature. The same relation is valid for relationship between the average interface concentration and the far field concentration in diffusional growth. The number density written in eq.(7.42) can further be evaluated to show that it varies hyperbolically with time:

$$n_V(t) = g(\tau) = e^{-3\tau} = \left(\frac{R_c(0)}{R_c}\right)^3 = \frac{27R_c(0)^3}{12K^*_{diff}t} \propto \frac{1}{t} \tag{7.50}$$

Note that these relations for $\langle R \rangle$ and n_V are valid only at large annealing times. In some case extremely long times are required to reach this asymptotic state [7].

This continuous decrease of the number density with time occurring in all coarsening systems makes numerical simulations difficult if they should cover extended periods of annealing times, since even starting with a large number of particles there might not be enough left at the end of a numerical simulation to yield sufficiently good statistics. A summary of the moments m^h_k of the LSW distribution is given in table 7.1.

k	m^h_k
-1	1.0775
0	1
1	1
2	1.1296

Table 7.1: Table of the moments of the size distribution of diffusional coarsening stated in eq.(7.35) with the normalization given in eq.(7.36).

The growth laws presented here are generally observed, i.e. the cube root dependence of time for the average particle size and the reduction of number density of particles with time. The absolute value of the rate constant given by this classical LSW treatment, however, has never been confirmed experimentally, a fact that could be due to the various assumptions made in deriving the results. This is discussed at the end of this chapter and especially in chapter 9.

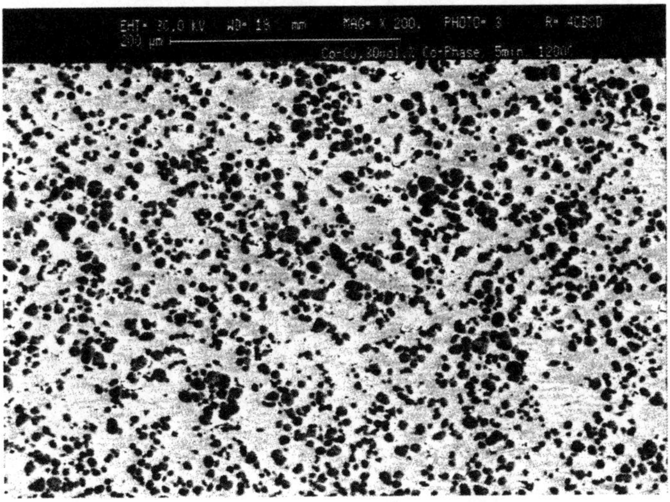

Fig. 7.3. Scanning Electron Micrographs (SEM) of a liquid phase sintered Cu-Co dispersion. The dark features are the cobalt particles embedded in a copper matrix that has dissolved approximately 8at.% Co. The sintering treatment was performed at 1473 K for 5 min. The volume content of solid cobalt particles at that temperature in the molten copper matrix was 30%.

Let us nevertheless show an example for a system undergoing ripening in a system of relevance to liquid phase sintering. Figs. 7.3 and 7.4 show micrographs of Co-particles dispersed in a Cu-Co matrix and sintered at high temperature under a reducing atmosphere (for details see [30]). It clearly can be seen that the Co particles have coarsened considerably. Evaluation of the particle sizes on such sections is done by measuring the intercept length they make with an array of lines and calculating then the average intercept length or measuring the area of each particle and calculating the radius of an area equivalent circle. Then one can plot for instance, the average area equivalent radius a as a function of coarsening time or the number density of particles. This is shown in figs. 7.5 and 7.6. They clearly reveal that the experimental observations are in agreement with the theoretical predictions for the exponents of the temporal power laws given in this chapter.

Let us summarize the LSW technique for determining the asymptotic state of a coarsening system:

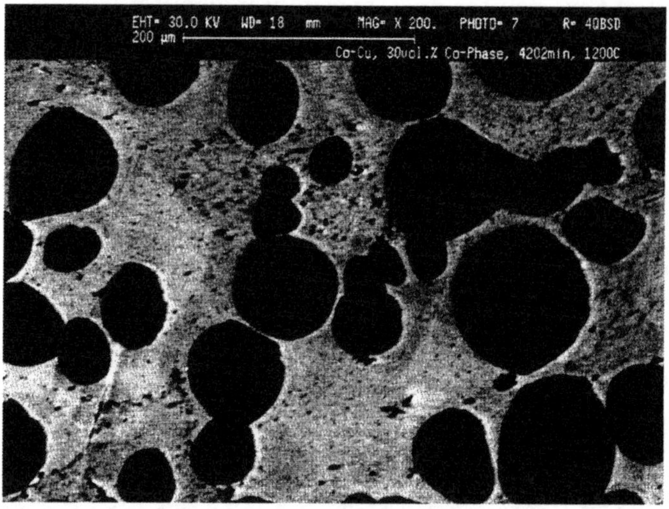

Fig. 7.4. SEM picture of the same material as shown in figure 7.3 but here after a sintering time of 4202 min.

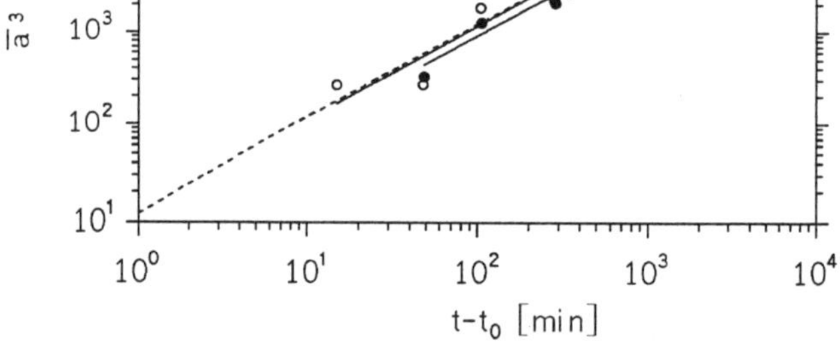

Fig. 7.5. Growth of the cube of the average particle radius of Co-particles in a Cu(Co) alloy melt. For both volume fractions of Co particles a linear increase of the cube radii with sintering time is observed experimentally. $\bar{a}_0$ is the experimentally determined particle radius in a section after five minutes sintering. Micrographs of these alloys are shown in figs.7.3 and 7.4.

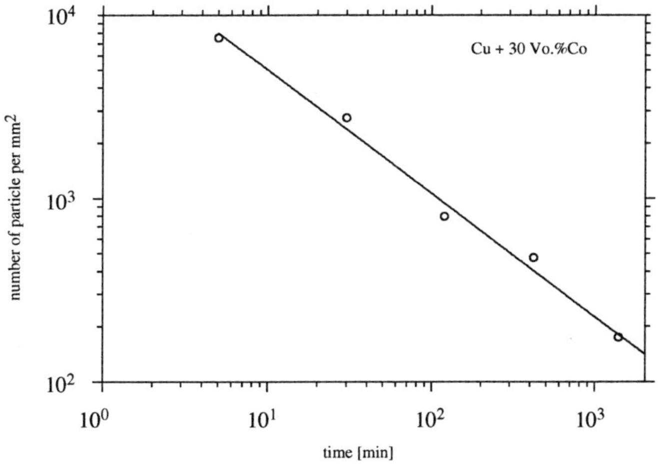

Fig. 7.6. Number density of Co particles measured on sections like those shown in figures 7.3 and 7.4. Notice that the number density decreases hyperbolically as predicted by the theory derived in this chapter.

1. First write down the growth rate equation using a suitable form of the Gibbs-Thomson equation.
2. Introducc the dimensionless coordinates of LSW and rewrite the growth rate law in these coordinates and establish a growth law for the critical radius, thus introducing a constant ν.
3. Solve the size distribution with a separation ansatz.
4. Determine the separation constant using the constancy of volume fraction and look for the asymptotic behavior to ensure that the distribution has a final cut-off (this step could be skipped as it does not determine ν).
5. Determine the cut-off radius and the constant ν for the growth of the critical radius. Determine the zero of the reduced growth rate (see step 2) and the zero of its first derivative with respect to reduced radius. Solve these two equations to yield ν and ρ_m.
6. Use these constants to determine the radius dependent portion of the size distribution and normalize it.
7. Determine the proportionality constant between the average radius and the critical radius, integrate the growth law for the critical radius, (obtained in step 2) with respect to time and transform it to the average radius.

Using this recipe a large number of studies on coarsening in dilute dispersions have been performed in the past (especially looking for the effect of volume fraction, influence of convective mass or heat transport, coarsening at grain boundaries or dislocations) and since the technique can be tedious but is simple, it always can be applied to new situations to get a first insight into the behavior of the physical system under investigation. If more precise information is needed or it is unclear if an asymptotic scaling exists at all, the technique of Marqusee and Ross, that will be explained in the next chapter, should be consulted.

Let us make some comments or interpretation on the results of the LSW theory.

- First, the theory predicts that in the asymptotic regime, after long annealing times, there exists a unique scaled distribution function. After sufficiently long coarsening a distribution will independent of the initial size distribution. In radius space the size distribution moves to larger and larger radii in as much as annealing proceeds. But scaling the measured radii at any instant of time with their average value at that time will always lead to the same distribution. Thus coarsening is self-similar in nature.
- Second, the average radius grows proportional to the cube root of time. The growth rate depends only on thermophysical properties of the system under investigation. Care must be taken, not to overstress the importance of such a relation. Finding in an experimental study the growth exponent 1/3 does not mean that coarsening occurred nor does it mean that LSW or better a self similar coarsening was observed [7]. The growth exponent is simply a matter of the driving force for the process under consideration. We showed that in diffusional growth the change of radius with time is proportional to the driving force (supercooling or supersaturation) and inversely proportional to the radius itself. If in a dispersion the curvature determines either the supersaturation or the supercooling, the driving force term depends itself on the radius due to the Gibbs-Thomson effect. Thus we always will have a relation $dR/dt \propto 1/R^2$ and thus find a growth exponent 1/3. There are other processes which also lead to the same growth exponent, e.g. a second order interface reaction or coalescence of particles by Brownian motion.
- We assumed that the dispersion is infinitely diluted. We have used the solution of the Laplace equation for temperature or concentration and prescribed boundary conditions at the particle surface and,

in particular, at infinity. Especially the last condition means that the solute or temperature fields of the particles do not influence each other. It is a single particle approximation.
- The particles are ideally spherical, which may not be the case, either in solid solutions or in solid-liquid systems.
- The analysis of LSW does not account for many effects observed in real systems, such as: finite volume fraction of particles, elastic stress fields around misfitting inclusions, particle motion, non spherical particle shapes, particle contacts (coalescence), combined transport of solute and heat, combined effect of diffusive transport and interface kinetics etc.. Some of these problems have been solved and the references to these studies can be found in the bibliography at the end of this book.

Finally we would like to mention that in the newest literature on coarsening the special solution of LSW was questioned. For instance Brown [43] showed that solutions that guarantee mass conservation are also possible for other values of ν than that derived above. Careful mathematical inspection by for instance Niethammer [225] showed that the LSW solution presented here is not the only possible one. This analysis agrees with the conclusions reached by Meerson [212, 213] and others [248]. However, these analyses show that the LSW solution the only particle size distribution that is stable whereas other possible solutions are unstable against fluctuations or noise. A treatment of these aspects of self-similar coarsening is far beyond the scope of this book. The interested reader is advised to look into the papers cited.

7.2 Coarsening with convective transport

The LSW analysis of convective diffusive mass transport at high Peclet numbers was performed by Ratke and Thieringer [242]. With the growth laws stated in section 6.6 and using the recipe of the LSW analysis given above one derives the following relations (the details of the calculations are omitted here, since there is nothing new to be learned in deriving them). All of these results employ the equations for the growth or coarsening rate of a particle derived previously, and assume that $\mathrm{Pe} \gg 1$.

Solid particles moving with Stokes velocity

In this case we obtain for the average radius the growth law:

$$\langle R\rangle^2 = K_{1S}t \tag{7.51}$$

with

$$K_{1S} = 1.576\left(\frac{2D^2g(\rho_p-\rho_m)}{9\eta}\right)^{1/3}\frac{\ell^\alpha}{x_B^{\beta,e}-x_B^{\alpha,e}} \tag{7.52}$$

here ρ_p, ρ_m are the densities of particle and matrix. The size distribution is identical to that of a first order reaction and will be presented in the next chapter using the Marqusee and Ross technique to analyse coarsening (see eq.(8.49 and figure 8.2).

Droplets moving with Marangoni velocity

In this case we obtain for the average radius principally the same growth law as before, but with another rate constant:

$$\langle R\rangle^2 = K_{1M}t \tag{7.53}$$

with

$$K_{1M} = 1.576\left(\frac{D|\frac{\partial\sigma}{\partial T}||\frac{dT}{dx}|(\eta+\eta_p)}{3\eta(2\eta+3\eta_p)(2+\lambda_p/\lambda}\right)^{1/2}\frac{\ell^\alpha}{x_B^{\beta,e}-x_B^{\alpha,e}} \tag{7.54}$$

Here λ, λ_p denote the heat diffusivities of the matrix and the drop, the η's denote the viscosities. The size distribution is identical to that of a first order reaction and will be presented in the next chapter (see eq.(8.49 and figure 8.2).

Droplets moving with Stokes velocity

In this case we obtain indeed a new growth law for the average radius. It reads:

$$\langle R\rangle^{3/2} = K_2t \tag{7.55}$$

with the rate constant given by:

$$K_2 = 0.837\left(\frac{Dg(\rho_p-\rho_m)(\eta+\eta_p)}{3\pi\eta(2\eta+3\eta_p)}\right)^{1/2}\frac{\ell^\alpha}{x_B^{\beta,e}-x_B^{\alpha,e}} \tag{7.56}$$

The size distribution can be calculated along the lines presented above and turns out to be (after lengthy calculations using the method of partial fractions to evaluate the integrals or after a short time using symbolic computations with Mathematica^{TM}):

$$
\begin{aligned}
h(\rho) &= 2 \cdot \frac{81}{2^{5/3}} \frac{1}{[\rho^{3/2} - \frac{3\sqrt{3}}{2}(\rho - 1)][\rho^{1/2} + \frac{\sqrt{3}}{2}]^{2/3}} \\
&+ \frac{\rho^{1/2}}{[\sqrt{3} - \rho^{1/2}]^{16/3}} \exp\left(-\frac{4\rho^{1/2}}{\sqrt{3} - \rho^{1/2}}\right) \qquad (7.57)
\end{aligned}
$$

It is interesting to note that for this case we obtain $\rho_m = 3$ and $\nu = 3\sqrt{3}/2$. The shape of the size distribution is shown in figure 7.7. Although the size distribution principally extends to range of $\rho = 3$,

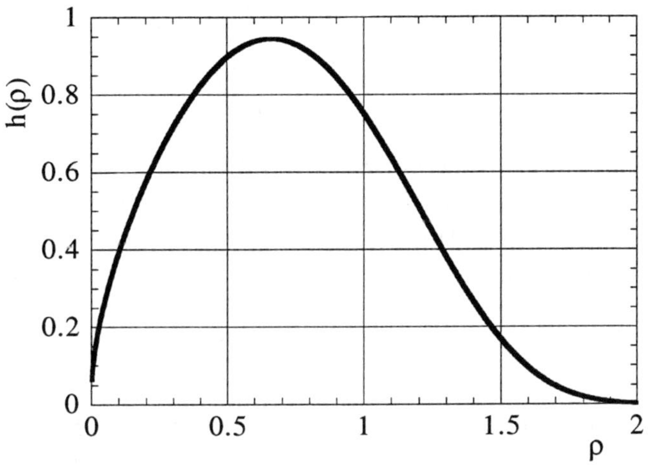

Fig. 7.7. Asymptotic size distribution for coarsening of drops mocing in an infinite matrix with Stokes velocity.

the probability is down to nearly zero close to a value of 2.

Solid particles moving at constant velocity

There are only two cases not being treated sofar. If the fluid flow velocity is constant and not a function of particle radius. Whereas this situation might be unreal for freely suspended drops or particles it is an interesting picture for solid spheres. Since fixing spheres or cylinders in space and set up a fluid flow of constant velocity (equal to pushing through a network of spheres or cylinders a flow with a constant pressure difference) is an idealization for a dendritic zone or mush in which fluid flow effects the rate of morphological changes. We therefore will give here the growth rate for that situation. In chapter 4 we derived in section 4.6.2 for the volume change of a particle if the fluid flow passing a sphere has a constant, radius independent flow velocity v_0 the equation:

$$\frac{dV}{dt} = 7.98(x_B^{\infty} - x_B^{I})D^{2/3}v_0^{1/3}R^{4/3} \tag{7.58}$$

The radius change is then obtained from a mass balance such that $dR/dt = dV/dt/(4\pi R^2(x_B^{\beta} - x_B^{\alpha}))$ yielding for the growth rate the equation:

$$\frac{dR}{dt} = K_{U=const}\frac{1}{R^{2/3}} \tag{7.59}$$

with

$$K_{U=const} = 7.98\frac{x_B^{\infty} - x_B^{\alpha}}{x_B^{\beta} - x_B^{\alpha}}D^{2/3}v_0^{1/3} \tag{7.60}$$

Integration would give for growth a relation $R^{5/3} \propto t$. Here we are interested in coarsening and therefore we identify as before x_B^{∞} with the far field matrix concentration and put in the Gibbs-Thomson relations to obtain the equation:

$$\frac{dR}{dt} \propto \frac{1}{R^{5/3}}(\frac{R}{R_c} - 1) \tag{7.61}$$

With the recipe given above one easily performs the LSW analysis to finds $\rho_m = 8/5$, $\nu = (8/3)(8/5)^{5/3}$ and thus for the average radius $\bar{R}$ the equation:

$$\langle R\rangle^{8/3} = K_{UCC} \cdot t \tag{7.62}$$

with

$$K_{UCC} = 0.215\frac{\ell^{\alpha}}{\pi(x_B^{\beta} - x_B^{\alpha})}D^{2/3}v_0^{1/3} \tag{7.63}$$

The size distribution is presented in figure 7.8. The growth exponent of 8/3 and the cut-off radius of 8/5 are close to the classical LSW values for diffusional growth and thus are hard to detect the deviation from LSW. Let us give an example to illustrate that the situation discussed here theoretically can be met in real systems. Akaiwa, Hardy and Voorhees [3] studied Ostwald ripening in solid-liquid mixtures using a Sn-Pb alloy over a wide range of volume fraction solid. They induced convection by a slow rotation of a disk-shaped sample with

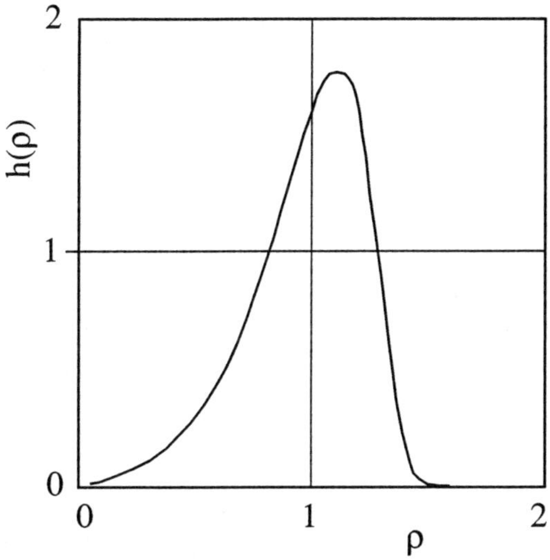

Fig. 7.8. Normalized particle size distribution for solid spheres moving with a constant radius independet velocity thorugh a liquid matrix.

the axis of rotation perpendicular to the gravity vector. At low volume fraction of solid (2 %) the experiments show that convection can alter the exponent of the temporal power law for the average intercept length from its classical value of 1/3 to 0.37. The result is shown in fig.7.9. This is close to the value predicted in eq.(7.62), which, however was derived for large Peclet numbers. At intermediate volume fractions solid they observed the temporal exponent to be 1/3 but the amplitude of the temporal power law increased with increasing rate of rotation. Let us close this section with some general comments on coarsening with convective mass or heat transport. The theory presented above employed an expression for mass or heat transport valid at large Peclet numbers. In coarsening, however, there are always particles or drops being dissolved, meaning that there are particles for which the Peclet number is close to zero. Therefore a complete theory would have to use an expression for transport covering the whole range of Peclet numbers, from diffusion to large values of the Peclet number, as outlined in chapter 4.6.4. This problem is unsolved. There were, however, numerical studies performed indicating that the distribution function and exponents of the temporal power laws gradually approach those predicted in this chapter [241]. Another aspect needs to be considered. The derivation employs a surfaced concentration gradient to determine the mass or heat transport by convection and thus

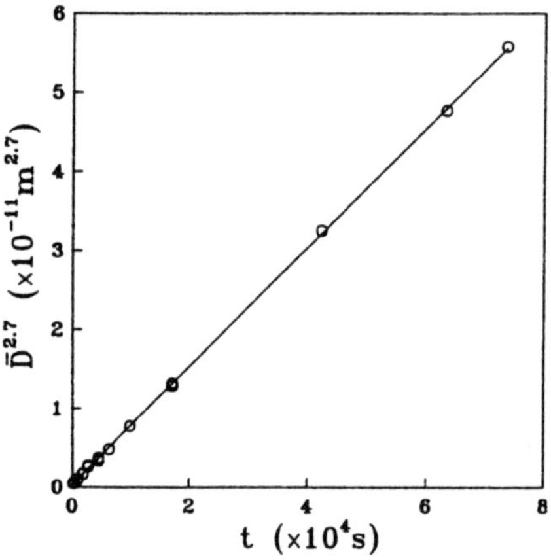

Fig. 7.9. The average particle diameter to the 2.7 power as a function of time observed by Akaiwa, Hardy and Voorhees [3] in a Pb-Sn alloy just above the eutectic temperature with a volume fraction solid of 0.02. Thin disk like samples were rotated with the axis of rotation perpendicular to the gravity vector. Thus a convective fluid flow was induced changing mass transport in the dispersion.

spherical symmetry is restored. In reality, however the asymmetry of the interfacial concentration gradient may induce significant changes in the morphology of solid particles from spheres. In liquid-liquid dispersion there will be no changes, since the interface tension will restore the sphere shape as long as the Reynolds number is small (see [182]). In another numerical study on coarsening of cylindrical arrays at constant temperature and a constant fluid flow velocity using a phase field model [23] the results shown in this section on convective effects agreed with these calculations. This new numerical method self-consistently includes finite volume fraction effects and morphological changes.

7.3 Second order reaction coarsening

The coarsening law was already presented in the last chapter. For convenience we write it down again here:

$$\frac{dR}{dt} = K^s_{2nd} \frac{1}{R^2} (\rho - 1)^2 \mathrm{sgn}(\rho - 1) \tag{7.64}$$

Transforming coordinates to the ρ, τ system of LSW we obtain:

$$\frac{d\rho}{d\tau} = \nu_2 \frac{(\rho - 1)^2}{\rho^2} \mathrm{sgn}(\rho - 1) - \rho \tag{7.65}$$

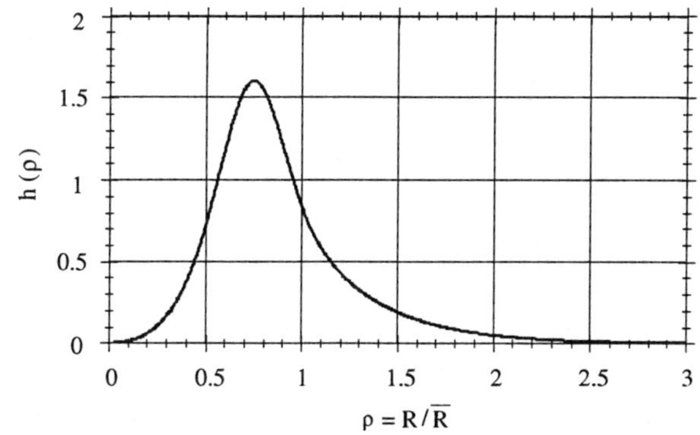

Fig. 7.10. Normalized particle size distribution for pure interface controlled coarsening due to a second-order reaction.

with

$$\nu_2 = \frac{K^s_{2nd}}{R_c^2 dR_c/dt} \tag{7.66}$$

Compared to all expression known up to now for various types of mass transport processes in the limit of a zero volume fraction of particles, this growth rate differs considerably in the dependence on the term $\rho - 1$. In all other expressions given in chapter 6 $d\rho/d\tau$ is directly proportional to $\rho - 1$. Here it is proportional to the square of this term. Furthermore, $d\rho/d\tau$ is a continuous function of ρ even though the sgn-function is present. If the term $\text{sgn}(\rho - 1)$ in eq.(7.65) were not present, there could only be particle growth and no dissolution. According to the recipe we use the two equations $d\rho/d\tau = 0$ and $d/d\rho(d\rho/d\tau) = 0$ to determine the values of ρ_m and ν_2, leading to $\nu_2 = 27/8$ and $\rho_m = 3$. The rate equation for the critical radius, eq.(7.66), can now be readily integrated to yield,

$$R_c^3 = \frac{8}{9} K^s_{2nd} t. \tag{7.67}$$

For a comparison with experiments one has to determine the relationship between the critical and the average radius. Using the distribution function as discussed above leads to $\langle R \rangle = 0.8877 R_c$. Thus the average radius is smaller than the critical. This is usual in all cases with the exception of pure diffusion controlled coarsening. Thus the growth law of eq.(7.67) is similar to that for pure diffusional controlled growth, with the same growth exponent n=3, but with different physical constants. A numerical evaluation of the equation for $h(\rho)$ leads to the time independent particle size distribution function shown in Fig.7.10.

The new distribution function differs strongly from that for diffusional growth or growth by a first-order interface reaction.

This difference is especially evident by the presence of many particles with larger normalized radii, where the distribution for a second-order reaction decreases to zero very slowly. In particular, the large cut-off radius of 3 is unusual. For diffusion controlled growth, in the limit of zero volume fraction, the cut-off radius is 3/2, for reaction controlled growth (first order) it is 2. A value of three is only known for convective-diffusive mass-transport for droplets moving by Stokes motion in a liquid matrix. In contrast to that case, however, for growth by second-order reaction treated here the probability finding a particle with a size larger than $\rho = 2$ is not negligible. Thus, if the long-time limit has been reached in an experimental situation, by examing the shape of the the particle size distribution, it should be possible to determine if coarsening via a second-order reaction is occurring. In addition the maximum of the distribution function for coarsening via a second-order reaction has a value that is rather small compared with that for a diffusion controlled coarsening process.

8. Ostwald ripening - Marqusee and Ross type analysis

The Wagner-analysis rests on the introduction of a critical radius and the assumption that the volume fraction is a conserved quantity. This is inconsistent since the small supersaturation decays during coarsening and thus the volume also varies with time. It is possible to remove this inconsistency using instead the more general mass or enthalpy conservation laws[1]. The asymptotic analysis employs a power series representation for the particle size distribution in time and shows that asymptotically a time independent state is reached under the appropriate scaling, which is unique and independent of the initial conditions. This analysis was performed by Marqusee and Ross and is especially suitable to incorporate effects beyond simple mass diffusion. Moreover, the analysis is self-consistent unlike the Wagner-analysis. We therefore present their approach in some detail. We first treat coarsening due to interface kinetics in a supersaturated matrix and then treat coarsening due to diffusional heat transfer.

8.1 Interface driven coarsening in a supersaturated matrix

We assume as before that we have spherical particles with a spectrum of sizes distributed at random in a matrix. The system shall be infinite in size such that the volume fraction of dispersed particles is close to zero. We assume a situation as described in the section on interface driven growth and coarsening (see section 6.4). Let us rewrite the

[1] Conservation of mass or energy are a quite more general physical principles than the conservation of the volume fraction of precipitates or dispersed phase in general and thus using such principles leads to more general answers. In the case of coarsening, when the supersaturation or the temperature in a dispersion are very close to their equilibrium values the differences are small, such that mass or enthalpy conservation and volume fraction conservation are are merely identical. From the general physical point of view there still is a difference.

growth rate derived in that section but this time avoiding the use of a critical radius,

$$\frac{dR}{dt} = k_s(x_B^\infty - x_B^I). \tag{8.1}$$

We assume that the initial concentration of B-atoms, x_B^0, is conserved throughout the coarsening process,

$$x_B^0 = x_B^\alpha(1 - \Phi(t)) + x_B^\beta \Phi(t) \tag{8.2}$$

$$= x_B^\alpha + (x_B^\beta - x_B^\alpha)\Phi(t), \tag{8.3}$$

where $\Phi(t)$ denotes the volume fraction of particles. Let

$$x_B^\alpha = x_B^\infty \tag{8.4}$$

with x_B^α the matrix concentration at ∞. Then,

$$x_B^0 = x_B^\infty + (x_B^\beta - x_B^\alpha)\Phi(t) \tag{8.5}$$

Since the curvature effect on concentration during coarsening is in general small compared to the difference of the concentration between the α- and the β-phase we assume

$$x_B^\beta - x_B^\alpha \approx x_B^{\beta,e} - x_B^{\alpha,e} \tag{8.6}$$

with $x_B^{\alpha,e}$ and $x_B^{\beta,e}$ the equilibrium concentrations of B in the matrix and in the particle assuming a flat interface. This gives

$$x_B^0 = x_B^\infty + (x_B^{\beta,e} - x_B^{\alpha,e})\Phi(t) \tag{8.7}$$

where the volume fraction $\Phi(t)$ can be calculated from

$$\Phi(t) = \frac{4\pi}{3}\int_0^\infty R^3 f(R,t)dR. \tag{8.8}$$

Using the Gibbs-Thomson relation for x_B^I, eq.(2.120), we obtain

$$\frac{dR}{dt} = k_s\left(x_B^0 - (x_B^{\beta,e} - x_B^{\alpha,e})\Phi(t) - x_B^{\alpha,e} - \frac{\ell^\alpha}{R}\right) \tag{8.9}$$

We introduce dimensionless coordinates,

$$r = \frac{R}{\ell^\alpha}, \tag{8.10}$$

$$\sigma(\tau) = x_B^0 - x_B^{\alpha,e} - (x_B^{\beta,e} - x_B^{\alpha,e})\Phi, \tag{8.11}$$

$$\tau = \frac{k_s}{\ell^\alpha}t. \tag{8.12}$$

Then, the growth law expressed in these coordinates becomes

$$\frac{dr}{d\tau} = \sigma(\tau) - \frac{1}{r}. \tag{8.13}$$

The continuity equation takes the form

$$\frac{\partial f}{\partial \tau} + \frac{\partial}{\partial r}\left(\frac{dr}{d\tau} f\right) = 0 \tag{8.14}$$

and the dimensionless supersaturation

$$\begin{aligned}\sigma(\tau) &= x_B^0 - x_B^{\alpha,e} - \frac{4\pi(\ell^\alpha)^3(x_B^{\beta,e} - x_B^{\alpha,e})}{3} \int_0^\infty r^3 f(r,\tau) dr \\ &= \sigma_0 - \sigma^* \int_0^\infty r^3 f(r,\tau) dr, \end{aligned} \tag{8.15}$$

with

$$\sigma_0 = x_B^0 - x_B^{\alpha,e} \tag{8.16}$$

as the initial supersaturation and σ^* as an obvious collection of terms. These three equations have to be solved together. In principle one could insert eqs.(8.13,8.15) into the continuity equation and would obtain a non-linear integro-differential equation to be solved. This generally has to be done numerically if one is interested to obtain information on the time evolution.

The Marqusee and Ross (MR) approach avoids this, as does the LSW analysis, and focuses on the late stage when the system reaches steady-state. We assume that the particle number as well as matrix supersaturation slowly decrease with time. These assumptions are reasonable for experimental situations. Some particles in the particle size distribution are smaller than the critical radius and will shrink and vanish, thus, decreasing the number of particles. The critical radius increases and, thus, the matrix supersaturation decreases. We use a scaling ansatz for the particle radius and expand the particle size distribution in a series of distributions F_0, F_1, $F_2 \ldots$, which decay with different temporal exponents,

$$z = r\tau^{-x} \tag{8.17}$$

$$f(r,\tau) = \tau^{-y}\left[F_0(z) + \tau^{-y_1} F_1(z) + \text{higher order terms}\right], \tag{8.18}$$

with $0 < y_1 < y_2 \ldots$ This expansion is based on the assumption of a general decay as discussed above. However, it should be noted that we assume that the particle size distribution at all times is composed of only slowly decaying distributions. This is certainly not always the case. Particularly in the early stages different behavior is possible. A detailed study on this was performed by Chen and Voorhees [63] who solved the set of equations numerically and studied the evolution of the particle size distributions. They found that despite the complicated initial evolution, at late times the system still approaches the same steady-state solution found using a procedure similar to that described below. Furthermore, our goal is just to find *one* and not every possible solution to the problem. As mentioned in the last chapter there are in fact more than one solution to the problem, although most of these solutions are unstable against noise.

We insert the expansion of the size distribution into the mass conservation condition,

$$\sigma(t) = \sigma_0 - \sigma^* \tau^{4x} \tau^{-y} \left[\int_0^\infty z^3 F_0(z) dz + \tau^{-y_1} \int_0^\infty z^3 F_1(z) dz + \ldots \right] \tag{8.19}$$

In the long-time limit only the term with the slowest decay rate will survive and the term with y_1 and all higher order terms vanish. Asymptotically the supersaturation $\sigma(\tau)$ should be zero, since then the system must be in equilibrium. This can only be achieved, if we choose $y = 4x$. Thus, in the long time limit only F_0 survives. In the short time F_0 would *not* exist since the overall matrix supersaturation is not zero but decreases. Again, we are only interested in the long time limit, where upon using $y = 4x$ we obtain

$$\sigma_0 = \sigma^* \int_0^\infty z^3 F_0(z) dz. \tag{8.20}$$

Using this equality, eq.(8.20), we see that the time dependent supersaturation has only one term left (if we neglect the higher order terms),

$$\sigma(\tau) = \tau^{-y_1} \int_0^\infty z^3 F_1(z) dz = \sigma_1 \tau^{-y_1}. \tag{8.21}$$

We will not derive an expression for $F_1(z)$, although this is possible, but only use the fact that that σ_1 is a constant and determine its value (in a similar manner as we did in the Wagner-analysis for ν). We will describe the evolution of the size distribution by the one factor which is left, namely the leading power in time,

$$f(r, \tau) = \tau^{-4x} F_0(z). \tag{8.22}$$

The exponent y_1 in eq.(8.20) is determined by rewriting the growth law in terms of the new coordinates,

$$\frac{dr}{d\tau} = \sigma_1 \tau^{-y_1} - \frac{\tau^{-x}}{z} \tag{8.23}$$

$$= \tau^{-x}\left(\sigma_1 \tau^{-y_1+x} - \frac{1}{z}\right) \tag{8.24}$$

$$= \tau^{-x} V(z; x - y_1) \tag{8.25}$$

Inserting eqs.(8.22,8.25) into the continuity equation, eq.(8.14), using $\partial/\partial r = \tau^{-x}\partial/\partial z$ leads to

$$\tau^{-4x-1}[4xF_0 + xzF_0'] = \tau^{-6x}\frac{\partial V}{\partial z}F_0 + \tau^{-6x}VF_0', \tag{8.26}$$

where we have used

$$\frac{\partial f}{\partial \tau} = -4x\tau^{-4x-1}F_0 - xz\tau^{-4x}F_0'\tau^{-1}. \tag{8.27}$$

The prime $'$ indicates differentiation with respect to z. eq.(8.26), which is an equation defining the size distribution, becomes independent of time τ, this was the intent in the expansion of the size distribution in terms of powers of time, if all powers in time sum to zero. We therefore choose

$$y_1 = x \quad \text{in} \quad V(z; x - y_1), \tag{8.28}$$

$$x = 1/2. \tag{8.29}$$

The last equality for x is a direct consequence of the first choice, since from eq.(8.26) the first choice leads to $-4x - 1 = -6x \;\rightarrow\; x = 1/2$. We are now able to directly give the time dependence of the number density,

$$n_V(\tau) = \int_0^\infty f(r,\tau)dr \propto \tau^{-y}\tau^{x}\int_0^\infty F_0(z)dz \propto \tau^{-3x} \propto \tau^{-3/2}. \tag{8.30}$$

For the average radius we get

$$\langle r\rangle = \frac{1}{n_V}\int_0^\infty r f(r,\tau)dr \propto \frac{\tau^{-4x}\tau^{2x}}{\tau^{-3/2}} \propto \tau^{1/2} \tag{8.31}$$

and the supersaturation goes to zero as

$$\sigma(t) \propto \tau^{-y_1} \propto \tau^{-x} \propto \tau^{-1/2}. \tag{8.32}$$

We are left with the problem to determine the size distribution $F_0(z)$. Eq.(8.26) reads after inserting the value for x and the function V

$$\left[2F_0 + \frac{z}{2}F_0'\right] = \frac{1}{z^2}F_0 + \left(\sigma_1 - \frac{1}{z}\right)F_0'. \tag{8.33}$$

Let us first state, that this distribution must also have a final cut-off radius z_{max}. We will prove this by the following evaluation: For large z eq.(8.33) becomes

$$2F_0 + z/2F_0' = 0 \qquad \text{for large } z \text{ only.} \tag{8.34}$$

which gives,

$$F_0 \propto \frac{1}{z^4}. \tag{8.35}$$

Inserting this into the expression for the initial supersaturation, eq.(8.20), yields

$$\sigma_0 = \sigma^* \int_0^\infty z^3 F_0(z)dz \propto \int_0^\infty z^{-1}dz \propto \ln(z) \quad \text{at large } z \tag{8.36}$$

This logarithmic divergence can only be avoided if $F_0(z)$ has a final cut-off z_{max}.

Thus, we are left with the same problem as in the LSW analysis: we have to determine the constant σ_1 and the cut-off value z_{max} together. The analysis can be made easier if we rewrite the differential equation for F_0 slightly. Introducing

$$W(z, \sigma_1) = \sigma_1 - \frac{1}{z} - \frac{z}{2} \tag{8.37}$$

we can write eq.(8.33) as

$$\frac{3}{2}F_0 = \frac{d}{dz}\left(W(z,\sigma_1)F_0\right) = F_0\frac{dW}{dz} + W\frac{dF_0}{dz}. \tag{8.38}$$

The functional behavior of W for different values of σ_1 is shown in fig.8.1. Its similarity to fig.7.1 is striking. W can be interpreted as a normalized growth rate, comparable to the $d\rho/d\tau$-function in the LSW theory.

The argument used to determine the values of the constants is quite similar to the manner we derived in the preceeding chapter the values of ν and ρ_m. The details are therefore omitted here. We again require that the distribution function shall be zero at z_{max}, $F_0(z_{max}) = 0$. The

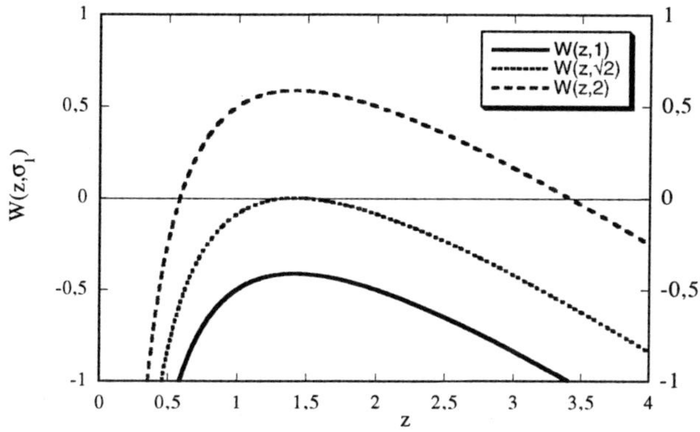

Fig. 8.1. Dependence of $W(z, \sigma_1)$ on z and σ_1.

first condition to determine both constants is given by the intercept of W with the axis $W = 0$ at z_{max},

$$W(z_{max}, \sigma_1) = 0 \quad \text{or} \quad \sigma_1 - \frac{1}{z_{max}} - \frac{z_{max}}{2} = 0. \tag{8.39}$$

The second condition states that at this point the function W has a maximum,

$$\left.\frac{d}{dz} W(z, \sigma_1)\right|_{z_{max}} = \frac{1}{z_{max}^2} - \frac{1}{2} = 0. \tag{8.40}$$

Both conditions together lead to

$$z_{max} = \sqrt{2}, \tag{8.41}$$
$$\sigma_1 = \sqrt{2}. \tag{8.42}$$

We insert this value into eq.(8.33) and obtain the final equation to be solved for the size distribution,

$$(2 - \frac{1}{z^2})F_0(z) = \left[\sqrt{2} - \frac{1}{z} - \frac{z}{2}\right] \frac{dF_0}{dz}. \tag{8.43}$$

which gives,

$$F_0(z) = F_N \frac{z}{(\sqrt{2} - z)^5} \exp\left(-\frac{3z}{\sqrt{2} - z}\right), \tag{8.44}$$

where F_N is a normalization constant. For comparison with experiments it is more convenient to transform this equation to radii, especially, since we will use LSW-like measures, i.e. $\rho = r/r_c = R/R_c$. In the LSW definition we have for the critical radius

$$\frac{1}{r_c} = \sigma(\tau) = \sigma_1 \tau^{-1/2}. \tag{8.45}$$

Therefore, we obtain from the definition of z

$$\frac{1}{z_c} = \sigma_1. \tag{8.46}$$

Using

$$\rho = \frac{r}{r_c} = \frac{z}{z_c} = z\sigma_1 \tag{8.47}$$

we obtain for the maximum radius in the ρ, τ space of LSW

$$\rho_{max} = \frac{z_{max}}{z_c} = 2. \tag{8.48}$$

Inserting eq.(8.47) into eq.(8.44) we obtain the classical result of Wagner,

$$h(\rho) = h_N \frac{\rho}{(2-\rho)^5} \exp\left(-\frac{3\rho}{2-\rho}\right). \tag{8.49}$$

with h_N a normalization constant that can be calculated directly by requiring $\int h(\rho)d\rho = 1$ yielding $h_N = 1/24$. The result is shown in Fig. 8.2. The temporal power law for the average particle radius is

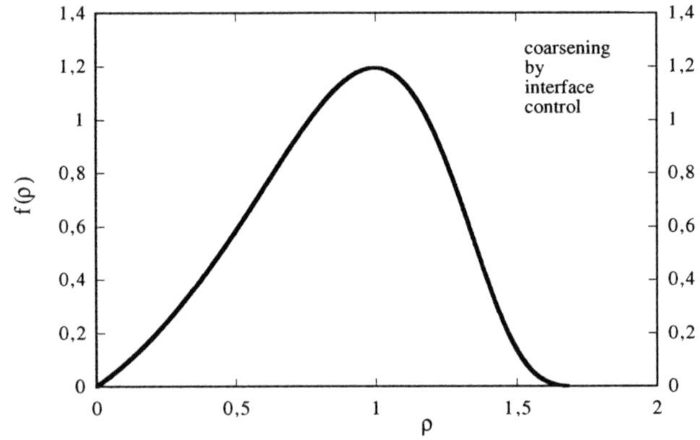

Fig. 8.2. Normalized size distribution for interface reaction of heat or mass.

obtained evaluating eq.(8.31) using the particle size distribution,

$$\langle r \rangle = \frac{4\sqrt{2}}{9} \tau^{1/2}. \tag{8.50}$$

For the dimensionless supersaturation we obtain

$$\sigma = \sigma_1 \tau^{y_1} = \sqrt{2}\tau^{-1/2} \tag{8.51}$$

With the definition given in eqs. (8.7) and (8.15) we can derive, how the volume fraction develops with time:

$$\Phi(t) = \frac{x_B^0 - x_B^{\alpha,e}}{x_B^{\beta,e} - x_B^{\alpha,e}} - \frac{\sqrt{2}}{\sqrt{\tau}(x_B^{\beta,e} - x_B^{\alpha,e})} \tag{8.52}$$

The volume fraction increases to its maximum level given by the lever rule as $\tau \to \infty$.

There is a simple relation between the average particle radius and the critical radius, namely,

$$\langle R \rangle = \frac{\int_0^\infty R f(R,t) dR}{\int_0^\infty f(R,t) dR} = R_c \int_0^2 \rho h(\rho) d\rho \tag{8.53}$$

leading to

$$\langle R \rangle = \frac{8}{9} R_c. \tag{8.54}$$

This shows that in interface driven coarsening the average and the critical radius are not identical, in contrast to the purely diffusive case. We rewrite the time law for the average radius in dimensional coordinates,

$$\langle R \rangle = \frac{4\sqrt{2} k_s \ell^\alpha}{9} t^{1/2}. \tag{8.55}$$

The result is identical to that obtained by Wagner with the 'critical radius' approach. The advantage of the Marqusee and Ross approach is that it was not necessary to assume that the volume fraction was constant. The asymptotic analysis shows that to the order of the expansion where $\langle r \rangle \propto t^{1/2}$, the volume fraction changes with time. Only after long annealing times the second term in eq.(8.52) could eventually become small compared to the first term. The Marqusee and Ross analysis clearly shows, that the size distribution and growth law obtained are valid in the asymptotic regime, i.e. after long annealing times, since it expands the size distribution into powers of time. This is somewhat unclear in the Wagner-analysis and the Wagner analysis turns out to be not self-consistent. The approach used in the MR analysis is to not make *one* separation ansatz but to use a series of separation ansatz's, see eq.(8.18), together with a suitable scaling (knowing already from the classical LSW analysis that a scaling is possible). In addition the

MR analysis shows that the mass conservation condition together with the growth law is necessary to determine the scaling behavior of the system at long coarsening times.

Let us finally calculate the moments of the size distribution given in eq.(8.49). The moments are defined as:

$$m_k^h = \int_0^{\rho_m} \rho^k h(\rho) d\rho \tag{8.56}$$

Evaluating numerically the moments yields the results given in table 8.1.

k	m_k^h
-1	13/9
0	1
1	8/9
2	8/9

Table 8.1: Table of the moments of the size distribution of interface controlled coarsening of first order given in eq.(8.49) with the normalization stated there.

The standard deviation defined in eq.7.45 is:

$$\delta\sigma^2 = 8/81 \quad \text{or} \quad \delta\sigma = 2\sqrt{2}/9 \tag{8.57}$$

To illustrate that this behaviour also is observed experimentally we show some results of Exner and coworkers [92, 93, 95, 96, 97]. They investigated the coarsening carbides dispersed in liquid cobalt during liquid phase sintering. For all types of carbides investigated (e.g. TaC, VC, WC) they observed that the square of the particle diameter increases linearly with time and they also developed tests to compare the theoretical distributions with experimental ones. One example of their size distributions is shown in fig.8.3. Fig.8.4 shows for three different sintering times the variation of the measured average intercept length with sintering time, clearly revealing that the result of the Marqusee and Ross approach described above is correct.

8.2 Diffusive coarsening in a supercooled melt

In this section we will show how the MR-approach can be used to analyze the coarsening of spherical particles if the process occurs by

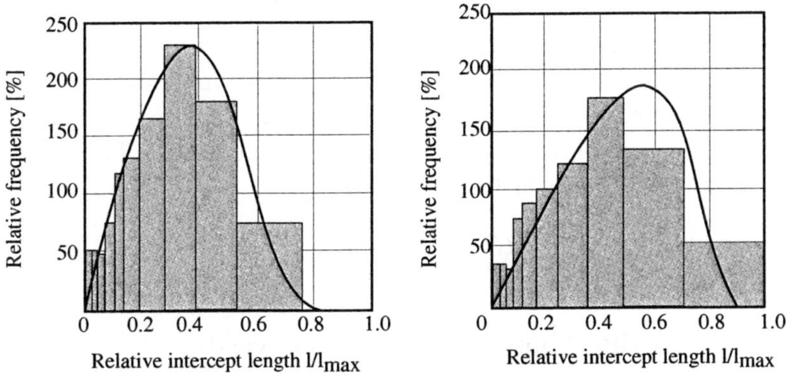

Fig. 8.3. Normalized size distribution of VC particles in liquid Cobalt sintered at 1500°C. Shown is the mean intercept length normalized by the maximum observed one and their relative frequency. The left figure shows the theoretical curve for an interface reaction of first order and the right diagram compares the experimental results with the prediction of diffusional coarsening.

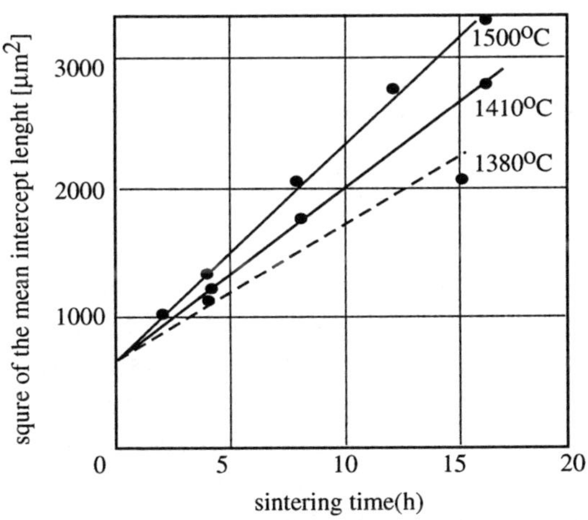

Fig. 8.4. Measured average intercept length of VC particles in liquid Cobalt sintered different temperatures. Shown is the square of the mean intercept length. It clearly varies linearly with time as predicted.

the diffusion of heat in a pure material. Instead of the mass conservation condition used previously, we conserve heat, the system is isoenthalpic. We will not go through the whole analysis, since no new results appear compared to those derived in section 7. The essential point is only to show, how to use the conservation of enthalpy in the context of coarsening theory.

For convenience let us rewrite the coarsening equation for one spherical particle given in eq.(6.1),

$$\frac{dR}{dt} = \frac{\lambda}{\rho L} \frac{T_I - T_\infty}{R}. \tag{8.58}$$

Inserting for T_I the Gibbs-Thomson relation yields

$$\frac{dR}{dt} = \frac{\lambda}{\rho L R} \left(T_m - \frac{2\Gamma}{R} - T_\infty \right). \tag{8.59}$$

Introducing the following reduced coordinates,

$$\Theta = \frac{T_m - T_\infty}{T_m}, \tag{8.60}$$

$$r = \frac{R T_m}{2\Gamma}, \tag{8.61}$$

$$\tau = \frac{\lambda T_m^3}{4\rho L \Gamma^2} t, \tag{8.62}$$

we can write the growth rate in dimensionless coordinates,

$$\frac{dr}{d\tau} = \frac{1}{r} \left(\Theta(\tau) - \frac{1}{r} \right). \tag{8.63}$$

$\Theta(\tau)$ represents the relative departure of the temperature from its equilibrium value T_m, i.e. the supercooling. We expect that after long coarsening times Θ tends to zero, since the system should reach equilibrium. We have now to establish a relation between this supercooling and the size distribution as we did in the preceding section, in which σ, a dimensionless supersaturation, took the place of Θ. There we related σ to the mass conservation equation. In this context we will examine how the enthalpy of the system evolves with time.

The enthalpy per unit volume can be expressed as

$$H_V(T) = H_V^e + \rho C_P (T - T^{eq}), \tag{8.64}$$

wit H_V^e the enthalpy per unit volume at equilibrium, if the actual temperature of the system T is not too far away from the equilibrium temperature $T^e = T_m$. Applying this relation to the system of liquid matrix and dispersed particles we obtain

$$H_V^{liq.}(T) = H_V^{liq.,e} + \rho C_P^{liq.}(\bar{T} - T_m), \tag{8.65}$$

where $\bar{T}$ is the volume averaged temperature of the liquid. The temperature of the liquid is non uniform due to the presence of temperature gradients that induce the heat flow that is necessary for coarsening. For the solid we have to account for the distribution of particle sizes and, thus, particles of different size are at different temperatures,

$$H_V^{sol.}(T) = H_V^{sol.,e} + \frac{4\pi}{3}\rho C_P^{sol.} \int_0^\infty (T_i - T_m) R^3 f(R,t) dR, \tag{8.66}$$

where T_i is a function of R through the Gibbs-Thomson equation. The overall enthalpy balance for the two-phase mixtures is

$$H_V^0 = (1 - \Phi) H_V^{liq.}(T) + \Phi H_V^{sol.}(T). \tag{8.67}$$

The total enthalpy per volume of the system, H_V^0, is a conserved quantity and thus constant. Inserting the expressions eqs.(8.65,8.66) and using $\rho L = H_V^{liq.,e} - H_V^{sol.,e}$, we obtain

$$H_V^0 = H_V^{liq.,e} - \Phi \rho L + \rho C_P^{liq.}(1 - \Phi)(\bar{T} - T_m) + \rho C_P^{sol.} \Phi J_1 \tag{8.68}$$

where

$$J_1 = \frac{4\pi}{3} \int_0^\infty (T_I - T_m) R^3 f(R,t) dR \tag{8.69}$$

If we assume that the deviation in the temperature of the particles from the equilibrium temperature is small, as is the case during coarsening in most systems, then $L \gg C_P^{sol.} J_1$ and the last term can be neglected. Thus, eq.(8.68) becomes,

$$\frac{H_V^0 - H_V^{liq.,e}}{\rho L} = -\Phi + \frac{C_P^{liq.}}{L}(1 - \Phi)(\bar{T} - T_m) \tag{8.70}$$

Since we are using eq.(8.58) to describe the growth rate of a particle, we are implicitly assuming that the volume fraction of solid is very small, $\Phi \ll 1$. In addition, the volume averaged temperature $\bar{T}$ must therefore also be very close to that at infinity, T_∞ and thus,

$$\frac{H_V^0 - H_V^{liq.,e}}{\rho L} = -\Phi + \frac{C_P^{liq.}}{L}(T_\infty - T_m) \tag{8.71}$$

Using now the nondimensionalisation given above we can rewrite this as

$$\Theta(t) = -\frac{H_V^0 - H_V^{liq.,e}}{\rho C_P^{liq.} T_m} + \frac{L}{C_P^{liq.} T_m}\Phi\,. \tag{8.72}$$

This finally can be transformed to

$$\Theta(t) = -\Theta_0 + \vartheta \int_0^\infty r^3 f(r,\tau)dr, \tag{8.73}$$

with an obvious definition of Θ_0 and the abbreviation

$$\vartheta = \frac{4\pi L}{3C_P^{liq.} T_m}\left(\frac{2\Gamma}{T_m}\right)^3 \tag{8.74}$$

Eq. 8.73 exactly is the expression we were looking for to be inserted into the growth rate of eq.(8.63). Using the transformation and series expansion outlined in the preceding section, eqs.(8.17,8.18) and inserting this into the expression for the relative supercooling, eq.(8.73) we obtain

$$\Theta(t) = -\Theta_0 + \vartheta\tau^{-y+4x}\left[\int_0^\infty z^3 F_0(z)dz + \tau^{-y_1}\int_0^\infty z^3 F_1(z)dz + \ldots\right]. \tag{8.75}$$

Since the supercooling should be zero as τ tends to infinity we must have $y = 4x$ and, thus,

$$\Theta_0 = \vartheta \int_0^\infty z^3 F_0(z)dz, \tag{8.76}$$

and thus the supercooling depends on time as

$$\Theta(t) = \vartheta\tau^{-y_1} \int_0^\infty z^3 F_1(z)dz = \Theta_1\tau^{-y_1}. \tag{8.77}$$

Θ_1 is a constant and we again neglect higher order terms as before. Rewriting the growth rate as

$$\frac{dr}{d\tau} = \frac{\tau^{-2x}}{z}\left(\Theta_1\tau^{-y_1+x} - \frac{1}{z}\right) = \tau^{-2x}V(z; -y_1 + x) \tag{8.78}$$

and inserting it into the continuity equation we obtain

$$\tau^{-4x-1}[4xF_0 + xzF_0'] = \tau^{-7x}\frac{\partial V}{\partial z}F_0 + \tau^{-7x}VF_0' \tag{8.79}$$

This equation becomes independent of time if we again choose $y_1 = x$ and evaluate the exponents $4x + 1 = 7x$ to obtain $x = 1/3$. Having calculated the exponent the rest of the story is mathematically similar to that for interface kinetics treated before. Therefore we list important results:

$$\Theta_1 = \left(\frac{9}{4}\right)^{1/3} \tag{8.80}$$

$$z_m = \left(\frac{3}{2}\right)^{1/3} \tag{8.81}$$

$$\langle r\rangle^3 \propto \tau \tag{8.82}$$

$$n_V \propto \tau^{-1} \tag{8.83}$$

$$T_\infty = T_m\left(1 - \frac{\Theta_1}{\tau^{1/3}}\right) \approx \bar{T} \tag{8.84}$$

Interesting to see that the volume averaged temperature $\bar{T}$ or the far field temperature continuously increase towards the equilibrium value given by the melting temperature. This result cannot be obtained from the Wagner analysis.

For a system undergoing coarsening by mass transfer we list the important results in dimensional form explicitly.

$$\langle R\rangle^3(t) = \frac{8\ell^\alpha D}{9(x_B^{\beta,e} - x_B^{\alpha,e})}t \tag{8.85}$$

$$x_B^\infty = x_B^{\alpha,e} + \frac{9(\ell^\alpha)^2(x_B^{\beta,e} - x_B^{\alpha,e})}{D}t^{-1/3} \tag{8.86}$$

$$n_V(t) = 1.99\left(\frac{3(x_B^0 - x_B^{\alpha,e})}{8\pi D\ell^\alpha}\right)t^{-1} \tag{8.87}$$

For a dilute solution, $x_B^{\beta,e} \approx 1$, $G_m'' = \mathcal{R}T/x_B^{\alpha,e}$ so that eq.(8.85) reduces to,

$$\langle R\rangle^3(t) = \frac{8\sigma x_B^{\alpha,e}D}{9\mathcal{R}T}t \tag{8.88}$$

So that the ripening rate of a two phase mixture is proportional to the diffusion coefficient in the solid, the solubility of the solute in the matrix and the magnitude of the interfacial energy.

9. Multiparticle diffusion analysis

The LSW theory for coarsening has a serious drawback: it does not account for diffusional interactions between particles. As a result the coarsening rate of a particle is a function only of the radius of a particle relative to the average particle size. This is not the case in nonzero volume fraction systems, for when a particle of a given size is surrounded by particles larger than itself it must coarsen in a different manner than when this particle is surrounded by particles that are smaller than itself. Diffusional interactions between particles also must lead to a coarsening rate of the system that is a function of the volume fraction. As the volume fraction increases the interparticle separation decreases but the concentration in the matrix at the interfaces of the particles is not a function of the interparticle separation, it set by the Gibbs-Thomson equation. Thus the concentration gradients in the matrix at the interfaces increases with increasing volume fraction and the coarsening rate of the system must therefore increase. Both the physics of local diffusional interactions and the dependence of the coarsening rate on volume fraction are missing from the LSW theory.

Diffusional interactions clearly act to increase the coarsening rate above that predicted by LSW. However, these diffusional interactions, unlike the fluid flow, do not alter the temporal power laws from that predicted by LSW. The increase in coarsening rate manifests itself as an increase in the amplitude of the temporal power laws. Thus, for example, at nonzero volume fractions $\langle R \rangle^3 = K(\phi)t$, where ϕ is the volume fraction of coarsening phase. The presence of interparticle diffusional interactions also does not alter the LSW prediction of a unique time-independent particle size distribution when the particle radii are scaled by the average particle size. However, the shape of the distribution becomes a function of the volume fraction.

There have been many theories proposed to account for the effects of diffusional interactions on the coarsening process. Since it is impossible to discuss each theory in detail we present representative examples of broad classes of these theories. We begin by developing a solution to the multiparticle diffusion problem that governs the coarsening process at nonzero volume fractions. This approach is, in some sense, a microscopic description of the coarsening process as all the detailed diffusional interactions are included in the theory. While this theory can be used to determine the coarsening rate of individual particles, it does not, by itself, give the statistically averaged properties of coarsening ensembles that are easy to measure experimentally. Thus, we follow this with a presentation of models that employ effective media of various designs to predict these statistically averaged properties without knowledge of the detailed diffusional interactions between particles. We close with a discussion of first-principles statistical mechanical theories and computer simulations. Both of these approaches are based upon the microscopic description of the coarsening process.

9.1 The Multiparticle Diffusion Problem

In general, the diffusion field satisfies the time dependent diffusion equation,

$$\frac{\partial x_B}{\partial t} = D\nabla^2 x_B. \tag{9.1}$$

However, during Ostwald ripening it is possible to describe the diffusion field using the quasi-stationary approximation. In this case, instead of eq.(9.1) we shall employ,

$$D\nabla^2 x_B = 0. \tag{9.2}$$

The validity of this approximation rests on two assumptions: (1) the rate of particle growth is slow on the time scale of diffusion. Thus, the actual time dependent diffusion field in the matrix is nearly relaxed to its time independent form. (2) The distance from a particle at which the concentration field has to be determined cannot be too large. It is clear physically, that a change in the concentration field propagates through the system at a finite speed that is related to the magnitude of the diffusion coefficient. In contrast, a diffusion field satisfying Laplace's equation gives, essentially, an infinitely fast communication rate. Thus, at a sufficiently large distance from a particle the concentration field given by the quasi-stationary approximation and time dependent diffusion equation must differ. Coriell and Parker derived the following condition under which the concentration field surrounding a growing spherical particle agrees with that found by the quasi-stationary approximation,

$$\left(\frac{S_0}{2}\right)^{1/2} \frac{r}{R} \ll 1, \tag{9.3}$$

where S is the supersaturation in the matrix,

$$S_0 = \frac{x_B^{\infty} - x_B^{\alpha}}{x_B^{\beta} - x_B^{\alpha}} \tag{9.4}$$

and r is the distance from the center of a particle where the concentration is evaluated. So, for small supersaturations, as is common during Ostwald ripening, one can employ the quasi-stationary approximation as long as the distance between two particles, given approximately by r, is not too large, or, the volume fraction is not too low. This is because, as we will see, an accurate description of the diffusion field on the order of multiples of the average interparticle separation is necessary to model the ripening process. This condition can also be rewritten in terms of the growth rate of a particle, since from eq.(4.32) $S_0 \sim dR/dt$. Thus, requiring small S_0 is equivalent to requiring slow particle growth rates. Equation(9.3) also implies that the zero volume fraction assumption of LSW is inconsistent with the quasi-stationary approximation that is used to describe particle growth in this theory.

We shall develop a solution to the diffusion problem that is consistent with the particles having a spherical morphology. This can be done in a self-consistent manner using multiple scattering theory [200], irreducible tensor [25] or using boundary integrals and multipole expansions [5]. Here we will employ a more straightforward approach. We shall represent the emission or adsorption of solute from shrinking or growing particles as point sources or sinks located at the center the particles. This point source/sink approximation was first employed in the context of a coarsening problem by Weins and Cahn [306]. The diffusion field for a system consisting of n point sources/sinks, or particles, located at positions $\boldsymbol{r}_i$ is given by

$$\nabla^2 x_B(\boldsymbol{r}) = -4\pi \sum_i^n B_i \delta(\boldsymbol{r} - \boldsymbol{r}_i), \tag{9.5}$$

where B_i is the strength of the point source/sink, and $\delta(\boldsymbol{r})$ is the Dirac delta function located at a position $\boldsymbol{r}_i$. The strength of the point source is determined by the boundary conditions: the Gibbs-Thomson equation, and global mass conservation.

To understand the physical significance of the point source/sink strength, integrate eq.(9.5) over a region containing only one particle,

$$\int_V \nabla^2 x_B(\boldsymbol{r})\, dv = -4\pi \int_V B_i \delta(\boldsymbol{r} - \boldsymbol{r}_i)\, dv. \tag{9.6}$$

The integral over the volume can be converted to an integral over the surface of the particle by using the divergence theorem for the flux $\boldsymbol{J}$

$$\int_V \nabla \cdot \boldsymbol{J} dv = \int_s \boldsymbol{J} \cdot \boldsymbol{n} ds. \tag{9.7}$$

$\boldsymbol{n}$ is the unit vector normal to the surface. Using Fick's first law in the LHS of eq.(9.7) gives,

$$-D \int_V \nabla^2 x_B(\boldsymbol{r})\, dv = \int_s \boldsymbol{J} \cdot \boldsymbol{n} ds. \tag{9.8}$$

Substituting eq.(9.5) with $n = 1$,

$$4\pi D \int_V B_1 \delta(\mathbf{r} - \mathbf{r}_1)\, dv = \int_s \mathbf{J} \cdot \mathbf{n} ds. \tag{9.9}$$

Since the integral is 0 for V not including particle 1 and 1 for V including particle 1,

$$B_1 = \frac{1}{4\pi D} \int_{s_1} \boldsymbol{J} \cdot \boldsymbol{n} ds. \tag{9.10}$$

Thus, B_1 is proportional to the total mass flowing into or out of a particle. If $B_1 < 0$ solute is flowing into the particle, since this implies that the solute flux and normal point in opposite directions. Conversely, if $B_1 > 0$ solute is flowing out of a particle since the solute flux and the normal point in the same direction. Once B is known for a given particle, the growth or shrinkage rate of the particle can be determined. To see this we use the mass balance condition at the interface for a spherical particle,

$$\left(x_B^\beta - x_B^\alpha\right) \frac{dR}{dt} = D \left.\frac{\partial x_B}{\partial r}\right|_{r=R}. \tag{9.11}$$

Since $\partial x_B/\partial r$ at the surface of a spherical particle is equal to $\nabla x_B \cdot \boldsymbol{n}$ and using Fick's first law, $\boldsymbol{J} = -(D/V_m)\nabla x_B$ where $/V_m$ is the density of solute (assumed equal in both phases), eq.(9.11)becomes

$$\frac{1}{V_m}\left(x_B^\beta - x_B^\alpha\right)\frac{dR}{dt} = -\boldsymbol{J}\cdot\boldsymbol{n}. \tag{9.12}$$

Using eq.(9.12) in eq.(9.10) we obtain

$$B_1 = -\frac{\left(x_B^\beta - x_B^\alpha\right)R_1^2}{DV_m}\frac{dR_1}{dt}. \tag{9.13}$$

Thus, if $B > 0$ the particles are growing and if $B < 0$ the particles are shrinking. Since $V(t) = 4\pi R(t)^3/3$, eq.(9.13) can be written as

$$B_1 = -\frac{\left(x_B^\beta - x_B^\alpha\right)}{4\pi DV_m}\frac{dV_1}{dt}. \tag{9.14}$$

So, B_i is directly proportional to the rate of change of the volume of particle i.

The diffusion field as shown in eq.(9.5) satisfies Poisson's equation, not the full time dependent diffusion equation. Thus, this equation does not conserve the total amount of solute in the system. As done by LSW this mass or solute conservation condition must be added explicitly. Here we shall neglect the small change in the total amount of solute in the matrix phase during coarsening, and assume the volume fraction of particle phase as constant. This approximation is valid when the average particle size is much larger than the critical radius for nucleation. The total volume of particle phase, V_T, is

$$V_T = \sum_i^n V_i. \tag{9.15}$$

If the total volume is to remain constant,

$$0 = \sum_i^n \frac{dV_i}{dt}. \tag{9.16}$$

Using eq.(9.14) for every particle i in eq.(9.16) we get

$$0 = \sum_i^n B_i. \tag{9.17}$$

Thus, the constraint of a constant volume fraction of particle phase implies that the source/sink strengths sum to zero.

A solution to eq.(9.5) is well known, see any text on electricity and magnetism (see for example [147]),

$$x_B(\boldsymbol{r}) = a_0 + \sum_i^n \frac{B_i}{|\boldsymbol{r} - \boldsymbol{r}_i|}, \tag{9.18}$$

where a_0 is a constant. The concentration diverges to infinity at the center of each particle. However, since the above holds only in the matrix, this is of no concern.

The $n+1$ constants, $a_0, B_i, i = 1 \dots n$ can be determined by applying the boundary conditions appropriate for the Ostwald ripening problem. In particular, the concentration in the matrix at the particle-matrix interface must be given by the Gibbs-Thomson equation,

$$x_B^\alpha(\boldsymbol{r}_i^s) = x_B^{\alpha,e} + \frac{\ell^\alpha}{R_i}, \tag{9.19}$$

where $\boldsymbol{r}_i^s$ is a vector to the surface of the ith particle. This vector can be expressed as $\boldsymbol{r}_i^s = \boldsymbol{r}_i + \boldsymbol{R}_i$, where $\boldsymbol{r}_i$ locates the center of particle i, and $\boldsymbol{R}_i = R_i\hat{\boldsymbol{r}}$, and $\hat{\boldsymbol{r}}$ is usual unit vector in the radial direction of a spherical coordinate system with an origin located at the center of the ith particle. Since the value of the diffusion field at the interface in the matrix must agree with the value given by the Gibbs-Thomson equation, for particle j where $j \neq i$, we set $\boldsymbol{r} = \boldsymbol{r}_j + \boldsymbol{R}_j$ in eq.(9.18),

$$a_0 + \frac{B_j}{R_j} + \sum_{i,i\neq j}^n \frac{B_i}{|\boldsymbol{r}_j + \boldsymbol{R}_j - \boldsymbol{r}_i|} = x_b^{\alpha,e} + \frac{\ell^\alpha}{R_j}. \tag{9.20}$$

We name the vector connecting the centers of particle j and i as $\boldsymbol{r}_{ji} = \boldsymbol{r}_j - \boldsymbol{r}_i$ and obtain

$$A + \frac{B_j}{R_j} + \sum_{i,i\neq j}^n \frac{B_i}{|\boldsymbol{R}_j - \boldsymbol{r}_{ji}|} = x_B^{\alpha,e} + \frac{\ell^\alpha}{R_j}. \tag{9.21}$$

We are now faced with a difficulty. The RHS of eq.(9.21) is a constant whereas the LHS of eq.(9.21) varies with the magnitude of $\boldsymbol{r}_{ji} + \boldsymbol{R}_j$, i.e. as the vector $\boldsymbol{R}_j$ sweeps over the surface of particle j. This implies that given the Gibbs-Thomson boundary condition the point source representation cannot be perfectly correct since the concentration produced by all particles $i \neq j$ on the surface of particle j will vary with position. Intuitively, however, this variation will be small if the separation between particles is large, $r_{ji} >> R_j$, where $r_{ji} = |\boldsymbol{r}_{ji}|$ and it may still serve as an adequate approximation in a certain limit. To make the extent of the approximation clear, we expand $1/|\boldsymbol{R}_j - \boldsymbol{r}_{ji}|$ in a series of spherical harmonics [19],

$$\frac{1}{|\boldsymbol{R}_j - \boldsymbol{r}_{ji}|} = 4\pi \sum_{k=0}^{\infty} \sum_{l=-k}^{k} \frac{1}{2k+1} Y_k^{l*}(\theta_R, \phi_R) Y_k^l(\theta_r, \phi_r) \frac{R^k}{r_{ji}^{k+1}}, \quad (9.22)$$

where $Y_k^{l*}(\theta_R, \phi_R)$ is the complex conjugate of the spherical harmonic as a function of the angles of the vector $\boldsymbol{R}_j$ of a spherical coordinate system located at the center of the jth particle, and $Y_k^l(\theta_r, \phi_r)$ is the spherical harmonic as a function of the angles of the vector $\boldsymbol{r}_{ji}$ of a spherical coordinate system located at the center of the jth particle (for a more detailed discussion of spherical harmonics see [147],[216]). See table 9.1 for a listing of the first few spherical harmonics. Note

$$\begin{aligned} Y_0^0(\theta, \phi) &= \frac{1}{\sqrt{4\pi}} \\ Y_1^1(\theta, \phi) &= -\sqrt{\frac{3}{8\pi}} \sin(\theta) e^{i\theta} \\ Y_0^1(\theta, \phi) &= \sqrt{\frac{3}{4\pi}} \cos(\theta) \\ Y_1^{-1}(\theta, \phi) &= +\sqrt{\frac{3}{8\pi}} \sin(\theta) e^{-i\theta} \end{aligned}$$

Table 9.1. Spherical harmonics

that each new term in the series is of smaller magnitude than the previous one since $R/r_{ji} < 1$. If the interparticle separation is large compared to the particle radii $R/r_{ji} \ll 1$ we take only the first term in the expansion. Thus eq.(9.22) becomes

$$\frac{1}{|\boldsymbol{R}_j - \boldsymbol{r}_{ji}|} = \frac{1}{r_{ji}} + O\left(R/r_{ji}^2\right). \quad (9.23)$$

Using this in eq.(9.21) we get

$$a_0 + \frac{B_j}{R_j} + \sum_{i,i \neq j}^{N} \frac{B_i}{r_{ji}} = x_B^{\alpha,e} + \frac{\ell^\alpha}{R_j} \quad (9.24)$$

This condition can be applied at all n particles in the system, yielding n equations for the $n+1$ unknowns, a_0 and $B_j, j = 1 \ldots n$. The final equation needed to determine all the unknown coefficients is the condition that the volume fraction of particle phase remain constant, eq.(9.17). Thus, once the particle locations and radii are given, solving this linear system of equations gives the particle source/sink strengths and hence the coarsening rate of each particle. It was found in numerical evaluations that this approximation is valid up to volume fractions of ca. 10%.

It is possible to include higher order spherical harmonics in the representation of the diffusion field [5, 196]. For example, dipolar terms can be included in the description of the diffusion field. These dipole terms provide a more accurate description of the diffusion field at small interparticle separations than the monopole approximation of eq.(9.5). Adding dipoles in the representation of the diffusion field, however, leads to a concentration gradient that will vary with position along the surface of the particle. This variation will lead to particle migration, but not change the shape of the particle from a sphere. The particle migrates because regions of the interface can grow or shrink faster than another region, due to the varying concentration gradient. This gives rise to a motion of the center-of-mass of the particle. Such motion has been observed experimentally in transparent system [291]. Adding dipoles allows the diffusion field to be described accurately to volume fractions not higher than approximately 30%. Adding higher multipole moments will provide a still more accurate representation of the diffusion field. However, to be consistent with these higher order terms the particles must also change shape from a sphere, greatly complicating the application of the Gibbs-Thomson boundary condition.

The relationship between the multiparticle diffusion solution and the classical LSW form for the growth rate of a particle of a size R can be seen by taking the limit $r_{ji} \to \infty$ in eq.(9.24), yielding.

$$a_0 + \frac{B_j}{R_j} = x_B^{\alpha,e} + \frac{\ell^\alpha}{R_j}. \tag{9.25}$$

Solving eq.(9.25) for B_j gives

$$B_j = \ell^\alpha + R_j \left(x_B^{\alpha,e} - a_0 \right), \tag{9.26}$$

and summing eq.(9.26) from 1 to n we get

$$\sum_{j=1}^{n} B_j = n\ell^\alpha + (x_B^{\alpha,e} - a_0) \sum_{j=1}^{n} R_j. \tag{9.27}$$

Since the LHS of eq.(9.27) is zero by the volume fraction conservation condition,

$$a_0 = x_B^{\alpha,e} + \frac{\ell^\alpha}{\langle R \rangle}. \tag{9.28}$$

Thus, the average composition in the matrix is given by the average particle size. Using eq.(9.28) in eq.(9.26) we obtain an expression for the rate of solute flow into or out of a particle,

$$B_i = \ell^\alpha \left(1 - R_i/\langle R \rangle\right). \tag{9.29}$$

So if the particle is larger than the average particle size, $B_i < 0$ and mass is flowing into a particle, but if the particle is smaller than the average particle size, $B_i > 0$ and mass is flowing out of a particle. Using eq.(9.29) in eq.(9.13) we obtain the LSW growth law, eq.(6.10). The point source approximation recovers the LSW kinetic equation in the limit of infinite particle separation.

9.2 Statistically averaged properties

If the volume fraction of particles is sufficiently small the above solution to the multiparticle diffusion problem is valid. Unfortunately, in most experiments the properties of importance that can be obtained are statistically averaged quantities, such as the growth rate of the average particle size and the shape of the time independent particle size distribution, and not the coarsening rate of an individual particle. Thus, the remaining challenge is to average the microscopic equations to arrive at the statistically averaged quantity of interest. The approaches to do this fall into are three categories: effective medium approaches, statistical mechanical theories, and direct computer simulations.

Before describing these theories, it is helpful to consider the physics underlying interparticle diffusional interactions. This will also establish the reasons why predicting the statistically averaged properties of an ensemble of coarsening particles is a very challenging problem. The first issue is the interaction distance between particles. In looking at eq.(9.24) it appears that one particle interacts over a distance $1/r_{ji}$.

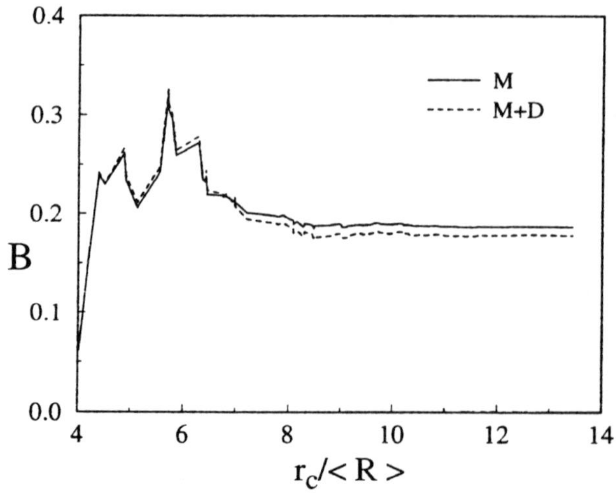

Fig. 9.1. The growth rate of a particle, or source/sink strength, of a particle as a function of the cut-off distance r_c relative to the average particle radius at a volume fraction of 0.1

This is an extremely long-ranged interaction. However, the actual interaction distance is much shorter as a result of diffusional screening. Consider the fate of the flux emanating from a shrinking particle. This flux is absorbed by neighboring growing particles. Thus, the distance over which the flux from this particle exists is smaller than one would expect from the $1/r_{ij}$ dependence of eq.(9.24). As a illustration of this, Figure(9.1) shows the variation in the source/sink strength of a particle located at the origin where only those particles within a spherical region of radius equal to the cut-off distance, r_c, have been considered in determining the source/sink strength B. The volume fraction of particles in this case is 10%. As the cut-off distance increases the coarsening rate of the particle varies with the size and locations of the particles that happen to be in the shell of that radius surrounding the particle. However, if the radius of shell, r_c, becomes sufficiently large, on the order of $8\langle R\rangle$, the growth rate of the particle becomes independent of r_c, which means it has converged to the value it would have in an infinite system. The screening length ℓ_s is defined as that distance beyond which a particle does not interact with other particles. For the 10% volume fraction system considered in Fig. 9.1 it is approximately $8\langle R\rangle$. This screening distance is a function of the volume fraction: As the volume fraction decreases, and the spacing between the particles increases, the screening length increases. Physically this is because the flux leaving a shrinking particle must travel longer distances before being absorbed by growing particles. Fig. (9.1) shows that the interaction distance is much larger than the average particle size but smaller than would be expected on the basis of the $1/r_{ji}$ dependence of eq.(9.24). A second consideration are the effects of interparticle spatial correlations. One would expect that during ripening particles are not distributed randomly in space. Consider the result of placing particles in a space in a random fashion. It is then possible for a small particle to be placed very close to a large particle. A large gradient will develop between the particles and the small particle will disappear quickly. Thus, one would expect that there will be a deficit of small particles near large particles. Thus the diffusional interactions, which are the cause of the change in the coarsening rate with volume fraction, are a function of the spatial correlations of the particles. These spatial correlations develop due to the diffusional interactions themselves. This coupling makes developing rigorous statistical mechanical theories extremely challenging.

9.3 Effective Medium Theories

Determining the statistically averaged growth rate or source/sink strength using eq.(9.24) is a challenging task. In an effort to avoid addressing the details of interparticle diffusional interactions embodied by eq.(9.24) many people have employed effective medium theories. The advantage of an effective medium theory is that it provides the statistically averaged properties of the ensemble of particles from the outset. The solution of the equations defining the average growth rate of a particle of a given size using an effective medium is straightforward. The central task, therefore, is to construct the effective medium in a fashion such that it yields the statistically averaged properties of a coarsening ensemble. We shall discuss two types of theories; 1) models that replace the coarsening medium by an averaging sphere with a particle of a given size at the center of the sphere [11, 20, 276, 202, 206], and 2) models that employ diffusion in a lossy medium [38, 39, 201].

9.3.1 Averaging sphere models

In this approach a particles of size R is placed in a sphere or radius r_c, where $R < r_c$. This particle is taken to represent all particles in the system of this size and thus once the coarsening rate of this particle is determined, the statistically averaged growth rate of a particle of size R is known. All of these models begin with the quasi-stationary approximation to the diffusion equation as presented above (here and in the following we use only the mass diffusion case). Due to the spherical symmetry, the general solution is (see eq.(4.45))

$$x_B(r) = A_0 + \frac{A_1}{r}. \tag{9.30}$$

The constants A_0, A_1 are fixed by boundary conditions. Instead of prescribing a boundary condition for the concentration field at infinity as we did in the LSW theory, the concentration is fixed at the external sphere, a distance r_c away from the center of the spherical particle. We thus have the boundary conditions

$$x_B(r_c) = x_{B,r_c}, \tag{9.31}$$

$$x_B(R) = x_B^{\alpha}(R). \tag{9.32}$$

Using the boundary conditions and eq.(9.30) the constants are

$$A_0 = x_B^{\alpha}(R) - (x_B^{\alpha}(R) - x_{B,r_c})/(1 - R/r_c) \tag{9.33}$$

and

$$A_1 = (x_B^{\alpha}(R) - x_{B,r_c})R/(1 - R/r_c). \tag{9.34}$$

The concentration field becomes

$$x_B(r) = \frac{x_{B,r_c} - x_B^{\alpha}(R)R/r_c}{1 - R/r_c} - \left(\frac{x_{B,r_c} - x_B^{\alpha}(R)}{1 - R/r_c}\right)\frac{R}{r}. \tag{9.35}$$

The concentration gradient at the particle interface is

$$\left.\frac{\partial x_B}{\partial r}\right|_{r=R} = \frac{x_{B,r_c} - x_B^{\alpha}(R)}{R(1 - R/r_c)}. \tag{9.36}$$

From the equation it is immediately clear that as r_c moves towards the particle surface at $r = R$, the concentration gradient becomes steeper. This, however, means a faster growth rate and thus faster coarsening. The growth rate of a particle is given as before by a mass balance at the interface eq.(9.11). The nearly ten different models discussed in the literature differ in the way they choose the radius of the enclosing sphere r_c and how they connect it to the volume fraction [11, 20, 276, 206].

As an illustration of the advantages and drawbacks of these models, we discuss in detail two very different approaches. The first one is a nearest neighbor model developed by Ardell (the first consistent model on the effect of finite volume fraction), and a more recent statistical model by Marsh and Glicksman.

In *Ardell's approach* the particles are assumed to be distributed randomly in space. If this is true, then the distance to the next nearest neighbor can be calculated. This was done prior to Ardell by Chandrasekhar in 1943 [58]. Using that result the average center to center distance between particles is given. If we assume that the interparticle diffusional interactions are on this length scale, then it is possible to assign the average center-to-center distance minus two times the average particle radius as a value of r_c. The relationship between r_c and the interparticle separation, as represented by $\Phi^{1/3}$ is [1]

$$r_c = R + R_c \frac{e^{8\Phi}\Gamma(\Phi)}{6\Phi^{1/3}} = R + R_c \psi(\Phi). \tag{9.37}$$

The most important assumption in determining the nearest neighbor distance is that all particles have the same size, and that this is just the critical size defined above using the Gibbs-Thomson relation. There are no physical arguments supporting that assumption, although it sounds reasonable to assume that the size of the enclosing sphere scales with the critical radius. In particular, it has not been shown that by choosing r_c as in eq.(9.37) will yield the same statistically averaged growth rate given by the multiparticle diffusion solution. Nevertheless, substituting the relation into eq.(9.36) we obtain after some rearrangement

$$\left.\frac{\partial x_B}{\partial r}\right|_{r=R} = \frac{x_{B,r_c} - x_B^\alpha(R)}{R}(1 + \rho/\psi(\Phi)). \tag{9.38}$$

where $\rho = R/R_c$. Using the Gibbs-Thomson relation and the definition of the critical radius to eliminate x_{B,r_c} the growth rate of eq.(6.10) then reads

$$\frac{dR}{dt} = K_{ss}^* \frac{1}{R^2}(\frac{R}{R_c} - 1)(1 + \rho/\psi(\Phi)) \tag{9.39}$$

One can now use the recipe for a LSW analysis (see page 139) to derive the growth equation for the critical and average radius and the size distribution. The technical details are skipped here. It is important to notice, however, that the cut-off radius of the size distribution, the

[1] For Details of the derivation for $\psi(\Phi)$ see the original article by Ardell from 1972.

rate constant describing the growth of a critical particle, ν, and the size distribution itself will depend on the function $\psi(\Phi)$ and thus on the volume fraction. The temporal exponents, as mentioned above, do not. The behavior of $\psi(\Phi)$, the cut-off $\rho_m(\Phi)$, $\nu(\Phi)$ and the growth rate constant for the average radius, $K(\Phi)$, in relation to the LSW-value for zero volume fraction is shown in figures 9.2 and 9.3. Some size distributions are shown in figure 9.4.

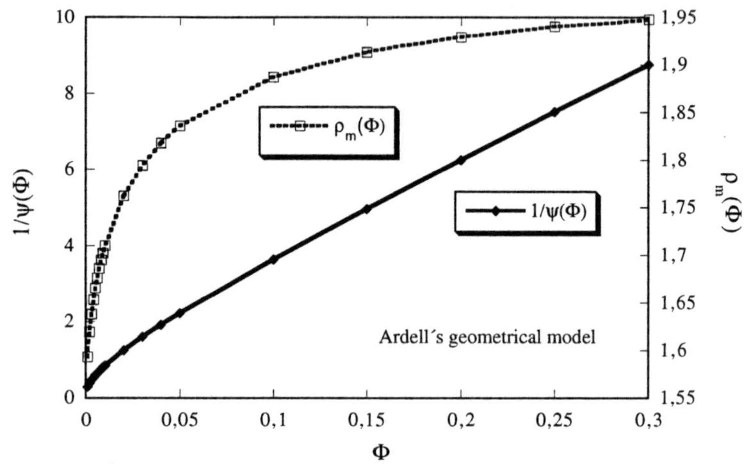

Fig. 9.2. Dependence of the function $\psi(\Phi)$ on volume fraction and the cut-off radius in Ardells model

Some comments are necessary. It was reasonably easy to determine the variation in the particle size distributions and coarsening rates as a function of volume fraction using this approach. Identifying the radius of the embedding sphere with the nearest neighbor distance, while perhaps a reasonable guess, is not correct given that the diffusional interaction distance is much larger than this, see Fig. 9.1. Thus, the effect of choosing this distance is rather drastic: the gradient at the interface is very steep and becomes steeper with increasing volume fraction. The result is that the rate constant depends strongly on volume fraction as shown in Fig.9.3. A nearest neighbor model may thus represent something like an upper bound for any volume fraction effect. It is possible to develop any number of these models by assigning r_c to other stereologically related parameters, such as the mean free path between particles, functions of the volume fraction, etc., see [276].

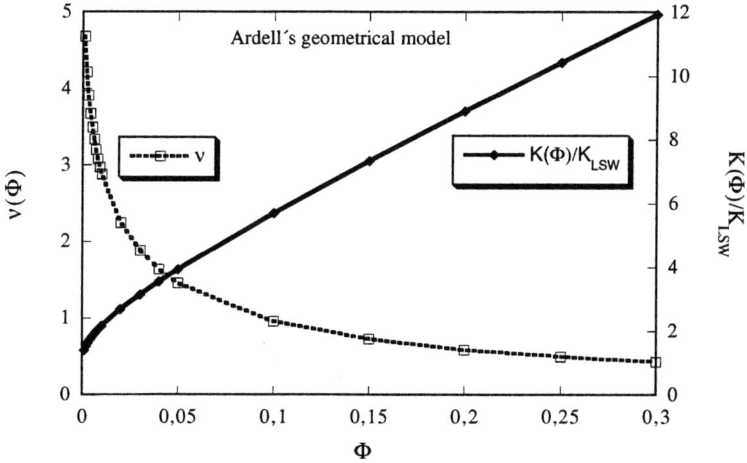

Fig. 9.3. Variation of the inverse growth rate of the critical radius ν with volume fraction and its effect on the proportionality constant in the growth law of the average particle radius.

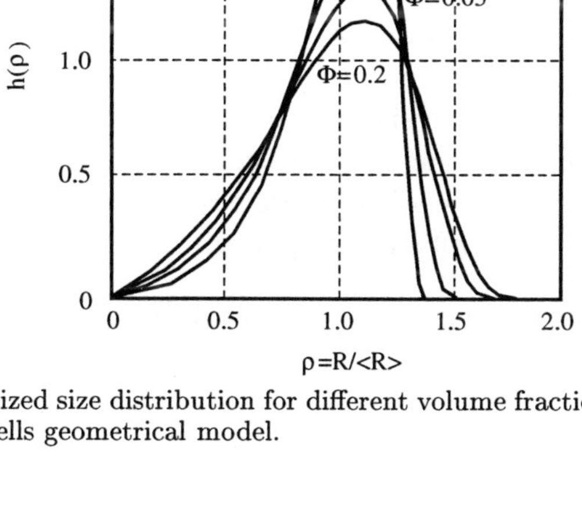

Fig. 9.4. Normalized size distribution for different volume fraction dispersed phase according to Ardells geometrical model.

The starting point of the *Marsh and Glicksman analysis* looks on a first glance like another geometrical model since they use eq.(9.35) and rewrite the concentration of solute in the matrix introducing

$$\chi(r) = \frac{x_B(r) - x_B^{\alpha,e}}{\ell^\alpha}, \tag{9.40}$$

$$\overline{B} = \frac{2(R - R_c)}{R_c(1 - R/r_c)}, \tag{9.41}$$

as

$$\chi(r) = \frac{\overline{B} + 2}{R} - \frac{\overline{B}}{r}, \tag{9.42}$$

where $\overline{B} = -B$. This means that they used the Gibbs-Thomson equation, eq.(2.124), the classical definition of the critical radius, eq.(6.9), and assumed that at the surface of the embedding sphere the concentration attains the value corresponding to the Gibbs-Thomson value at the critical radius. The essential difference between this approach and any geometrical model is, however, that they do not assign the radius of the embedding sphere to some stereological quantity, but determine it in a statistical consistent way using certain assumptions. The reasoning behind it and its meaning will now be explained in some detail.

First their quantity $\overline{B}$ is the analog of the source/sink strength of the multiparticle diffusion discussed above. In the classical LSW approximation of zero volume fraction, i.e. $r_c \to \infty$, $\overline{B}$ reduces to $\overline{B} = 2(\rho - 1)$, see eq.(9.29), with the usual definition of the reduced radius. Thus, $\overline{B} < 0$ characterizes a shrinking particle and $\overline{B} > 0$ a growing one. In the finite volume fraction, the value of $\overline{B}$ varies with two parameters, the critical size R_c and the size of the embedding sphere r_c.

The key to their analysis is the idea that the embedding sphere grows and shrinks with the particle size. At the same time a minimum size that depends on the critical particle radius is maintained. We write for the volume of the embedding sphere

$$V_{embed} = V_{embed}^{I} + V_{embed}^{II} \tag{9.43}$$

with

$$V_{embed}^{I} = PV_{particle} \tag{9.44}$$

being simply proportional to the particle volume and

$$V_{embed}^{II} = EV_c. \tag{9.45}$$

with V_c the volume of a critical particle. P and E are two proportionality constants. This guarantees that the V_{embed} grows and shrinks with the particle size but never becomes smaller than V_{embed}^2. This is even more obvious if we normalize the volume with the critical volume (as done before with radii in the LSW analysis),

$$v_{embed} = Pv_{particle} + E. \tag{9.46}$$

It should be noted that the total volume of the embedding spheres is chosen to be conserved and equal to the total volume of matrix plus particles. Furthermore, the division of the matrix into embedding spheres leaves some free from particles: $v_{p.} = 0$, but $v_{embed} = E$. This equation can not be applied readily to the coarsening of a single particle, but the quantities entering are statistical averages over the system. Thus, it cannot be used to determine directly the growth rate of an individual particle. Equation (9.46) leaves us with two unknowns, P and E. They are determined from two constraints on the growth.

The first constraint is simply that the volume fraction always shall be constant. Let us denote $h(v_{p.} = V_{particle}/V_c)$ as the normalized size distribution (instead of using radii, one can simply switch to volume using the standard procedures to express probability distributions in different coordinates). Then, the volume fraction is given by

$$\Phi = \frac{\int_0^{v_m} v_{p.} h(v_{p.}) dv_{p.}}{\int_0^{v_m} (Pv_{p.} + E) h(v_{p.}) dv_{p.}}, \tag{9.47}$$

which is nothing but the ratio of a all spheres divided by that of all embedding spheres. The second constraint is that at the surface of all embedding spheres the concentration of solute shall exactly match the value corresponding to the critical radius. This also means that the average concentration inside the matrix, i.e. in the swiss cheese in which the holes represent the particles, has to be equal to that concentration value. Performing the averaging using the normalized size distribution gives the second constraint.

This leaves us with the problem: how is the size distribution determined in this approach? The continuity equation is written in terms of volume instead of radius,

$$\frac{\partial f(V_{p.})}{\partial t} + \frac{\partial}{\partial V_{p.}} \left(\frac{dV}{dt} f(V_{p.}) \right) = 0. \tag{9.48}$$

We use again, as in the LSW analysis, reduced coordinates but only transform the volume with $v_{p.} = V_{p.}/V_c$. Using

$$\frac{\partial}{\partial V_{p.}} = \frac{1}{V_c} \frac{\partial}{\partial v_{p.}} \tag{9.49}$$

we get

$$\frac{\partial f(v_{p.})}{\partial t} + \frac{\partial f(v_{p.})}{\partial v_{p.}} \frac{\partial v_{p.}}{\partial t} + \frac{1}{V_c} \frac{\partial}{\partial v_{p.}} \left[\frac{dV_{p.}}{dt} f(v_{p.}) \right] = 0 \tag{9.50}$$

The second term arises from the from the change of $f(V)$ to $f(v)$. Using $\frac{dv}{dt} = (1/V_c)dV_{p.}/dt + V_{p.}/(V_c)^2 dV_c/dt$ we can rearrange eq.(9.50) to become

$$V_c \frac{\partial f(v_{p.})}{\partial t} + \left(2\frac{dV_{p.}}{dt} - v_{p.} \frac{dV_c}{dt} \right) \frac{\partial f(v_{p.})}{\partial v} + f(v_{p.}) \frac{d}{dt} \left(\frac{dV_{p.}}{dt} \right) = 0 \tag{9.51}$$

With the LSW result that the critical radius changes as the cube root of time or $dV_c/dt = K = const.$ and writing $\overline{B} = dV_{p.}/dt$ we get

$$(2\overline{B} - v_{p.} K) \frac{\partial f(v_{p.})}{\partial v} + f(v_{p.}) \frac{d\overline{B}}{dt} + V_c \frac{\partial f(v_{p.})}{\partial t} = 0 \tag{9.52}$$

As in the LSW analysis we are looking for a separation solution $f = g(t) \cdot h(v_{p.})$. Inserting this expression and using standard procedures leads to

$$-V^* \frac{1}{g} \frac{dg}{dt} = \beta, \tag{9.53}$$

$$(2\overline{B} - vK) \frac{1}{h} \frac{dh}{dv} + \frac{d\overline{B}(v_{p.})}{dv_{p.}} = \beta. \tag{9.54}$$

It remains to determine the constant β. This is done in the same way as in the Wagner analysis. The constancy of volume fraction allows to relate β with the rate constant K following the same procedure as outlined above in the LSW analysis. This gives $\beta = 2K$. The general solution of the equation for h is then

$$h(v_{p.}) = \frac{K}{2\overline{B}(v_{p.}) - vK} \exp \left(\int_0^{v_{p.}} \frac{K dz}{2\overline{B} - zK} \right). \tag{9.55}$$

This general solution together with the two equations concerning the constraints for the embedding matrix spheres are solved iteratively to obtain the size distribution, the average growth rate, and of course the two proportionality constants P and E. Some results of this analysis are shown in figs. 9.5, 9.6, and 9.7.

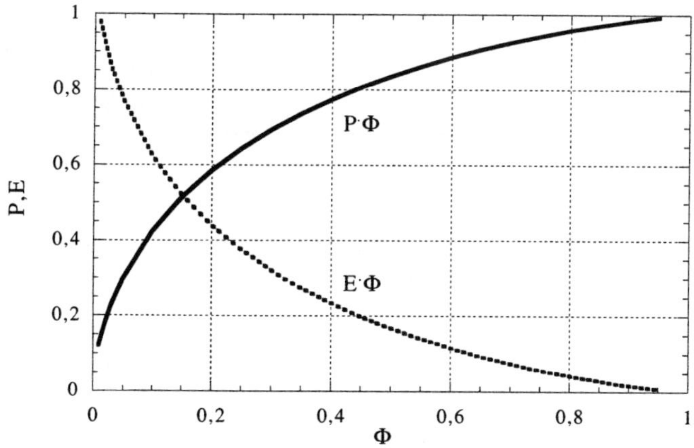

Fig. 9.5. Variation of the proportionality constants with volume fraction.

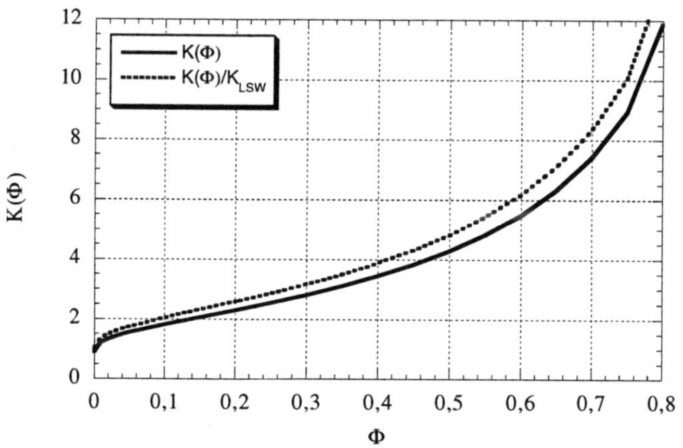

Fig. 9.6. Rate constant for the average radius and in relation to the LSW rate constant as predicted by the Marsh and Glicksman approach.

Finally some comments to this approach. A particularly nice aspect of the Marsh and Glicksman approach is that it is straightforward to determine the rate constants and particle size distributions at volume

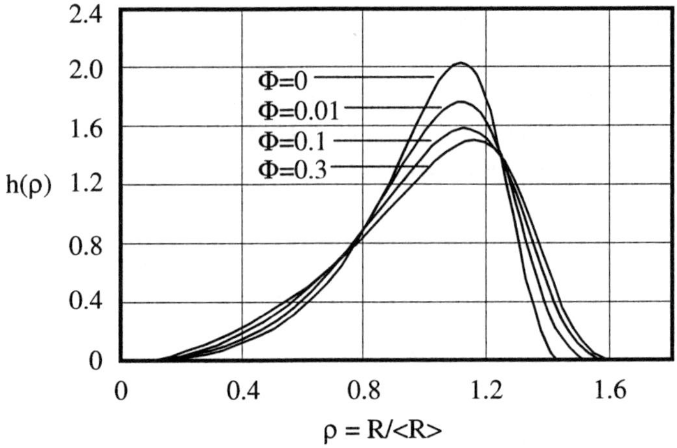

Fig. 9.7. Normalized size distribution for different volume fraction dispersed phase according to the Marsh and Glicksman model.

fractions above which solutions to the multiparticle diffusion problem are not possible using a spherical geometry. It is, however, not a fundamental approach to the problem, since numerous assumptions are used, which sound reasonable, but are not much better justified than in the geometrical models. Such assumptions are made in eqs.(9.44,9.45). These assumptions form the basis of the model. The assumptions that the volume of particles of critical size grows at a constant rate and that volume fraction is constant are very reasonable, but would have to be proved seriously using a statistical mechanical approach or simulations. Especially at high volume fractions coarsening is confined to thin channels between the particles and thus coarsening could even change its time dependence. A transition from one power law to another cannot be obtained with the assumptions made, but needs a more fundamental analysis.

9.3.2 Brailsford and Wynblatt approach

Brailsford and Wynblatt employ a different effective medium where the emission or adsorption of solute from the surrounding array of coarsening particles is represented as distributed sources and sinks within the matrix. These sources and sinks are taken to represent a sea of randomly distributed particles. One important difference between this approach and the previous approaches is that Brailsford has established that at low volume fractions a similar effective medium yields the same results for statistically averaged quantities as does the solution to the multiparticle diffusion problem, assuming that the particles are distributed randomly in space. Thus, at least in the limit of small volume fractions and by neglecting spatial correlations, the results of a multiparticle diffusion solution and this theory are in agreement. This approach has also been used by Marquesee to describe coarsening in two dimensions [201].

This effective medium is based upon a homogeneous source of solute, P_{total}, in units of atoms per sec, and sink $Dk_{total}^2 x_B^\infty$ of solute in the matrix, where k_{total}^2 is the total sink strength [2]. These homogeneous sources and sinks represent, in a statistically averaged sense, the transport of solute in the matrix between growing and shrinking particles. Solute conservation demands that the total sink and source strengths balance, $P_{total} = Dk_{total}^2 x_B^\infty$.

We assume an underlying number density distribution $f(R,t)dR$ and define $Dk_R^2 x_B^\infty dR$ to be the combined loss rate to all particles in the size range between R and $R + dR$ and $P_R dR$ to be the combined emission rate from these particles. Both, k_R and P_R are a function of particle size R and time t. Then,

$$k_{total}^2 (t) = \int_0^\infty k_R^2 (R,t)\, dR \tag{9.56}$$

$$P_{total} (t) = \int_0^\infty P_R (R,t)\, dR \tag{9.57}$$

The solute adsorbed or emitted by particles of a given size class R,$R+dR$ is a result of the particles growing or shrinking, thus

$$4\pi R^2 \frac{dR}{dt} f (R,t) = Dk_R^2 x_B^\infty - P_R \tag{9.58}$$

Once the combined source and sink strengths for the different particle size classes are determined, then the statistically averaged growth rate of a particle of a given size is known.

The basic approach used by Brailsford and Wynblatt to construct their effective medium dates back to Maxwell in his work on the electrical resistivity of a composite consisting of domains of different resistivity. In our case it is assumed that the particles are distributed randomly and thus an equivalent homogeneous medium is assumed to

[2] One might have expected a certain symbol for the sink strength as was done for the source strength. The definition given here later simplifies drastically algebra.

exist where there is a source of solute P_{total} and a sink $Dk_{total}x_B^\infty$. Again we assume a shell of radius R_{embed} surrounding each particle in which there are no other sources and sinks. The radius of this shell is chosen so that within the shell the volume fraction equals that of the whole system, $R^3/R_{embed}{}^3 = \phi$. Outside this shell the matrix contains sources and sinks according to P_{total} and $Dk_{total}x_B^\infty$. One has to be careful to not conclude that since sources and sinks are balanced the outside medium is in effect empty. Here is where the analogy to Maxwell's work gives the right inside: a medium in which *on average* sources and sinks balance will actually provide a shielding effect. Mathematically this becomes clear when we look at the solutions to the diffusion field. Within the spherical shell diffusion field is given by the Laplace's equation

$$\frac{1}{r^2}\frac{d}{dr}\left(r^2\frac{dx_B}{dr}\right) = 0 \quad \text{for} \quad R \leq r \leq R_{embed} \tag{9.59}$$

while external to this sphere, due to the source of solute, Poisson's equation holds,

$$\frac{1}{r^2}\frac{d}{dr}\left(r^2\frac{dx_B}{dr}\right) = Dk_{total}^2 x_B - P_{total} \quad \text{for} \quad R_{embed} \leq r < \infty \tag{9.60}$$

The boundary conditions are that $x_B(r) \to x_B^\infty$ as $r \to \infty$, that x_B is given by the Gibbs-Thomson equation at $r = R$, and that concentration and its derivative must be continuous at $r = R_{embed}$.

The solution to eq.(9.59) is

$$x_B(R) = A_0 + A_1/r. \tag{9.61}$$

The solution to eq.(9.60) using the far-field condition is

$$x_B = x_B^\infty - \frac{A_2 R_{embed}}{r} e^{-k_{total}(r - R_{embed})}, \tag{9.62}$$

where A_0, A_1 and A_2 are constants to be determined by the boundary conditions.

Here we encounter the first indication that the diffusion field in this system does not decay as $1/r$ at long distances from the particle. Rather in this case it decays exponentially with a characteristic distance k_{total}, which is related to the screening effect of the matrix discussed above. This screened potential occurs in many other contexts as well, such as Debye-Hückle theory of ionic solutions and diffusion limited reactions, the Yukawa potential in quantum mechanics, screening potential in plasmas etc.. Applying the boundary conditions yields,

$$\begin{aligned} A_0 &= x_B^\alpha(R) - \frac{A_1}{R} \\ A_2 &= -\frac{A_1}{R_{embed}\left(1 + k_T R_{embed}\right)} \\ A_1 &= -\left(x_B^\infty - x_B^\alpha(R)\right) R\, W\left(R, R_{embed}\right) \end{aligned} \tag{9.63}$$

where $W(R, R_{embed}) = \left(1 + k_{total} R_{embed}\right) / \left[1 + k_{total}\left(R_{embed} - R\right)\right]$. The rate of solute flow to the test particle is $4\pi R_{embed}^2 D A_1/\Omega$, where Ω is an atomic volume. The rate of flow to $f(R)dR$ particles per unit volume is $4\pi R_{embed}^2 D A_1 f dR$. This must be equal to the net flow $\left(D k_R^2 x_B^\infty - P_R\right) dR$. Thus,

$$4\pi R_{embed}^2 D A_1 f = D k_R^2 x_B^\infty - P_R \tag{9.64}$$

Using eq.(9.62) in eq.(9.64) yields,

$$-4\pi R_{embed}^2 D f \left(x_B^\infty - x_B^\alpha(R)\right) R W\left(R, R_{embed}\right) = D k_R^2 x_B^\infty - P_R \tag{9.65}$$

Sorting the terms leads to,

$$\begin{aligned} & x_B^\infty \left[D k_R^2 + 4\pi R_{embed}^2 D f R W\left(R, R_{embed}\right)\right] - \\ P_R \quad & -4\pi R_{embed}^2 D f x_B^\alpha(R) R W\left(R, R_{embed}\right) = 0 \end{aligned} \tag{9.66}$$

Using that this relation must hold for any x_B^∞ the first term must be identical to zero and thus we obtain

$$k_R^2 = -4\pi R_{embed}^2 R f\left(R, t\right) W\left(R, R_{embed}\right) \tag{9.67}$$

and

$$P_R = x_B^\alpha(R) D k_R^2. \tag{9.68}$$

The total sink strength can now be determined by substituting eq.(9.67) in eq.(9.56),

$$k_{total}^2\left(t\right) = -4\pi R_{embed}^2 \int_0^\infty R f(R, t)\, W\left(R, R_{embed}\right) dR. \tag{9.69}$$

Thus, P_{total} is also specified.

The growth rate is given by using the relationship between the source/sink strength and the coarsening rate of a particle. This was derived previously in the section on the multiparticle diffusion problem, as the diffusion field in this case satisfies Laplace's equation as does the field near the particle in the effective medium. Letting $B_1 = A_1$ and $R_1 = R$ in eq.(9.13)

$$\frac{dR}{dt} = -\frac{DA_1}{\left(x_B^\beta - x_B^\alpha\right) R^2}. \tag{9.70}$$

Using eq.(9.62) in eq.(9.70) yields

$$\frac{dR}{dt} = \frac{DW\left(R, R_{embed}\right)\left(x_B^\infty - x_B^\alpha(R)\right)}{R\left(x_B^\beta - x_B^\alpha\right)}. \tag{9.71}$$

Using that $P_{total} = Dk_{total}^2 x_B^\infty$ together with eq.(9.57), eq.(9.68),eq.(9.67) and eq.(9.69) we get

$$x_B^\infty = \frac{\int_0^\infty x_B^\alpha(R) R f\left(R,t\right) W\left(R, R_{embed}\right) dR}{\int_0^\infty R f\left(R,t\right) W\left(R, R_{embed}\right) dR}. \tag{9.72}$$

Substituting eq.(9.72) in eq.(9.71) yields the desired equation for the statistically averaged growth rate of a particle of a given size,

$$\frac{dR}{dt} = \frac{DW}{R\left(x_B^\beta - x_B^\alpha\right)}\left(\frac{\int_0^\infty x_B^\alpha R f W dR}{\int_0^\infty R f W dR} - x_B^\alpha\right) \tag{9.73}$$

where the dependence of the various quantities on R and t have been suppressed in the above. Expressing x_B^α as a function of R using the Gibbs-Thomson equation, the kinetic equation is exactly in the form required for the asymptotic analysis of the type used to determine the coarsening rate and time-independent particle size distributions.

In the limit $\phi = 0$ there is no production of solute in the region exterior to the shell, and thus $k_{total} = 0$. Using that $W\left(R, R_{embed}\right) = 1$ in this limit and the Gibbs-Thomson equation eq.(9.72) yields,

$$x_B^\infty = x_B^{\alpha,e} + \ell^\alpha \frac{\int_0^\infty f\left(R,t\right) dR}{\int_0^\infty R f\left(R,t\right)}. \tag{9.74}$$

Thus, the concentration in the matrix is set by the average particle size and the kinetic equation reduces to that of LSW.

To simplify the computations, instead of using the self-consistent expression for $W(R, R_{embed})$ Brailsford and Wynblatt employ an interpolation function for $W(R, R_{embed})$ that past experience indicates is a reasonable approximation,

$$W(R, R_{embed}) = 1 + k_{total} R. \tag{9.75}$$

Using this in eq.(9.69) gives a quadratic equation for k_{total} with the solution

$$k_{total} = \frac{1}{2}\left[M_2 + \left(M_2^2 + 4M_1\right)^{1/2}\right], \tag{9.76}$$

where

$$M_n = 4\pi \int_0^{\infty} R^n f(R,t)\, dR. \tag{9.77}$$

Thus, using this simple approximation allows the total sink strength, and thus the kinetic equation, to be expressed as a function of moments of the distribution function.

The asymptotic analysis yields a rate constant for the average particles size that increases with volume fraction as shown in Fig. 9.8. The increase in K with volume fraction is much weaker than with Ardell's theory, as expected. The particle size distributions are broader than the predictions of LSW, becoming broader and more symmetric with increasing volume fraction. The predictions of the theory are easily extendable to higher volume fractions and, as we shall see, are close to the predictions of theories that are based upon averaging directly the multiparticle diffusion solution.

More recently Yao, Elder, Guo and Grant developed a mean-field theory for coarsening that has much of the same elements as the Brailsford and Wynblatt theory [315]. Namely, the presence of screening and a diffusion field that is given by a solution to Poisson's equation with a uniform distribution of sinks and sources. The predictions of this theory are nearly indistinguishable from those of the statistical mechanical theories of Marqusee and Ross and Tokuyama, Kawasaki and Enomoto mentioned below to $O\left(\phi^{1/2}\right)$.

9.4 Statistical Mechanical Theories

There are many theories that employ statistical mechanical methods developed in other contexts to directly average the microscopic equations for the source/sink strength in an effort to determine the statistically averaged coarsening rate of a particle of a given size. The

Marqusee and Ross theory assumes that the particles are randomly distributed in space [200]. Theories that account for the non-random spatial distribution that develops as a result of the ripening process itself are Tokuyama and Kawasaki [270, 271, 272] and Enomoto et al. (soft collisions in their treatment) [85, 87, 88] and Marder [197, 198]. All of these theories require that the mathematically divergent terms resulting from averaging the equations be treated in a self-consistent manner. Treating the effects of a non-random spatial distribution of particles is even more challenging. As a result, while all these theories begin with eq.(9.24), all the theories make different predictions on rate constant as a function of volume fraction. A detailed derivation of the results of any of these theories is beyond the scope of this book. Thus, we only present the highlights of the results of a few these theories.

The first theory to employ first-principles statistical mechanical methods to find the statistically averaged growth rate was the theory by Marqusee and Ross. They employed statistical methods developed for the related diffusion limited reaction problem. They found the statistically averaged growth rate of a particle,

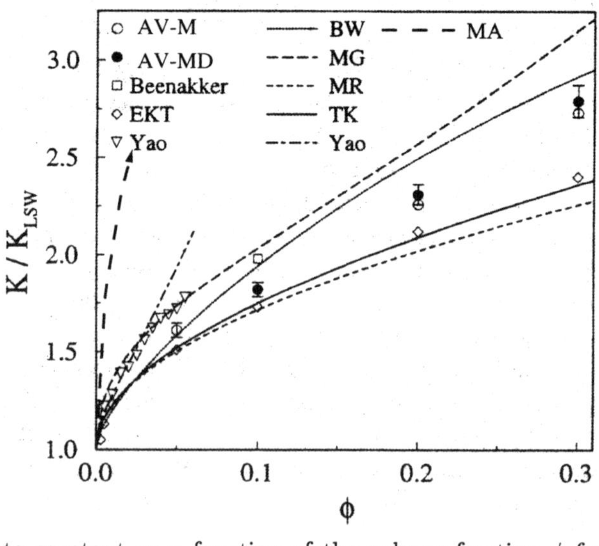

Fig. 9.8. Rate constant as a function of the volume fraction ϕ for the theories of Marder (M), Brailsford and Wynblatt (BW), Tokuyama and Kawasaki and Enomoto et al. (TK), Marqusee and Ross (MR), Marsh and Glicksman (MG), and the computer simulations of Akaiwa and Voorhees in the monopole approximation (AV-M) and monopole+dipole approximation (AV-MD), and Yao et al (Y).

$$\frac{dR}{dt} = \frac{D}{R\left(x_B^\beta - x_B^\alpha\right)} \left[\langle x_B \rangle - x_B^\alpha(R)\right] \left[1 + R\sqrt{4\pi n_V \langle R \rangle}\right] \qquad (9.78)$$

where $n_v(t)$ is the number of particles per unit volume. Thus, for any non-zero volume fraction the growth rate is above that of LSW. Both, $\langle x_B \rangle$ and of course $\langle R \rangle$, are functions of the particle size distribution and hence volume fraction. Beyond the detailed predictions of the theory is the qualitative dependence of the growth rate of a particle on the volume fraction. Consider the simple limit of a monodispersion. Thus, $R = \langle R \rangle$ and the last term in eq.(9.78) becomes $1 + \sqrt{3\phi}$. The increase in growth rate of a particle scales as $\sqrt{\phi}$. A more detailed analysis shows that this is indeed the case for a polydisperse system in the limit $\phi \to 0$ [201]. The theory of Brailsford and Wynblatt also yields this result in the limit $\phi \to 0$. This is something of a surprise as one would expect that the increase in the rate constant would scale with the interparticle separation distance, and hence as $\phi^{1/3}$. This is in fact the case when the particles are arranged on a regular periodic lattice. The reason for the difference is that the particles are distributed randomly. The random nature of the spatial distributions changes the dependence from $K \sim \phi^{1/3}$ to $K \sim \phi^{1/2}$ in the limit $\phi \to 0$. The predictions of the rate constant are shown in Fig. 9.8. The time independent particle size distributions are broader and more symmetric than that predicted by LSW, see Fig. 9.9.

There have been a number of other first principles theories. These theories do not assume that the particles are randomly distributed in space and thus attempt to determine the spatial correlation function as well as the rate constant and particle size distribution. Tokuyama and Kawasaki and Enomoto, Tokuyama and Kawasaki consider the development of correlations – "soft collisions" in their terminology. Although they show that though correlations develop, they have a rather weak effect on the rate constant, note the small difference between the predictions of their theory and that of Marquesee and Ross, where correlations are neglected, in Fig. 9.8. The particle size distribution depends on volume fraction see Fig. 9.10. While Marder's theory also predicts that correlations exist, the rate constant is much higher than that predicted by any other theory or first-principles simulation and is quite similar to that of Ardell's theory. Che and Hoyt have reexamined Marder's theory and have identified one of the assumptions employed by Marder in the calculation of the two-point correlation

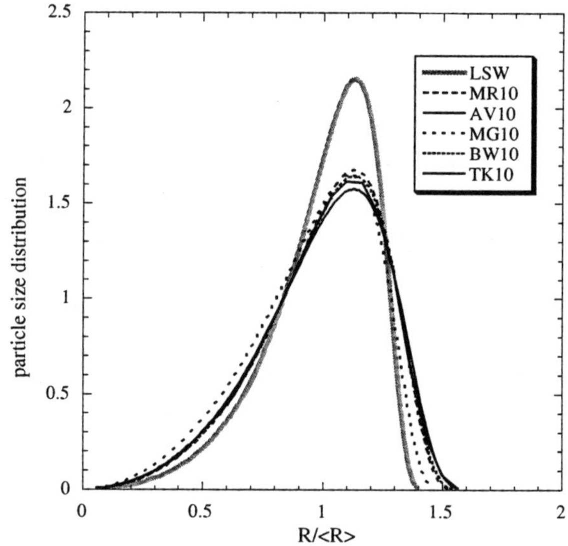

Fig. 9.9. The scaled particle size distributions at a volume fraction of 0.1 for the theories of Marquess and Ross (MR), Tokuyama and Kawasaki (TK), Brailsford and Wynblatt (BW), Marsh and Glicksman (MG), Akaiwa and Voorhees (AV) as compared to the LSW theory (LSW).

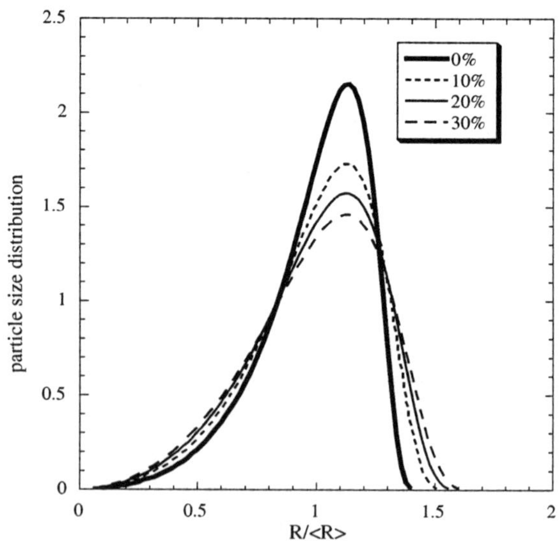

Fig. 9.10. Steady-state particle size distributions predicted by the Tokuyama and Kawasaki theory at various volume fractions.

function as being too drastic [61]. When this assumption is dropped the predictions agree well with the other theories.

9.5 Computer Simulations of Coarsening

Since the equations defining the solution to the multiparticle diffusion problem determine the coarsening rate of a particle at a given location, one can employ computer simulations to determine the statistically averaged properties of these coarsening ensembles. These equations were used by Weins and Cahn to study the effects of local arrangements of particles, typically less than 10, on their coarsening rates. They showed clearly that the coarsening rate of a particle can be a strong function of its local environment [306]. Voorhees and Glicksman were

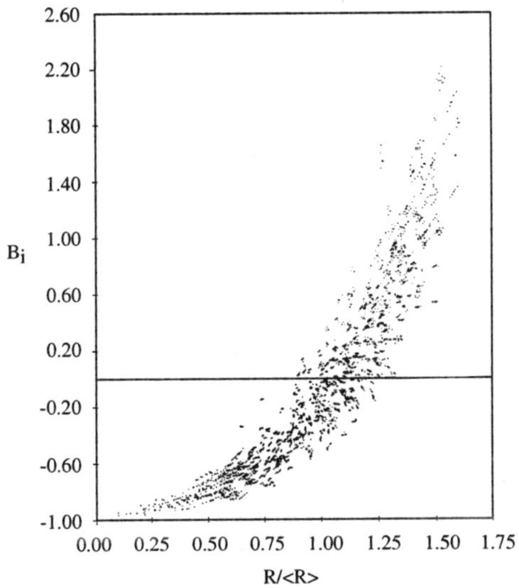

Fig. 9.11. Steady-state dispersion of $\overline{B}_i$ at a volume fraction of 0.35. Particles with $\overline{B} > 0$ grow while particles with $\overline{B} < 0$ shrink.

the first to employ computer simulation techniques on these equations to address the statistically averaged properties of coarsening mixtures. Particles were placed at random locations in a box and the box was periodically repeated to fill all space. The effects from these periodically repeated boxes was accounted for by using the Ewald method. This method guarantees a reasonably fast convergence rate of the summation appearing in eq.(9.24). Thus, the interaction of one particle with every other particle in the system was considered. Once the B_i were determined the particle radii were updated in time using eq.(9.13) and taking a small timestep. The asymptotic distributions and rate constants were determined by using the distribution obtained at the end of one calculation as the initial condition of another. If the distributions at the beginning and end of the calculation were the same, then it was deemed that it was the correct time independent distribution. This approach is acceptable as long as spatial correlations do not affect the particle size distribution or rate constant, since there was no attempt to determine the correlations at the end of a run and use these as an initial condition for another run. Thus, the approach, essentially, assumed that the particles were distributed at random. The effect of diffusional interactions on the coarsening rate is evident by the scatter in the source/sink strength as shown in Fig. 9.11. The coarsening rate of a particle of a given size is not unique, but is a strong function of its local environment. However, note that as the particle becomes smaller, the magnitude of the scatter decreases. This is a result of the localization of the diffusion field surrounding shrinking particles. As a particle shrinks to zero, the diffusion field created by this particle approaches a delta function and becomes extremely localized. This leads to weak interactions with neighboring particles. More recently Yao *et al.* employed the same approach to compare with the predictions of their mean-field treatment [315]. Beenaker employed this idea to measure the spatial correlation function during coarsening [24].

The most complete study of coarsening using simulations has been done by Akaiwa and Voorhees. They include monopoles and dipolar terms, and thus the particles are permitted to migrate through the matrix as a result of diffusional interactions. To avoid considering the interaction of one particle with all the other particles in the system, the screening distance was computed at a given volume fraction and only particles within this screening distance about a particle of interest were included in calculating its coarsening rate. Thus, although 50,000 to 20,000 particles were employed at $t = 0$ the size of system of linear equations, eq.(9.24), was limited to $O(100)$. Finally, by comparing the results of calculations with 20,000 and 50,000 particles, it was determined that 20,000 was sufficient to allow the steady-state non-random spatial correlations to develop.

The predictions for the rate constant and particle size distributions are shown in Fig. 9.8 and Fig. 9.9. All the theories and simulations shown agree fairly close on the particle size distribution at $\phi = 0.1$. The rate constants computed using both the monopole (M) and monopole+dipole (M+D) representation are also shown. Above volume fractions of 0.1, the dipolar terms increase the value of the rate constant above that given by the monopole approximation. Above a volume fraction of 0.3, it is clear that even with the dipole terms, and the ensuing particle migration, significant particle overlap occurs. Thus, to proceed above volume fractions of approximately 0.3, it is necessary to consider particle shape changes. Similar conclusions have been reached by Mandyam *et al.* using a slightly different numerical approach [196].

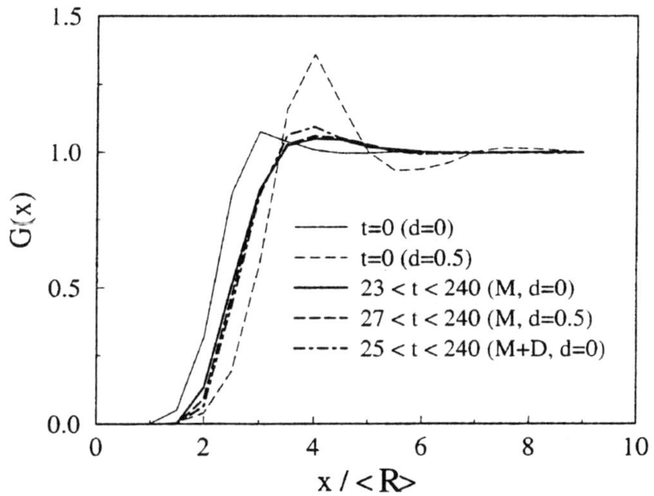

Fig. 9.12. The radial distribution functions after considerable coarsening at a volume fraction $\phi = 0.1$ along with the initial radial distribution functions employed at $t = 0$.

Akaiwa-Voorhees also monitored the development of spatial correlations during the coarsening process. This was measured by examining the number density of the particles $G(x)$, where $G(x)$ = (Number of particles in a spherical shell at x)/(Number of particles in the shell for a uniform dispersion). The evolution of this radial distribution function with time is shown in Fig. 9.12. Shown are two different initial spatial distributions of particles, $d = 0$ corresponds to a random distribution and $d = 0.5$ corresponds to a spatial distribution where there are depletion zones near each particle where no other particle exists.

After sufficient time both coarsen to a unique time independent radial distribution function when the distance x is scaled by the average particle size.

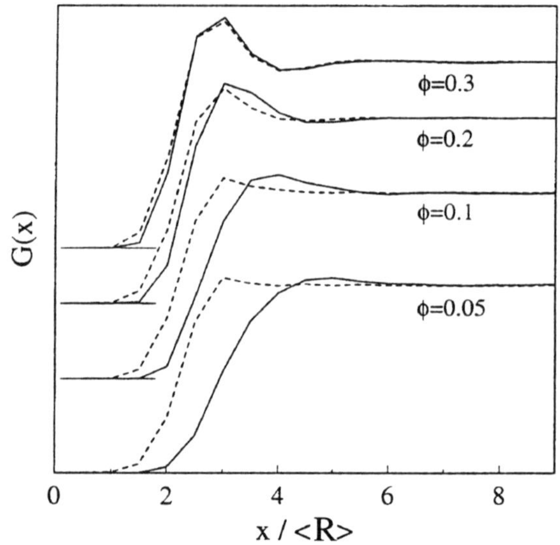

Fig. 9.13. Steady-state radial distribution functions and the initial random spatial distribution functions for various volume fractions.

The steady-state radial distribution function can be much different from that of a random distribution at low volume fractions, see Fig. 9.13. As the volume fraction decreases, the difference between the steady-state distribution and that of a random distribution increases. The similarity between the random distribution and the steady-state distribution at $\phi = 0.3$ likely reflects the strong effect of packing at these high volume fractions. Interestingly, the strongly non-random nature of the correlations at low volume fractions does not have a strong influence on the value of the rate constant as compared to the Marquesee-Ross theory that assumes a random spatial distribution.

Not only does the ripening process affect the overall spatial distribution of particles, but also size-size correlations. To assess this the size-range of the particle size distribution was divided into three classes: small, medium and large. Shown in Fig. 9.14 is the probability of finding a small particle near a large particle. $G(\text{small}, \text{large}, r) = 1$ implies that the probability of finding a small particle at a distance

r from a large particle is equal to that for a uniform distribution of particles. As was the case with the radial distribution function, the small-large correlation function becomes unique after a certain coarsening time. It is clearly different from that of a random distribution ($d = 0$), implying that the small particles are farther from the large particles as compared to a random distribution. This is because, as mentioned above, if the particles are distributed randomly those small particles which may be placed very close to large particles disappear very quickly. At larger distances, however, there is a peak in the small-large correlation function indicating an enhanced probability of finding a small particles at this distance from a large particle. These are the small particles that are feeding solute to the growing large particles.

Particle spatial correlations can also be represented using the structure function. The structure function is related to the Fourier transform of the concentration field,

$$S(\boldsymbol{k},t) = \frac{1}{V}\int_V u(\boldsymbol{r}) \exp(i\boldsymbol{k}\cdot\boldsymbol{r})\, d\boldsymbol{r}, \tag{9.79}$$

where $u(\boldsymbol{r}) = x_B(\boldsymbol{r}) - \langle x_B \rangle$ and $\boldsymbol{k}$ is the wavevector in reciprocal space, $k = |\boldsymbol{k}|$. Assuming that there are no preferred directions in the two-phase mixture, it is only necessary to consider the spherical average of $S(\boldsymbol{k},t)$ which we will denote as the structure function $S(k,t)$,

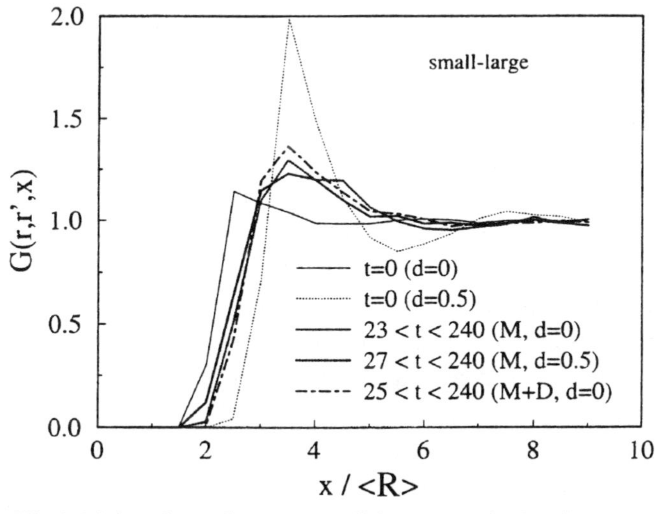

Fig. 9.14. The initial and steady-state small-large correlation function at a volume fraction of 0.1

$$S(k,t) = \frac{\int S(\boldsymbol{k},t)d\Omega_k}{\int d\Omega_k} \tag{9.80}$$

where Ω_k is solid spherical angle in k-space. The structure function is easily measured using small angle neutron or x-ray scattering. Thus, predictions on the shape and evolution of the structure function can be tested experimentally.

To compute the structure function in our case, the composition field in the matrix and particle will be taken as a constant. This is quite reasonable during Ostwald ripening due to the very small composition gradients that are present. In addition, although the absolute values of the composition in the matrix and particles affect the magnitude of the structure function they do not affect the variation in the structure function with wavenumber. Thus, we shall not be concerned with the particular alloy system chosen for the experiment. Assuming that the particles are spherical and taking the spherical average yields,

$$S(k,t) = \sum_j^N \left(\frac{4}{3}\pi R_j^3 \Psi\left(kR_j\right)\right)^2 + \sum_j^N \sum_{i \neq j}^N \frac{4}{3}\pi R_j^3 \frac{4}{3}\pi R_i^3 \Psi\left(kR_j\right)\Psi\left(kR_i\right)\frac{\sin\left(kr_{ji}\right)}{kr_{ji}} \tag{9.81}$$

where $\Psi\left(kR_j\right)$ is the Fourier transform of a sphere with uniform concentration,

$$\Psi\left(kR_j\right) = 3\frac{\sin\left(kr_j\right) - kR_j\cos\left(kR_j\right)}{\left(kR_j\right)^3} \tag{9.82}$$

While the details of the derivation of the above are not important, there are a number of important insights that can be drawn from eq.(9.81). First, it is clear that the structure function depends on the particle size distribution – see, in particular, the first term. This term is independent of the particle locations. Indeed, in the limit of large k, it is the only surviving term. Thus, the large k limit of the structure function depends solely on the particle size distribution and Ψ. Secondly, the structure function is also a function of the separation between particles, r_{ji}, see the second term in eq.(9.82). This term dominates in the limit of small k. Thus, the structure function reflects spatial correlations between particles. However, due to the weighting of the second term with R^3, it does not provide a direct measure of the correlations. It is clearly dependent of the particle size distribution. For example, it must be a function of the size-size correlation function mentioned previously as well as the radial distribution function.

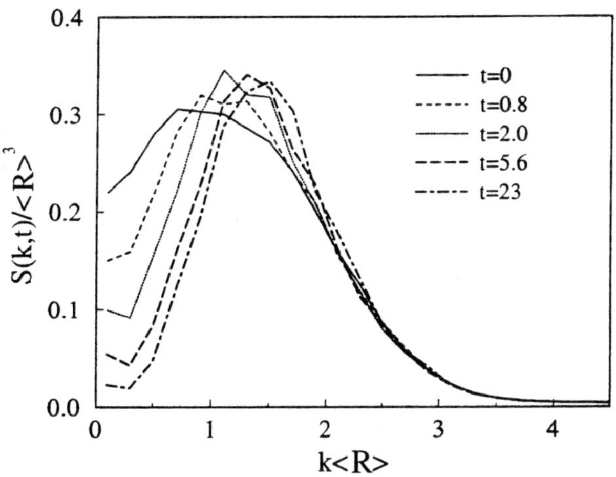

Fig. 9.15. Time evolution of the scaled structure function using the monopole approximation when particles were placed initially at random locations at a volume fraction of 0.1

Nevertheless, the evolution of the microstructure from a random arrangement to the correlated arrangement is shown quite clearly in Fig. 9.15. Both the structure function and wavevector in k space are scaled by the average particle size, and, thus, if steady-state coarsening is present, the structure function should be time invariant. At $t = 0$ the particles are distributed randomly, and thus the structure function does not decay to zero at the smallest k sampled. This is a result of the lack of interparticle correlations. As the coarsening process proceeds, the correlations develop and the amplitude of the structure function at small k decays. The initial particle size distribution used in this calculation was the steady-state distribution. Thus, the evolution of the scaled structure function is due solely to the development of interparticle spatial correlations. After sufficiently long coarsening the structure function does become time independent, see Fig. 9.16. Yao *et al.* also calculated structure functions from their simulations. However, due to the small system sizes employed, the scaled structure function at small k was not time independent. This structure function can be compared to experiments with no adjustable parameters.

The comparison between theory and experiment for $0.05 \leq \phi \leq 0.3$ shows reasonable agreement for the full width at half maximum of the structure function.

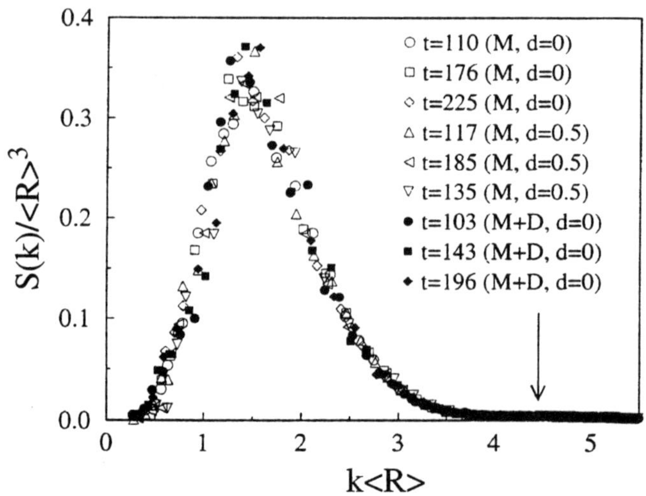

Fig. 9.16. The scaled time independent structure function in the late stages of coarsening using both monopole (M) or monopole and dipole (M+D) approximations. Two initial spatial arrangements of the particles were employed, random (d=0) and depletion zones (d=0.5).

Mandyam *et al.* have performed simulations using the multiparticle diffusion solution given above and they extended their treatment to include dipolar terms as well. However, instead of employing periodic displacements of a computational cell to fill all space, the coarsening kinetics of a finite cluster of particles was determined. Furthermore, the temporal evolution of the particle size distribution was not computed, but simply the growth rates of the particles for a given configuration. Many clusters were created and the growth rates of the particles in these clusters were determined. Using this approach it was possible to collect statistically significant information on the deviation of the coarsening rates of particles from that predicted by LSW. This deviation is defined as,

$$\Delta = \frac{\langle dR/dt\rangle - \langle (dR/dt)_{LSW}\rangle}{\langle (dR/dt)_{LSW}} \tag{9.83}$$

where $\langle\rangle$ in this case denote averages over the particle size distribution, and $(dR/dt)_{LSW}$ is the growth rate of a particles of size R as given by the LSW theory. A nonzero Δ is a result of the diffusional interactions between particles. Since this deviation clearly scales with the interparticle separation, $\Delta = \Delta(\phi)$. Through this approach it was possible to determine the scaling of the average diffusional interaction with the volume fraction. If these interactions are screened as predicted by that for a random array of particles, $\Delta \sim \phi^{1/2}$ as mentioned above. The screening length increases as the volume fraction decreases as the flux emitted from a shrinking particle must travel larger and larger distances before being absorbed by a growing particle. If the screening length exceeds the size of the cluster, then screening is impossible and deviation in the $\phi^{1/2}$ scaling should be observable. This is precisely the case, see Fig. 9.17. At volume fractions above $\phi \sim 10^{-2}$ $\phi^{1/2}$ scaling is observed. However, at extremely low volume fractions, the scaling changes to $\phi^{1/3}$. In this limit the deviation scales like the interparticle separation. These volume fractions are far below those accessible experimentally and such systems likely do not satisfy the requirement that the quasi-stationary approximation is valid (unless the coarsening rates of the particles is extremely small). Nevertheless, the calculations illustrate quite nicely the manner in which diffusional interactions scale with volume fraction during coarsening.

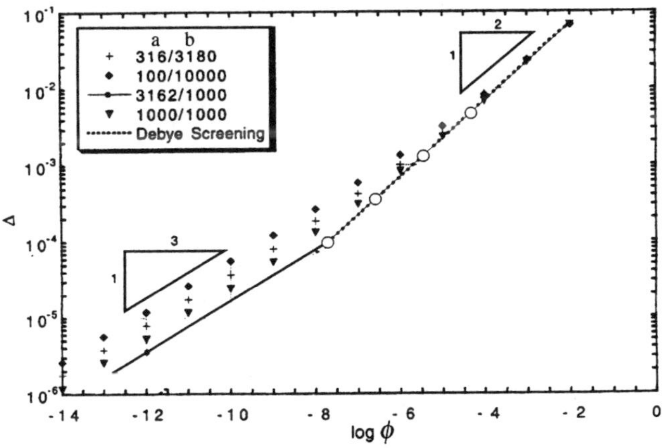

Fig. 9.17. Deviations in the coarsening rate from the LSW predictions at low volume fractions. The numbers a/b denote the number of particles in a cluster, a and the number of realizations employed in the calculation b. The open circles indicate the region where the scaling changes from $\phi^{1/2}$ to $\phi^{1/3}$

10. Nucleation, Growth and Coarsening

In the preceding chapters we treated growth and coarsening as separate processes. As sketched in the introduction (see chapter 1) there are several ways to prepare a two- or multi-phase system:

1. Nucleation from a supersaturated or supercooled liquid
2. Spinodal decomposition
3. Artificial mixing of powders

In the simplest picture transformations that proceed by nucleation can be divided into several steps: the nucleation process itself, followed by growth of the nuclei and finished by coarsening of the precipitates or particles or drops (in solid state transformations it is common to name the new second phase appearing in the matrix as precipitates, in transformations from a liquid to a solid they are called particles or grains and in liquid-liquid decomposition they are termed droplets). A prerequisite for the nucleation type of phase transformation is the existence of a limited miscibility in the phase diagram, see for example the eutectic phase diagram, fig.2.11, where a limited solubility exists in the solid state below the eutectic temperature. The miscibility gap in a eutectic can be described by, for example, the regular solution model. This gives the miscibility gap shown in Fig. 2.6 where the critical point does not exist since it is terminated by the eutectic reaction. Outside a miscibility gap the components are completely dissolvable into each other, inside the system decomposes into two equilibrium solid or liquid phases. When a thermal quench is performed from a single phase region (either solid or liquid) to a lower temperature the alloy decomposes into a two phases. The transformation or decomposition process does not occur spontaneously on reaching the miscibility gap boundary but generally some undercooling or supersaturation is necessary. From a thermodynamic standpoint a driving force to form precipitate nuclei of the second phase is necessary. Once nuclei are formed they can grow and later coarsen. All three processes (nucleation, growth and coarsening) do not occur in a strict sequence but overlap in time. The concurrent action of nucleation and growth was described in detail starting from first principles by Binder and Stauffer [35]. Their approach was put into a form suitable to compare it with experiments by Langer and Schwartz [184, 185]. A full numerical model that removes many of the approximations employed by Langer and Schwartz was developed by Kampmann and Wagner [163]. In this chapter we present a fully analytical treatment of the problem of concurrent nucleation, growth, and coarsening of solid particles or droplets based upon the theory of Langer and Schwartz. A brief review on the basic ideas of nucleation will be given in the first section of this chapter and a more detailed discussion is given in Appendix B.

There is another way transforming a single phase material into a multi-phase one. Spinodal decomposition occurs if one quenches a material from a single phase field to a temperature below the spinodal line (see fig.2.6). Then, the single phase system is unstable to composition fluctuations. Any concentration fluctuation occurring in the system grows by mass transport without the need for a nucleation step. The theory of spinodal decomposition was described first by Cahn and Hilliard [51, 52] and extended by for instance by Binder and coworkers [33, 34] and many others.

10.1 Nucleation and growth

There are two situations in which the nucleation of a phase is the first step in a phase transformation from a single to a multi-phase mixture. One can either quench a system from a single phase field into a two-phase field of a phase diagram and observe nucleation (this is typical for liquid-liquid systems) or in solid state transformations the single phase state is conserved by a rapid quench to a low temperature and the two-phase state is reached by subsequently annealing after the quench at higher temperatures.

The first process can be envisaged as follows: Consider a binary liquid mixture of A and B atoms having a phase diagram as depicted in fig.10.1 schematically. The liquid of alloy composition x_B^0 is initially at a temperature well above the coexistence or binodal line. Then, an instantaneous quench to a temperature T_q is performed. In equilibrium the two liquids coexisting at this temperature would have the compositions $x_B^{\alpha,e}$ and $x_B^{\beta,e}$. The decomposition of the liquid with the initial concentration x_B^0 just after the quench will lead to a continuous change of the supersaturation $S = (x_B(t) - x_B^{\alpha,e})/(x_B^{\beta,e} - x_B^{\alpha,e})$, where $x_B(t)$ is the average concentration of B-atoms in the matrix surrounding the nucleated and growing droplets of the β-phase at time t. The mean matrix concentration has initially a value very close to x_B^0 and decreases to $x_B^{\alpha,e}$ at equilibrium.

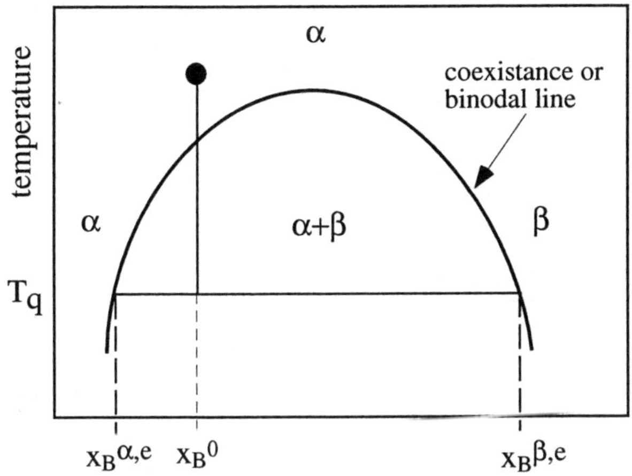

Fig. 10.1. Scheme of a phase diagram with a liquid miscibility gap.

In solid state transformations (like precipitation from a supersaturated solution in order to age harden a material) one rapidly quenches a material from a single-phase field to a low temperature and then anneals it at higher temperatures. In equilibrium the system would be in the two-phase region. The process of transformation occurs slowly due to the slow solid state diffusion (compared to liquid diffusion) and it is possible for metastable phases to nucleate and grow. It is not unusual in age hardenable alloys that, depending on annealing temperature and time, a whole sequence of metastable phase appears. The situation for precipitation of the stable phases is depicted schematically in fig.10.2. The initial alloy composition is x_B^0. The alloy is annealed in the single phase region above the solubility line and below the solidus line of the solid-liquid two-phase region (homogenization treatment). Then, a rapid quench to a temperature T_q is performed (typically placing the material into water, iced water or oil). The decomposition of the solid does not occur directly after the quench, since diffusion at room temperature is usually too slow to lead to nucleation and growth. A thermal treatment at T_{anneal} as indicated in the figure is needed to activate diffusion. The supersaturation, $S = (x_B(t) - x_B^{\alpha,e})/(x_B^{\beta,e} - x_B^{\alpha,e})$, as well as the mean matrix concentration change continuously during this anneal.

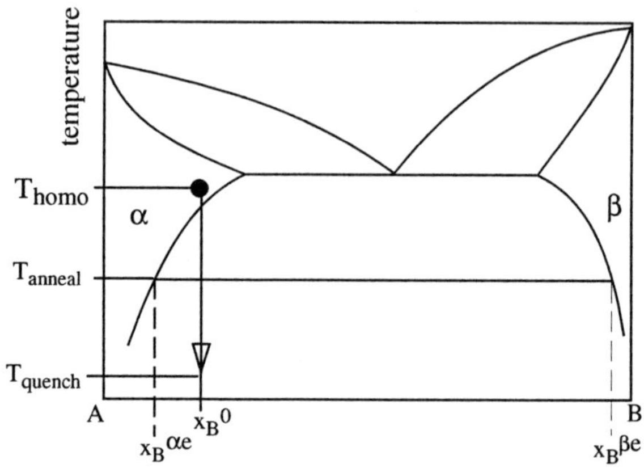

Fig. 10.2. Scheme of a phase diagram for a solid state transformation or precipitation process.

In appendix B we derive for the steady-state nucleation rate, I_{SS},

$$I_{SS} = I_0 \exp(-\frac{\Delta W_R^*}{k_B T}), \tag{10.1}$$

where the prefactor I_0 is a collection of several terms, like the diffusion coefficient, concentration of solute in the matrix etc. It depends mainly on temperature. ΔW_R^* is the work of formation for the critical nucleus defined as $\Delta W_R^* = 16\pi\sigma^3/(3\Delta G_V^2)$. ΔG_V is the change in Gibbs free energy per unit volume on nucleation (which is a function of the time dependent level of supersaturation or supercooling [35, 163, 184, 307]).

The nuclei grow if they are larger than the critical radius which is

$$R^* = \frac{-2\sigma V_m}{\Delta G_m}. \tag{10.2}$$

The critical radius of a nucleus is a function of the interfacial energy and the volume Gibbs free energy change. During growth and coarsening the far field composition will change. The energy or work of formation of the critical nucleus thus will depend on the supersaturation. Using eq.(4.63) and setting the alloy composition equal to x_B^∞ yields an analytic expression for ΔG_m and W_R^* becomes,

$$W_R^* = \frac{16\pi\sigma^3}{3\left[\frac{\partial^2 G_m^*}{\partial x_B^2}(x_B^\infty - x_B^{\alpha,e})(x_B^{\beta,e} - x_B^{\alpha,e})\right]^2}, \tag{10.3}$$

in the small supersaturation limit. Here $(\partial^2 G_m^* / \partial x_B^2)$ denotes the second derivative of the free energy evaluated at the composition of the critical nucleus. The critical radius, also in the small supersaturation limit, is

$$R^* = \frac{2\sigma}{-(\partial^2 G_m^* / \partial x_B^2)(x_B^\infty - x_B^{\alpha,e})(x_B^{\beta,e} - x_B^{\alpha,e})}. \tag{10.4}$$

Thus, as the transformation proceeds and $x_B^\infty \to x_B^{\alpha,e}$ $W_R^* \to \infty$ and by eq.(10.1) $I_{SS} \to 0$. The critical radius for nucleation diverges to infinity as $x_B^\infty \to x_B^{\alpha,e}$. Thus, both, I_{SS} and R^* will not be constant during the transformation.

The droplets nucleate with a certain size spectrum around the critical size[1] which is denoted here by $j(r)dR$ and it is connected to the nucleation rate by

$$I(t) = \int_{R^*}^{\infty} j(R,t)dR. \tag{10.5}$$

Thus $j(R,t)$ is the number density of nuclei entering a size class $R, R+dR$ in a given time interval. The nucleated particles are surrounded by a matrix which is supersaturated. Thus, there could be diffusional growth, growth by an interfacial reaction or any other type of growth as described in chapter 4.

In the chapter on the statistics of growth we developed a continuity equation for the size distribution of particles,

$$\frac{\partial f}{\partial t} + \frac{\partial}{\partial R}\left(\frac{dR}{dt} f\right) = 0, \tag{10.6}$$

where $f(R,t)dR$ denoted the number of particles per volume in the size range $R, R+dR$. In the following we will consider only diffusional growth described by the well known growth law,

$$\frac{dR}{dt} = \frac{DS}{R}, \tag{10.7}$$

[1] As discussed later in this chapter and in the appendix, all particles or clusters of atoms being smaller than the critical size are not stable and not able to grow but disappear. Therefore, the spectrum of sizes should have a lower cut-off at the critical radius.

where D denotes the diffusion coefficient and S the supersaturation (for further details see chapter 4). Note that as the transformation proceeds $x_B^\infty \to x_B^{\alpha,e}$ and $S \to 0$. We restrict ourselves to diffusional growth since it is the one that is most often encountered in metallurgical processes and the description of nucleation treated in the appendix is also restricted to diffusional growth.

Now we will consider nucleation and growth as occurring simultaneously. Since nucleation acts as a source of new particles, we simply have to add a source term to the above continuity equation,

$$\frac{\partial f}{\partial t} + \frac{\partial}{\partial R}\left(\frac{dR}{dt} f\right) = j(R,t). \tag{10.8}$$

The first term describes the time dependent change of the size distribution, the second term is the change due to growth and the term on the right side is the source term; particles appear only due to nucleation. This equation has to be solved using an expression for the growth rate like those described in chapter 4, the nucleation rate (see chapter B or eq.(10.1)) and mass conservation in order to describe measurable quantities like the mean particle radius and the volume fraction of particles as a function of time.

10.1.1 General solution: nucleation and growth

Assuming that there is only nucleation and growth and specifically that S_0 is constant, eq.(10.8) can be solved similar to procedures described in chapter 5, see also [151]. We thus neglect the change in supersaturation that can result from nucleation or growth. Although in physical metallurgy this looks like an unphysical assumption, it is can be realized growing crystals from solutions. During growth of Cu_2SO_4 crystals from aqueous solutions it is easily achievable to control the supersaturation by controlling the evaporation of the solvent and thus keep the concentration of it Cu_2SO_4 in the solvent constant). We rewrite eq.10.8 using eq.10.7

$$\frac{\partial f}{\partial t} - \frac{DS}{R^2} f + \frac{DS}{R}\frac{\partial f}{\partial t} = j(R,t) \tag{10.9}$$

We can solve this equation with the method of characteristics bearing in mind that any particle appearing at this stage is nucleated. Thus, before a time t_0 after the quench (the incubation time for nucleation) there are no particles nucleated and thus $f(R, t < t_0) = 0$. In contrast to the solutions described in the chapter on the growth statistics, Chapter 5, we have no initial size distribution and thus there is no Cauchy problem to solve (or it can be envisaged as a degenerate Cauchy problem with a initial distribution being zero everywhere). The first characteristic equation and its solution is as follows:

$$\begin{aligned} dt &= \frac{dR}{DS/R} \\ \Rightarrow \qquad RdR &= DSdt \\ \Rightarrow \qquad \int_{R_0}^{R} RdR &= DS\int_{t_0}^{t} dt \\ \frac{1}{2}(R^2 - R_0^2) &= DS(t-t_0) \\ R_0^2 &= R^2 - 2DS(t-t_0) \\ R^2 &= R_0^2 + 2DS(t-t_0) \end{aligned} \tag{10.10}$$

The second characteristic equation reads

$$\frac{df}{dt} = \frac{DS}{R^2} f + j(R,t). \tag{10.11}$$

Inserting the solutions obtained from the first characteristic equation leads to

$$\frac{df}{dt} = \frac{DS}{R_0^2 + 2DS(t-t_0)} f + j(\sqrt{R_0^2 + 2DS(t-t_0)}, t) \tag{10.12}$$

This equation is an ordinary differential equation of first order with one variable, namely t, and can easily be integrated leading to

$$f = \exp(\omega)\int_{t_0}^{t} j(\sqrt{R_0^2 + 2DS(\theta - t_0)}, \theta)\exp(-\omega)d\theta \tag{10.13}$$

with

$$\begin{aligned} \omega &= \int_{t_0}^{t} \frac{d\theta}{R_0^2 + 2DS(\theta - t_0)} \\ &= \frac{1}{2}\ln(R_0^2 + 2DS(\theta - t_0))\Big|_{t_0}^{t} \\ &= \ln\frac{R_0^2 + 2DS(t-t_0)}{R_0}. \end{aligned} \tag{10.14}$$

Inserting ω back into eq.(10.13) gives

$$f = \frac{\sqrt{R_0^2 + 2DS(t-t_0)}}{R_0} \int_{t_0}^{t} \frac{R_0 j(\sqrt{R_0^2 + 2DS(\theta - t_0)}, \theta)}{\sqrt{R_0^2 + 2DS(t-t_0)}} d\theta. \quad (10.15)$$

Inserting the expression for R_0 from eq.(10.10) we obtain the final solution of the problem of concurrent growth and nucleation in its most general form, denoted now as $f_{NG}(R,t)$

$$f_{NG}(R,t) = R \int_{t_0}^{t} \frac{j(\sqrt{R^2 - 2DS(t-\theta)}, \theta)}{\sqrt{R^2 - 2DS(t-\theta)}} d\theta. \quad (10.16)$$

The evolution of the size distribution depends strongly on the nucleation rate. Let us consider a special case.

10.1.2 Constant nucleation rate density

First we consider the case of nucleation and diffusional growth assuming that the nucleation rate $I(t) = I_{SS}^0$ is a constant independent of time. Let us further assume that any nucleus created shall have just the critical size R^* and that there shall be no incubation time for nucleation, i.e. $t_0 = 0$. Then $j(R,t) = I_{SS}^0 \delta(R - R^*)$, with $\delta(x)$ the Dirac δ-function. Performing the integral in eq.(10.16) and keeping in mind the definition of integrals over δ-functions[2] yields

$$f(R,t) = \frac{I_{SS}^0}{DS} \cdot R \quad \text{if} \quad R \leq \sqrt{2DSt} \quad \text{else} \quad f(R,t) = 0. \quad (10.17)$$

Note that $f(R,t)$ is a triangle with a maximum that varies with time and that the maximum particle size is given by $R_{max} = \sqrt{2DSt}$. The number density $n_V(t) = \int f(R,t)dR$ is

$$n_V(t) = I_{SS}^0 t \quad (10.18)$$

as expected, since the nucleation rate is constant the number density must increase linearly with time. The mean particle/droplet radius is

$$\langle R \rangle = \frac{\int R f(R,t) dR}{n_V(t)} = \frac{(I_{SS}^0 R_{max}^3 / 3DS)}{I_{SS}^0 t} = \frac{2}{3}\sqrt{2DSt}. \quad (10.19)$$

[2] The following relation holds for δ-functions: $\delta(x) = 0$ for all $x \neq 0$ and $\int_{-\infty}^{\infty} \delta(x)dx = 1$ For integrals over a finite interval: $\int_a^b \psi(x)\delta(f(x))dx = \psi(x_i)/|f'(x_i)|$ where the x_i are zeros of the function $f(x_i) = 0$. f must be a continuous function of x with a finite number of zeros and the interval $[a,b]$ contains just the zero x_i. In addition the derivative of f with respect to x shall not vanish at the zero $f'(x_i) \neq 0$.

Interestingly, the average radius does not depend on the nucleation rate explicitly. It is, however, smaller than the value calculated from the growth rate itself, which would be $\langle R\rangle = \sqrt{2DSt}$. Thus, the continuous input of very small particles into the volume reduces the average size by a constant factor 2/3.

There is no problem to drop the δ-function behavior and put, for example, a Gaussian distribution for the nucleation rate density $j(R)$, for example

$$j(R) = \frac{1}{s\sqrt{2\pi}} \exp\left(\frac{-(R - a_0R^*)^2}{2s^2}\right). \tag{10.20}$$

with a_0 an arbitrary constant and s the standard deviation. Then, eq.(10.16) would read

$$f_{NG}(R,t) = \frac{R}{s\sqrt{2\pi}} \int_{t_0}^{t} \frac{\exp\left(-\frac{(\sqrt{R^2-2DS(t-\theta)}-a_0)^2}{2s^2}\right)}{\sqrt{R^2 - 2DS(t-\theta)}} d\theta. \tag{10.21}$$

Introducing the coordinate transformation $\xi = \sqrt{R^2 - 2DS(t-\theta)}$ allows to calculate the integral [4] leading to,

$$f_{NG}(R,t) = \frac{Rs}{DS} \left[\operatorname{erf}\left(\frac{R - a_0R^*}{s\sqrt{2}}\right) - \operatorname{erf}\left(\frac{\sqrt{R^2 - 2DS(t-t_0)} - a_0R^*}{s\sqrt{2}}\right)\right] \tag{10.22}$$

The change introduced by such a Gauss distribution is shown in fig.10.3 using arbitrary values for D,S,a_0 and R^* since the purpose is just to show the differences to the triangle shaped distribution presented in eq.(10.17). In contrast to the triangle shaped distribution presented in eq.(10.17) which was obtained for the case of a δ-function type nucleation rate density, these distributions start at very small radii not linearly, proceed then linearly and tail off slowly after a sharp maximum. The shift in the maximum corresponds to the time dependent behavior of diffusive growth. The number density, i.e. the area under the distribution curve increases, as long as nucleation continuous. Although this model is more realistic than the δ-function approach, a more realistic model would introduce time dependent nucleation, not in the classical sense by introducing an incubation time, but accounting for the time dependent supersaturation level. Then the growth rate depends not only on radius but also on time

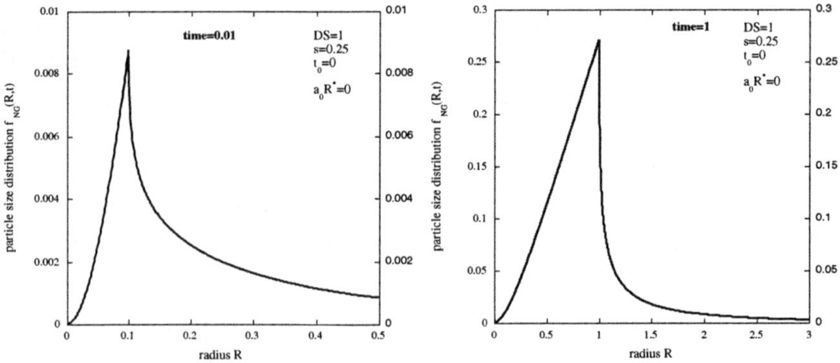

Fig. 10.3. If the nucleation rate density could be described by a Gauss distribution around the critical nucleus size the particle size distribution changes due to concurrent nucleation and diffusive growth at constant supersaturation as shown in these figures. In contrast to the triangle shaped distribution presented in eq.(10.17) which was obtained for the case of a δ-function type nucleation rate density, these distributions start at very small radii not linearly, proceed then linearly and tail off slowly after a sharp maximum.

and the nucleation rate changes since the energy barrier for nucleation varies (if the level of supersaturation decreases it increases). Doing this generally leads to a set of equations that cannot be solved analytically. An approach to solve that problem analytically is described in the next section.

We will, however, first illustrate briefly a numerical solution to the nucleation and growth problem. As explained above, curvature effects can be neglected if the supersaturation much larger during processing than any curvature induced change in the supersaturation due to the Gibbs-Thomson effect. Zhao and Ratke [318] investigated the rapid cooling through the liquid miscibility gap of low alloyed Al-Pb alloys (the phase diagram looks similar to that shown schematically in fig.2.6, but with a much steeper binodal line). Cooling a sample from the single phase field through the miscibility gap with cooling rates of 50 to 150 K/s leads on passing the liquid two-phase boundary to a rapid increase in supersaturation until nucleation sets in. The nuclei appear, grow by mass diffusion and consume the supersaturation. They calculated the time dependent nucleation rate using the expression given above and in appendix B, used the known functional form of the free energy of Al-Pb alloys, the known temperature dependence of the interfacial tension and also simulated the temperature dependence of the diffusion coefficient (for details see [318]). A typical microstructure of such a fast cooled sample is shown in fig.1.3 (the interested reader is also referred to [238]. The variation of the nucleation rate and the supersaturation is shown in fig.10.4. Figure 10.4 demonstrates that the

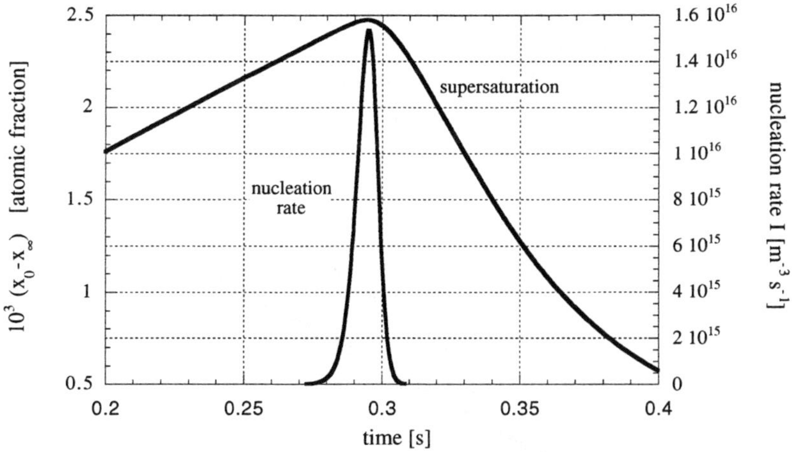

Fig. 10.4. Nucleation rate and supersaturation as developing with time of cooling an Al+7 wt%Pb alloy through the miscibility gap with 150 K/s. The supersaturation increases on passing the miscibility gap boundary and once nucleation has started it tends to decrease. The calculations were made with a numerical simulation based upon the theory outlined in this chapter, see [318].

first effect of cooling into the miscibility gap is a continuous increase of the supersaturation until it reaches a critical amount such that nucleation starts. Then the supersaturation does not directly decrease but increases further for a short period of time before it reaches it maximum value. During the early period of liquid-liquid decomposition the supersaturation of the matrix does not decrease but increases during cooling because both the nucleation rate I and the number density of droplets n_V are too small. The enhancement of supersaturation leads to a rapid increase of the nucleation rate and therefore also the number density of droplets rapidly increases (see figs. 10.4 and 10.5). A decrease of the level of supersaturation leads to an increase of the activation barrier for nucleation (to first order the barrier height is inversely proportional to the square of supersaturation, see appendix B). Since the nucleation rate depends exponentially on the barrier height it rapidly falls below a certain numerical detection value set arbitrarily to 1 nuclei per cubic meter and second. During this period of time the change in average droplet radius is a result of the common action of nucleation and growth of droplets being nucleated. Initially the average droplet radius increases rapidly with time because the very small droplets nucleated have a very high growth rate. A little later the growth rate slows down because very small droplets are nucleated continuously at a high rate, leading to the above mentioned slower increase in average radius compared with that of a single drop. After a short period of time, the nucleation rate decreases to a very low level and the average droplet radius increases by the diffusional growth only.

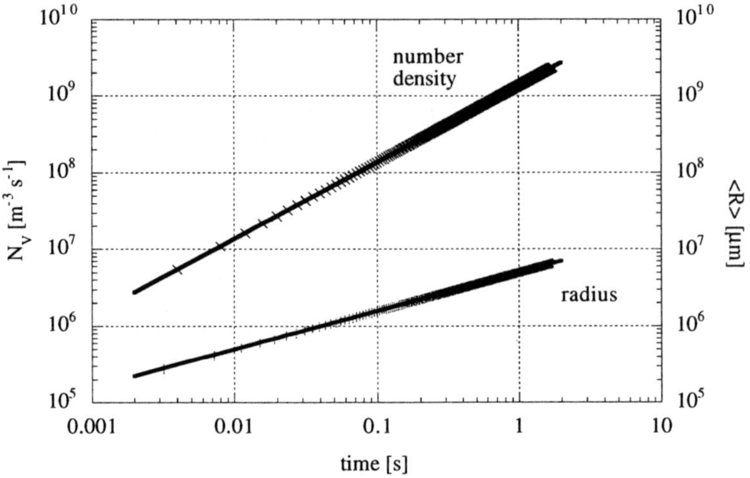

Fig. 10.5. Number density and average droplet as developing with time of cooling an Al+7 wt%Pb alloy through the miscibility gap with 150 K/s. The calculations were made with a numerical simulation based upon the theory outlined in this chapter, see [318].

10.2 Nucleation, growth and coarsening

As mentioned above one can basically divide the general sequence of forming a two phase alloy from a single phase one into three steps,

1. nucleation,
2. growth of the nuclei or particles,
3. coarsening (Ostwald ripening).

We have described theories of all three steps in Appendix B, Chapter 4 and 6 together with Chapter 7. We have mentioned several times that such a separation into three distinct steps is artificial. In the last ten years some unifying theories have been developed which calculate the evolution of the mean droplet size and their size distribution from the very beginning of the decomposition process to its final stage linking all steps self-consistently together [163, 185]. This unified approach shall be presented here briefly.

Starting point of our treatment is the continuity equation with a source term as already presented in eq.(10.8). For convenience we write it here again,

$$\frac{\partial f}{\partial t} + \frac{\partial}{\partial R}\left(\frac{dR}{dt} f\right) = j(R,t), \tag{10.23}$$

with $j(R)$ the number density of nuclei entering a size class $R, R+dR$ in a given time interval. It is related to the nucleation rate by

$$I(t) = \int_{R^*}^{\infty} j(R,t)dR. \tag{10.24}$$

where only particles with $R > R^*$ are considered part of the precipitate phase. We will only treat diffusional growth and coarsening of nuclei as done in the last section. Then, the growth rate of a particle in a supersaturated solution is (see eq.(4.47))

$$\frac{dR}{dt} = \left(\frac{x_B^\infty - x_B^\alpha}{x_B^\beta - x_B^\alpha}\right)\frac{D}{R}, \tag{10.25}$$

with x_B^∞ the matrix or far field concentration which is a function of time. We now use directly the Gibbs-Thomson relation (see eq.(2.124)),

$$\begin{aligned} x_B^\infty(t) &= x_B^{\alpha,e} + \frac{\ell^\alpha}{R^*(t)}\ , \\ x_B^\alpha(R) &= x_B^{\alpha,e} + \frac{\ell^\alpha}{R}\ , \end{aligned} \tag{10.26}$$

and we again neglect as in chapter 6 the effect of curvature on the difference $x_B^\beta - x_B^\alpha \approx x_B^{\beta,e} - x_B^{\alpha,e}$. The use of the Gibbs-Thomson equation as derived in the chapter on thermodynamics is an approximation. We there derived it under the assumption that ℓ^α/R^* is a small quantity, i.e. $\ell^\alpha/R^* \ll 1$. In the case of nucleation the critical radius generally is a small quantity and easily of the order of ℓ^α itself. Therefore one should revisit the derivation of the Gibbs-Thomson equation and not employ the series expansion used to derive eq.(10.26). Kampmann and Wagner extended the model described here to include the non-linear variation of the concentration with curvature for very small nuclei in the dilute solution approximation [163]. Inserting these equations into eq.(10.25) leads to

$$\frac{dR}{dt} = \frac{D\ell^{\alpha}}{R(x_B^{\beta,e} - x_B^{\alpha,e})}\left(\frac{1}{R^*} - \frac{1}{R}\right). \tag{10.27}$$

Unfortunately, $j(R)$ in eq.(10.23) cannot be calculated from first principles. We can only make assumptions on its form and functional dependence. In contrast to the section on nucleation and growth above, but similar to the situation dealt with in the chapters on coarsening, we have a growth rate that is time dependent via the critical radius and, thus, depends on the evolution of the size distribution. Therefore, we cannot present a general analytical solution. We would, however, at least like to have an expression for the time change of the average radius, the number density of particles and the average matrix concentration, since these quantities can be measured and a theoretical prediction could be compared with them. The average radius can be calculated without a detailed knowledge of the time evolution of the size distribution if we make an assumption on the size distribution of the nuclei.

Let us calculate the particle number density from

$$n_V(t) = \int_{R^*}^{\infty} f(R,t)dR. \tag{10.28}$$

The lower bound is set by the critical radius of nucleation because we assume that all particles smaller than R^* disappear and only particles larger can grow. The time change of the particle number density due to nucleation, growth and coarsening can be calculated from the above expression for $n_V(t)$,

$$\frac{dn_V(t)}{dt} = \frac{d}{dt}\int_{R^*}^{\infty} f(R,t)dR = \int_{R^*}^{\infty} \frac{\partial f}{\partial t}dR - f(R^*)\frac{dR^*}{dt}. \tag{10.29}$$

The last term originates from the fact that the critical radius is a time dependent function assuming that the supersaturation decays over time. Thus, the total derivative of the integral over the size distribution has two terms. The first term in this equation can be rearranged using the continuity equation,

$$\int_{R^*}^{\infty} \frac{\partial f}{\partial t} dR = -\int_{R^*}^{\infty} \frac{\partial}{\partial R}\left(\frac{dR}{dt} f\right) dR + \int_{R^*}^{\infty} j(R,t) dR. \tag{10.30}$$

The first integral on the right side of eq.(10.30) can be calculated directly,

$$\int_{R^*}^{\infty} \frac{\partial}{\partial R}\left(\frac{dR}{dt} f\right) dR = \left.\frac{dR}{dt} f\right|_{R=R^*}^{\infty} = \lim_{R\to\infty} \frac{dR}{dt} f - \left.\frac{dR}{dt} f\right|_{R=R^*} = 0. \tag{10.31}$$

The first term equals zero since the distribution function shall be zero at infinite radius. The second term is zero since the growth rate is zero at $R = R^*$, see eq.(10.27). The second integral on the right hand side of eq.(10.30) is just the definition of the nucleation rate. We thus obtain for the time change of the number density the expression

$$\frac{dn_V}{dt} = I(t) - f(R^*)\frac{dR^*}{dt}. \tag{10.32}$$

Let us now calculate the average particle size, defined by

$$\langle R\rangle = \frac{1}{n_V(t)} \int_{R^*}^{\infty} R f(R,t) dR. \tag{10.33}$$

Its time rate of change is calculated from

$$\frac{d\langle R\rangle}{dt} = -\frac{1}{n_V^2}\frac{dn_V}{dt} \int_{R^*}^{\infty} R f(R,t) dR + \frac{1}{n_V}\frac{d}{dt} \int_{R^*}^{\infty} R f(R,t) dR. \tag{10.34}$$

The evaluation of the derivative gives

$$\frac{d\langle R\rangle}{dt} = -\frac{1}{n_V}\frac{dn_V}{dt}\langle R\rangle + \frac{1}{n_V}\left(-R^*\frac{dR^*}{dt} f(R^*,t) + A_0\right) \tag{10.35}$$

with the term A_0 being calculated from

$$\begin{aligned} A_0 &= \int_{R^*}^{\infty} R\frac{\partial f}{\partial t} dR \\ &= -\int_{R^*}^{\infty} R\frac{\partial}{\partial R}\left(\frac{dR}{dt} f\right) dR + \int_{R^*}^{\infty} R j(R,t) dR \\ &= -\left. R\frac{dR}{dt} f\right|_{R^*}^{\infty} + \int_{R^*}^{\infty} \frac{dR}{dt} f(R,t) dR + \int_{R^*}^{\infty} R j(R,t) dR. \end{aligned} \tag{10.36}$$

The first term on the right side is zero. Thus, the average radius changes as

$$\frac{d\langle R\rangle}{dt} = \left\langle \frac{dR}{dt} \right\rangle - \frac{1}{n_V}\frac{dn_V}{dt}\langle R\rangle - \frac{1}{n_V}R^*\frac{dR^*}{dt}f(R^*,t) + \frac{1}{n_V}\int_{R^*}^{\infty} Rj(R)dR, \tag{10.37}$$

where we have used that the average of the growth rate is defined as

$$\left\langle \frac{dR}{dt} \right\rangle = \frac{1}{n_V}\int_{R^*}^{\infty} \frac{dR}{dt}f(R,t)dR. \tag{10.38}$$

Inserting eq.(10.32) into eq.(10.37) yields

$$\begin{aligned}\frac{d\langle R\rangle}{dt} &= \left\langle \frac{dR}{dt} \right\rangle - \frac{1}{n_V}\left(I(t) - f(R^*)\frac{dR^*}{dt}\right)\langle R\rangle \\ &\quad - \frac{1}{n_V}\left(R^*\frac{dR^*}{dt}f(R^*,t)\right) + \frac{1}{n_V}\int_{R^*}^{\infty} Rj(R,t)dR. \end{aligned} \tag{10.39}$$

Collecting terms and using eq.(10.24) leads to

$$\frac{d\langle R\rangle}{dt} = \left\langle \frac{dR}{dt} \right\rangle + \frac{1}{n_V}\int_{R^*}^{\infty} (R - \langle R\rangle)j(R)dR + (\langle R\rangle - R^*)\frac{f(R^*)}{n_V(t)}\frac{dR^*}{dt} \tag{10.40}$$

The three terms in this equation have the following meaning: The average radius changes by growth and coarsening (first term). The second term states that the nucleation rate or the stream of particle entering the system changes the average radius and the third term is a correction term to the second term, accounting for the changing nucleation which changes the average radius. The two equations for the change of number density of particles and the average radius cannot be solved in their current form. In order to solve these equations we now make a list of further approximations.

1. approximation:
The first term in eq.(10.40) is approximated as

$$\left\langle \frac{dR}{dt} \right\rangle \approx \left.\frac{dR}{dt}\right|_{R=\langle R\rangle} = \frac{D}{\langle R\rangle}\frac{\ell^\alpha}{x_B^{\beta,e} - x_B^{\alpha,e}}\left(\frac{1}{R^*} - \frac{1}{\langle R\rangle}\right). \tag{10.41}$$

2. approximation:
To deal with the second term we assume that the current density of nuclei is a Gaussian distribution around the critical radius with a variance of $(\Delta R^*)^2$.

$$j(R,t) = I(t)\frac{1}{\Delta R^*\sqrt{2\pi}}\exp\left(-\frac{(R-R^*)^2}{2(\Delta R^*)^2}\right) \tag{10.42}$$

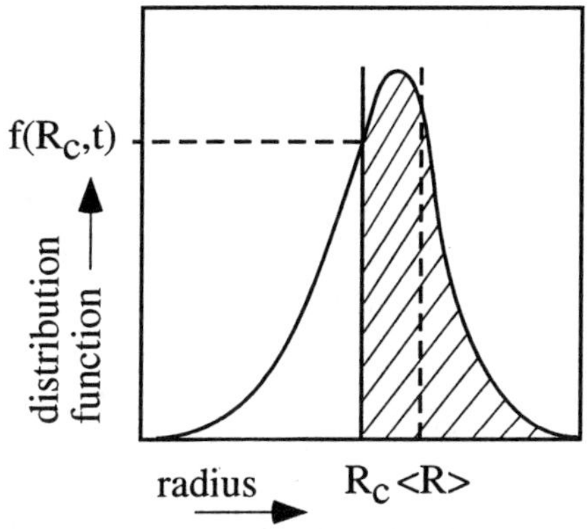

Fig. 10.6. Illustration of the relation between the full continuous size distribution of clusters (or nuclei). In this theory we assume that only nuclei being larger than the critical radius R^* contribute to the evolution of particle in radius and time.

A schematic sketch of the distribution is shown in Fig. 10.6. This allows us to calculate the second term in eq.(10.40),

$$\frac{\int_{R^*}^{\infty}(R - \langle R\rangle)j(R)dR}{n_V} = \frac{I_{SS}}{n_V \Delta R^* \sqrt{2\pi}} \int_{R^*}^{\infty} (R - \langle R\rangle) e^{\left(-\frac{(R-R^*)^2}{2(\Delta R^*)^2}\right)} dR. \tag{10.43}$$

Transforming the variables in the integral by defining a new variable $y = (R - R^*)/\sqrt{2}\Delta R^*$ and using the result that the integral over $1/\sqrt{2\pi} \cdot \exp(-y^2)$ is 1 and using the identity $\int y \exp -y^2 dy = -1/2 \exp(-y^2)$ one can evaluate the integral and obtain for the second term in eq.(10.40) the expression

$$\text{2nd term} = \frac{I_{SS}}{n_V}(R^* - \langle R\rangle + \Delta R^*). \tag{10.44}$$

3. approximation

For the third term we need an expression for $f(R^*)$. The integral over the size distribution defined in eq.(10.28) is approximated by a rectangular box of height $f(R^*)$ and width $\langle R\rangle - R^*$ and since we are not integrating to infinity we include a factor called a_1,

$$n_V(t) = \int_{R^*}^{\infty} f(R,t)dR \approx f(R^*)((\langle R\rangle - R^*)/a_1 \tag{10.45}$$

We thus have for the size distribution at the critical radius the value

$$f(R^*) = \frac{n_V a_1}{\langle R\rangle - R^*} \tag{10.46}$$

The factor a_1 is set by Langer and Schwartz arbitrarily as

$$a_1 = 0.317 \quad \text{for} \quad \langle R \rangle < 3/2R^*, \qquad a_1 = 0 \quad \text{elsewhere}, \tag{10.47}$$

to insure that the LSW result on coarsening is reproduced for long annealing times. We can now collect the terms and give a final equation for the time change of the average radius,

$$\frac{d\langle R \rangle}{dt} = \frac{D\ell^\alpha}{R(x_B^{\beta,e} - x_B^{\alpha,e})} \left(\frac{1}{R^*} - \frac{1}{\langle R \rangle} \right) + \frac{I_{SS}}{n_V}(R^* - \langle R \rangle + \Delta R^*) + a_1 \frac{dR^*}{dt}. \tag{10.48}$$

This equation has to be solved together with eq.(10.32) which, using the approximations given above, becomes

$$\frac{dn_V}{dt} = I(t) - \frac{n_V(t) a_1}{\langle R \rangle - R^*} \frac{dR^*}{dt}. \tag{10.49}$$

In addition we need the nucleation rate that depends not only on temperature but especially on supersaturation (or the critical radius). The system of equations is closed by a conservation equation for species,

$$\begin{aligned} x_B^\alpha(t) &= x_B^0 - \frac{4\pi}{3}(x_B^{\beta,e} - x_B^{\alpha,e}) \int_{R^*}^{\infty} R^3 f(R,t) dR \\ &\approx x_B^0 - \frac{4\pi}{3}(x_B^{\beta,e} - x_B^{\alpha,e}) \langle R \rangle^3 n_V(t). \end{aligned} \tag{10.50}$$

Note that in eq.(10.50) we approximated $\langle R^3 \rangle$ with $\langle R \rangle^3$. The particle size distribution is not calculated but only average values. The set of equations (10.45,10.48,10.49,10.50) together with an appropriate expression for the nucleation rate has to be solved numerically. The unknown quantities that can now be obtained numerically are the nucleation rate, the number density, the average matrix concentration and the average particle size. To illustrate this approach we present two experimental results which were compared by the authors with numerical calculations based on the above theory. The first is taken from an experimental investigation on the nucleation, growth and coarsening in Ni-Al alloys made by Wendt and Haasen [307]. They investigated the precipitation using atom probe field-ion microscopy and could thus measure particle sizes as small as one nanometer. The details of their investigation is given in the literature cited and shall not be repeated here. In the context of this monograph it is only important to notice that they measured the average precipitate size in

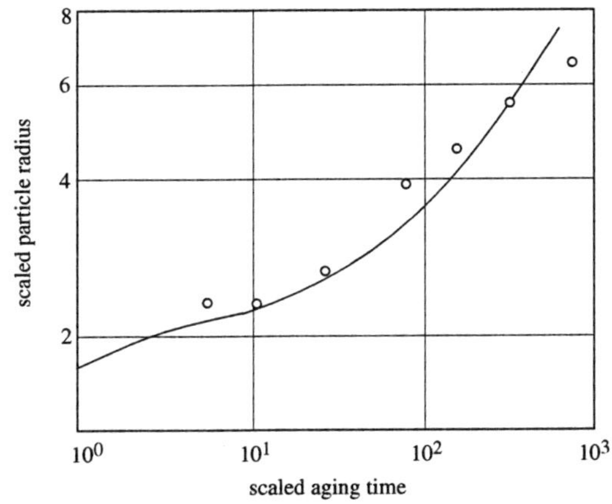

Fig. 10.7. Evolution of the average precipitate radius with time in a Ni-Al alloy according to Wendt and Haasen [307]. The theoretical curve was calculated with the theory outlined above using for the nucleation rate the expression given in the appendix and describing the solid solution of Al in Ni by a regular solution model. Details can be found in the original paper cited.

the very early stages of nucleation and growth and compared those with the theory presented here (using of course some minor modifications). In these alloys precipitates of the γ'-phase form from a supersaturated solid solution of Al in Ni, the γ-phase. One of their results for an Ni-14at.% Al alloys aged at 823K for various times is shown in Fig. 10.7. In order to compare the experimental results with theoretical ones they used scaled variables. The scaled radius was calculated from $\rho = \sqrt{16\pi\sigma/3k_BT}\langle R\rangle/2$ and the scaled time from $\tau = D\sqrt{16\pi\sigma/3k_BT}^3\ell^\alpha t/4$. A second example comes from a series of experiments by Kampmann and Wagner using Cu-1.9at.% Ti alloys [163]. They used the same experimental method as Wendt and Haasen to study the precipitation of Ti-rich precipitates in copper. They modified slightly the theory of Langer and Schwartz sketched above by using the non-linearized version of the Gibbs-Thomson relation and using the time dependent nucleation rate theory (see Appendix B). They also implemented a fully numerical solution to the continuity equation, and thus did not have to employ many of the approximations discussed above. From their results we only show the time evolution of the average particle radius in Fig. 10.8. In both cases the agreement between the theoretical predictions and the experimental results is reasonable, which shows the usefulness of the theoretical approach developed by Langer and Schwartz.

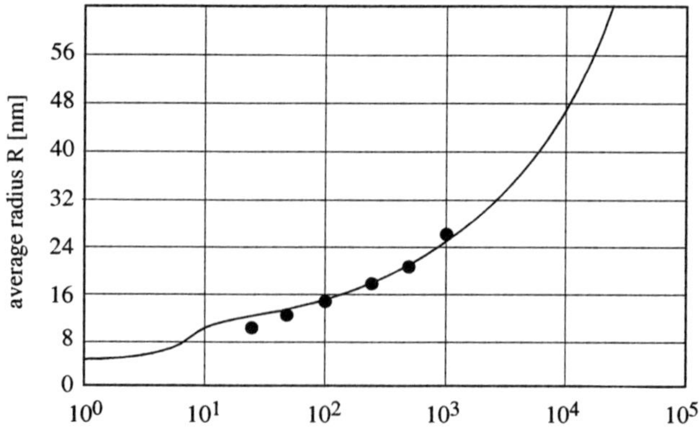

Fig. 10.8. Evolution of the average particle size in dilute Cu-Ti alloys according to Kampmann and Wagner [163]. The dots are the experimental results. The solid line is a numerical solution of the theory described in this section.

Let us finally make a critical comment to the approach of Langer and Schwartz. The whole procedure of approximating the size distribution as a rectangular just on the interval $\langle R \rangle - R^*$, as done in the 3rd approximation (see 221) is somehow arbitrary and not without problems. Since after long annealing times we know from coarsening theory that $\langle R \rangle$ comes intimately close to R^* the size distribution would converge towards a δ-function and definitively not to the LSW distribution. Fixing the constant a_1 might lead to the correct answer for the average radius as depending on time, but definitively not to a correct answer for the size distribution. There is no analytical solution known to this problem. Kampmann and Wagner [163] as well as others [35, 318, 319, 320] have developed numerical models that solve the problem of concurrent nucleation, growth and coarsening. The numerical analysis always leads to solutions converging to the well known LSW one under suitable conditions.

Appendix

A. Droplet and particle motion

In this chapter we will very briefly review some basic ideas of fluid dynamics and then derive in detail the equations of droplet and particle motion in a fluid matrix. These are useful to understand the effects of convective heat and mass transport on growth and coarsening or generally on microstructural evolution during solid-liquid phase transformations. There are several standard textbooks on fluid dynamics that discuss all aspects in detail. The reader may look into books from Batchelor [22], Landau and Lifshitz [182], Platten and Legros [232], Chandrasekhar [57], Sommerfeld [259] or even contact the classical textbook on theoretical physics from Morse and Feshbach [216].

A.1 Some basic fluid dynamic equations

Fluid dynamics describes the behavior of liquids and gases under the action of external and internal forces leading to flow or convection. The importance of fluid flow in some types of phase transformations became especially clear during the last two decades due to research under reduced gravity conditions. Especially in solidification and crystal growth it turned out to be of utmost importance to consider the effect of fluid flow of various origins on mass and heat transport and the microstructural evolution.

All types of fluid flow in gases or liquids must conserve mass. The fundamental equation describing the conservation of mass is derived as follows. Let us consider the mass of a volume V. Mass is the volume integral over the density,

$$M = \int_V \rho dV. \tag{A.1}$$

The flow through an infinitesimal surface element $\boldsymbol{df}$ per unit of time can be calculated from the flux density. Any flow is in general

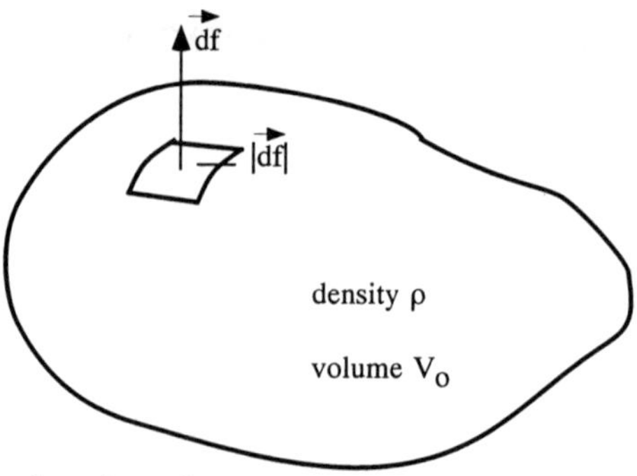

Fig. A.1. Volume and surface elements

the product of a density times a velocity. Here the density is that of the fluid and the velocity is a vector that can depend on space and time. The change of the flux through the surface element can therefore be expressed as

$$dI = \rho \boldsymbol{v} \cdot \boldsymbol{df} = \boldsymbol{j} \cdot \boldsymbol{df} \tag{A.2}$$

The variation of mass due to fluid flow into the volume can be calculated as the integral of the flux through the surface of the volume,

$$\begin{aligned} \frac{dM}{dt} &= \oint_{surface} \rho\, \boldsymbol{v} \cdot \boldsymbol{df} \\ &= -\frac{d}{dt} \int_V \rho dV \end{aligned} \tag{A.3}$$

or

$$\frac{d}{dt} \int_V \rho dV + \oint_{surface} \rho \boldsymbol{v} \cdot \boldsymbol{df} = 0. \tag{A.4}$$

Using Gauss theorem we can convert the integral over the surface into a volume integral and can then express eq.(A.4) as one single integral over the volume. This integral, however, is only zero if the integrand itself is zero,

$$\frac{d\rho}{dt} + \nabla \cdot (\rho \boldsymbol{v}) = 0. \tag{A.5}$$

This is the differential formulation of mass conservation. If there are no sources or sinks in a volume mass can only change due to fluxes into and out of the volume. eq.(A.5) can be written in a different form expressing the total time derivative of the density in a slightly different way,

$$\frac{d\rho}{dt} = \frac{\partial \rho}{\partial t} + \boldsymbol{v} \cdot \nabla \rho.$$

Using this in eq.(A.5) we obtain

$$\frac{d\rho}{dt} + \rho \nabla \cdot \boldsymbol{v} = 0. \tag{A.6}$$

Metallic melts are *incompressible*, whereas gases are compressible. A fluid is called incompressible if the mass density does not change in space and time, i.e. due to changing pressure. This is the case if

$$\frac{d\rho}{dt} = 0. \tag{A.7}$$

Then, we can simplify the continuity equation, eq.(A.6), to

$$\nabla \cdot \boldsymbol{v} = 0 \tag{A.8}$$

This equation shows a remarkable physical specialty of incompressible fluids: The flow is source free and the flow lines are always closed loops. One may compare that with Maxwell's equation in electromagnetism. In this case for the magnetic induction $\boldsymbol{B}$ we have $\nabla \cdot \boldsymbol{B} = 0$, meaning all magnetic flux lines are closed. For fluids eq.(A.8) this means that whenever a flow emerges in a fluid it must come back to its origin.

Let us consider the forces acting on a fluid element,

$$\begin{aligned} \boldsymbol{F}_V &= - \oint P d\boldsymbol{f} \\ &= - \int \nabla P dV. \end{aligned} \tag{A.9}$$

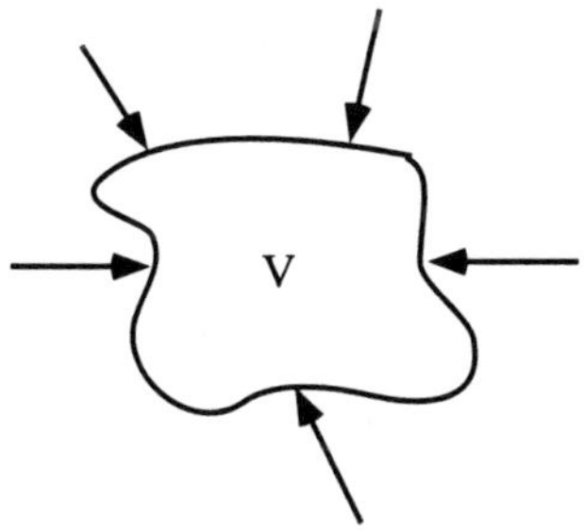

Fig. A.2. Forces on a liquid volume

The right term is negative since the normal vector to the surface surrounding the volume V is directed outward and, thus, commonly counted positive. The force acting is realized by a scalar P, the pressure, multiplied by the normal at a point on the surface. The normal vector gives the direction and the scalar P gives the amplitude. ∇P is a force density. According to Newton's law $\boldsymbol{F} = M\boldsymbol{a}$, with $\boldsymbol{a}$ as the acceleration, we can write

$$\rho \frac{d\boldsymbol{v}}{dt} = -\nabla P. \tag{A.10}$$

With

$$\frac{d\boldsymbol{v}}{dt} = \frac{\partial \boldsymbol{v}}{\partial t} + \boldsymbol{v} \cdot \nabla \boldsymbol{v}$$

we obtain the so-called *Euler equation*,

$$\frac{\partial \boldsymbol{v}}{\partial t} + \boldsymbol{v} \cdot \nabla \boldsymbol{v} = -\frac{1}{\rho} \nabla P. \tag{A.11}$$

Here $\partial \boldsymbol{v}/\partial t$ is the time dependent change of the fluid flow velocity at a fixed point in space and $\boldsymbol{v} \cdot \nabla \boldsymbol{v}$ is the pure spacial variation of the velocity being independent of time[1]. It exits even if there is a stationary flow pattern. Consider for example the flow through a pipe whose cross section changes at a certain point. Then, there are local variations in fluid velocity and thus accelerations although the flow pattern does not change with time. A stationary fluid flow field is therefore characterized by $\partial \boldsymbol{v}/\partial t = 0$.

There are special situation in which a flow is a *potential flow*, meaning that the flow field originates from a potential, ϕ, such that we have a flow velocity field described by

$$\boldsymbol{v} = \nabla \phi. \tag{A.12}$$

This is exactly the case when the flow is circulation free or *irrotational*, i.e.

$$\nabla \times \boldsymbol{v} = 0. \tag{A.13}$$

In the case of a potential flow the continuity equation reads

$$\nabla \cdot \boldsymbol{v} = \nabla \cdot (\nabla \phi) = \nabla^2 \phi = 0. \tag{A.14}$$

[1] Note that $\nabla \boldsymbol{v}$ is a tensor or a dyadic product. Scalar multiplication with $\boldsymbol{v}$ gives a vector.

This is the classical equation for a potential called *Laplace's equation.* Another special situation exists in two dimensional flows. The continuity equation can then be fulfilled automatically if the components of the velocity vector are defined the following way.

In rectangular coordinates the continuity equation reads,

$$\frac{\partial v_x}{\partial x} + \frac{\partial v_y}{\partial y} = 0. \tag{A.15}$$

If we define the velocity components as derivatives of a so-called stream function Ψ via

$$v_x = -\frac{\partial \Psi}{\partial y} \tag{A.16}$$

$$v_y = \frac{\partial \Psi}{\partial x} \tag{A.17}$$

we see that the continuity equation is fulfilled automatically with the unknown function $\Psi(x, y)$,

$$-\frac{\partial}{\partial x}\frac{\partial \Psi}{\partial y} + \frac{\partial}{\partial y}\frac{\partial \Psi}{\partial x} = 0 \tag{A.18}$$

This relation states that the order of differentiation can be interchanged and thus Ψ has a total differential. For a series of problems the stream functions are well known and can be found in standard textbooks on fluid dynamics given in the bibliography. Stream functions can also be defined in axisymmetric flow problems using cylindrical coordinates or spherical polar coordinates (where one axis is an axis of rotational symmetry) as

$$v_r = -\frac{1}{r^2 \sin^2\theta}\frac{\partial \Psi}{\partial \theta} \tag{A.19}$$

$$v_\theta = \frac{1}{r \sin^2\theta}\frac{\partial \Psi}{\partial r} \tag{A.20}$$

with r measured in the radial direction and θ the polar angle. We make use of this definitions in this monograph when we treat mass and heat transport towards moving spheres.

So far we have treated so-called *ideal fluids*, meaning there is no friction between fluid elements and thus shear stresses are not transferred between layers of the fluid. All real fluids, however, are non-ideal, they are viscous. The viscosity is phenomenologically a measure of the fluidity or the frictional forces transferred between adjacent layers in a fluid. If the viscosity is high, like in honey, the flow is drastically damped by the viscous forces whereas if the viscosity is low, like in water, the effect is comparatively small especially at low fluid flow velocities. The effect of viscosity can be taken into account using the well-known linear relation given by Newton for the relation between the fluid flow velocity and the viscous shear stress τ, namely

$$\tau = \eta \nabla \boldsymbol{v} \tag{A.21}$$

with similar relations valid for the other flow components[2]. We will not go into any details (see footnote 2) but show for the force balance on a fluid element. This leads to the well-known *Navier-Stokes equation* in incompressible isotropic fluids in which stress and flow velocity obey Newton's relation given above,

$$\rho[\frac{\partial \boldsymbol{v}}{\partial t} + \boldsymbol{v} \cdot \nabla \cdot \boldsymbol{v}] = -\nabla P + \eta \nabla^2 \boldsymbol{v} \tag{A.22}$$

This is the fundamental equation of fluid motion in viscous fluids. It describes momentum conservation. One can eliminate the pressure from the Navier Stokes equation by taking the rotational operator, rot $= \nabla\times$, on both sides of the eq.(A.22) to obtain the so-called vorticity formulation of the Navier Stokes equation for viscous fluids,

$$\frac{\partial}{\partial t}(\mathrm{rot}\boldsymbol{v}) = \mathrm{rot}(\boldsymbol{v}\,\mathrm{rot}\boldsymbol{v}) + \nu \Delta(\mathrm{rot}\boldsymbol{v}) \tag{A.23}$$

where we have defined the kinematic viscosity as $\nu = \eta/\rho$. In two-dimensional problems one can again define a stream function as done above for ideal fluids and solve the Navier Stokes equation for stream functions. Note, that in incompressible fluids the continuity equation given above is still valid. It is not changed by the appearance of the viscosity since it expresses mass conservation. In the following sections

[2] In eq.(A.21) the expression $\nabla \boldsymbol{v}$ means a dyadic product such that the shear stress τ is really a second rank tensor. In viscous fluids there is not only hydrostatic pressure as defined above for ideal fluids. Shear stresses are translated between adjacent fluid layers leading to a situation in which the scalar pressure is replaced by a second rank tensor of stresses. The off-diagonal elements are the shear stresses, whereas the diagonal elements are the hydrostatic pressure. In principle, the viscosity is not a scalar but a 4th rank tensor similar to the elasticity tensor in solid mechanics. For further details see the standard textbooks on fluid dynamics cited in the bibliography.

we will derive in detail the equations for particles - solid or liquid - moving in a gravity field. This type of particle motion is called *Stokes motion.* We consider only so-called *creeping flow*, meaning that the Reynolds number

$$\mathrm{Re} = \frac{vl}{\nu} \tag{A.24}$$

is small compared to 1. In the case of Stokes motion the length l is the particle radius, v the constant particle velocity that develops after a short period of acceleration when the particles buoyancy is compensated by the friction forces exerted by the surrounding fluid onto the particle. The condition $\mathrm{Re} \ll 1$ can be fulfilled by either having a small particle, a small settling speed or a highly viscous fluid. The problem of a moving sphere in a fluid is equivalent to a sphere fixed in space and the fluid moving towards it.

A.2 Solid sphere moving in a gravity field

Our starting point is the Navier Stokes equation (A.22) which reduces to a quite simple form, since small Reynolds numbers imply that the terms on the left hand side can be neglected. We thus have to solve the equation

$$0 = -\nabla P + \eta \nabla^2 \boldsymbol{v} \tag{A.25}$$

together with the continuity equation,

$$\nabla \cdot \boldsymbol{v} = 0. \tag{A.26}$$

Applying the divergence operator $(\nabla\cdot)$ to both equations leads to a simple equation for the pressure,

$$\nabla^2 P = 0 \tag{A.27}$$

The pressure satisfies a Laplace type equation. Before we proceed with solving eq.(A.27) we choose an appropriate coordinate system. The sphere center is taken as the origin of the coordinate system. We count the positive z-axis as being opposite to the sphere motion. Fixing the sphere at the origin then means the fluid moves from minus infinity to plus infinity in z-direction. Since we have a spherical problem we employ spherical polar coordinates r, θ, ϕ. In these coordinates we have for example

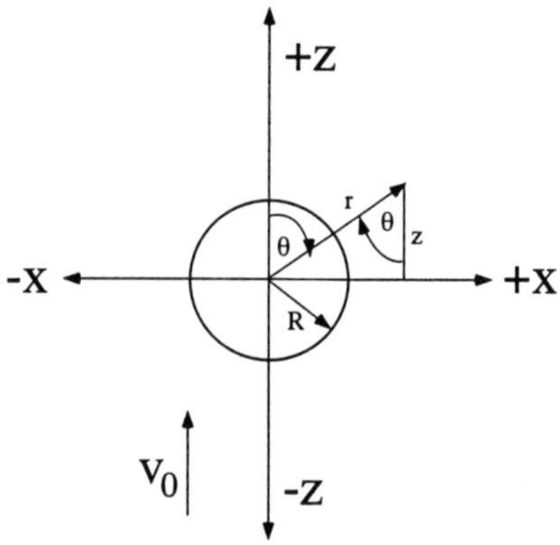

Fig. A.3. Illustration of the coordinate system used to calculate the Stokes flow around a spherical particle. The particle is fixed in the coordinate system and the fluid flows from minus infinity towards the sphere with a constant velocity.

$$\cos\theta = \frac{z}{r}. \tag{A.28}$$

The coordinate system is illustrated in Fig. A.2. One possible solution of the Laplace equation for the pressure would be

$$P = \frac{A_0}{r}. \tag{A.29}$$

This solution, however, is spherically symmetric and thus cannot account for the axial asymmetry along the z-axes. Another solution is

$$P = A_0 \frac{\partial}{\partial z}\frac{1}{r}. \tag{A.30}$$

which satisfies eq.(A.27) as can be shown by insertion. If we perform the differentiation in eq.(A.30) using eq.(A.28) we obtain for the pressure the expression

$$P = -\frac{A_0}{r^2}\cos\theta. \tag{A.31}$$

assuming P at is zero at $r \to \infty$. Instead of making these guess on the pressure field we could have started with the most general solution of Laplace's equation in spherical polar coordinates using spherical harmonics. This would be much more work to calculate but would lead to the same result.

Knowing the pressure field – besides the constant A_0 – we can calculate the velocity field solving the Navier-Stokes equation, eq.(A.25).

Since the problem has rotational symmetry around the z-axis the velocity vector $\boldsymbol{v} = (v_r, v_\theta, v_\phi)$ has only two components, i.e. v_ϕ and all derivatives with respect to ϕ are zero. The Laplace operator in spherical polar coordinates simplifies. We obtain the Navier-Stokes equation for the radial velocity component and will calculate the θ-component later using the continuity equation.

$$\frac{\eta}{r^2}\left[\frac{\partial^2(r^2 v_r)}{\partial r^2} + \frac{1}{\sin\theta}\frac{\partial}{\partial\theta}\left(\sin\theta\frac{\partial v_r}{\partial\theta}\right)\right] = \frac{\partial P}{\partial r} \tag{A.32}$$

To solve this equation we first calculate the pressure derivative with respect to the radius using eq.(A.31),

$$\frac{\partial P}{\partial r} = \frac{2A_0}{r^3}\cos\theta. \tag{A.33}$$

In order to eliminate of the cosine term in the equation for the radial velocity component the radial velocity itself should be proportional to the cosine. We therefore make the following ansatz:

$$v_r = \frac{w(r)}{r}\cos\theta \tag{A.34}$$

Inserting this into eq.(A.32) using eq.(A.33) we obtain after a short calculation an ordinary differential equation for the unknown function $w(r)$,

$$\frac{\eta}{r^2}\left(\frac{d^2}{dr^2}(rw(r)) - \frac{2}{r}w(r)\right) = \frac{2A_0}{r^3} \tag{A.35}$$

This is an inhomogeneous ordinary differential equation of second order. Its solution can be obtained by looking for a particular integral and the general integral of the homogeneous equation. A particular integral is easily found as the constant

$$w_p(r) = -\frac{A_0}{\eta} \tag{A.36}$$

and the general integral of the homogeneous equation (this means the right hand term in eq.(A.35) is set equal to zero) is

$$w_h(r) = A_1 r + \frac{A_2}{r^2}. \tag{A.37}$$

Therefore, the general solution of eq.(A.35) is the sum of both

$$w(r) = w_p(r) + w_h(r) = -\frac{A_0}{\eta} + A_1 r + \frac{A_2}{r^2}. \tag{A.38}$$

The radial velocity component is then

$$v_r(r,\theta) = \left(-\frac{A_0}{\eta r} + A_1 + \frac{A_2}{r^3}\right)\cos\theta. \tag{A.39}$$

The continuity equation reads in spherical polar coordinates

$$\nabla \cdot \boldsymbol{v} = \frac{1}{r^2}\frac{\partial}{\partial r}(r^2 v_r) + \frac{1}{r\sin\theta}\frac{\partial}{\partial\theta}(\sin\theta\, v_\theta) = 0. \tag{A.40}$$

Inserting eq.(A.39) suggests we write for the polar component of r the equation

$$v_\theta = w_1(r)\sin\theta. \tag{A.41}$$

We, then, obtain after an easy but lengthy calculation the result,

$$w_1(r) = \frac{A}{2\eta r} - B + \frac{C}{2r^3}, \tag{A.42}$$

and thus for the velocity component

$$v_\theta = \left(\frac{A_0}{2\eta r} - A_1 + \frac{A_2}{2r^3}\right)\sin\theta. \tag{A.43}$$

We are left with the problem to determine the unknown coefficients A_0, A_1, A_2. This is done as usual using boundary conditions. The boundary conditions are

$$v_r = v_\theta = 0 \quad \text{at} \quad r = R, \tag{A.44}$$
$$v_r = v_0\cos\theta \quad \text{at} \quad r \to \infty, \tag{A.45}$$
$$v_\theta = -v_0\sin\theta \quad \text{at} \quad r \to \infty. \tag{A.46}$$

The first condition is the so-called no-slip condition, meaning at a solid interface the flow velocity is zero. The two other conditions just state that the flow at infinity goes into the positive z-direction at constant velocity v_0. Determining the constants is a lengthy calculation task and omitted here. The result for the velocity components in the end is

$$v_r = v_0\left(1 - \frac{3R}{2r} + \frac{R^3}{2r^3}\right)\cos\theta, \tag{A.47}$$

$$v_\theta = v_0\left(-1 + \frac{3R}{4r} + \frac{R^3}{4r^3}\right)\sin\theta. \tag{A.48}$$

The pressure reads now

$$P = -\frac{3}{2}\eta v_0 R \frac{\cos\theta}{r^2} \tag{A.49}$$

These equations require a comment. The decay of fluid flow velocity components around the spherical particle have a lowest order of $1/r$, i.e. they behave like a Coulomb potential. This means in a dispersion of spheres the flow fields will interact even at small volume concentrations quite similar to the diffusional interaction we have treated in the context of finite volume fraction effects on coarsening. Here, however, the problem would be much more difficult to evaluate since we first have a vector field instead of a scalar one and second besides the Coulomb term we have a dipole term, and third, the problem has only axial symmetry or in other words no spherical symmetry.

In order to calculate the particle settling speed we have to calculate the drag exerted by the flow onto the sphere and balance it with the buoyant forces. The drag can be calculated from the pressure and the gradient of the flow field using Newton's law are explained in the preceding section. The calculation goes as follows.

The hydrodynamic pressure acts in radial direction. Its component in the positive z direction at $r = R$ is

$$-P\cos\theta = \frac{3}{2}\eta v_0 \frac{\cos^2\theta}{R}. \tag{A.50}$$

Integration over the particle surface using the surface element $2\pi R^2 \sin\theta d\theta$ gives

$$3\pi\eta R v_0 \int_0^{\pi} \cos^2\theta \sin\theta d\theta = 2\pi\eta R v_0 \tag{A.51}$$

The viscous fluid flow exerts additional drag onto the sphere. We had written down the simple Newton ansatz for the relation between velocity gradient and shear stress which reads now

$$\tau_{r\theta} = -\eta \frac{1}{r}\frac{\partial v_r}{\partial\theta} + \frac{\partial v_\theta}{\partial r} - \frac{1}{r}v_\theta. \tag{A.52}$$

We skip a derivation of this relation and suggest the interested reader to consult books like those from Batchelor [22] or Landau and Lifshitz

[182]). Insertion of the velocity components and evaluating the result at $r = R$ yields

$$\tau_{r\theta} = \frac{3}{2}\frac{\eta v_0}{R}\sin\theta. \tag{A.53}$$

The component of the viscous shear stress into the positive z-direction is

$$\tau_{r\theta}\sin\theta = \frac{3}{2}\frac{\eta v_0}{R}\sin^2\theta \tag{A.54}$$

and the integral over the surface is given by

$$3\pi\eta R v_0 \int_0^{\pi} \sin^3\theta d\theta = 4\pi\eta R v_0. \tag{A.55}$$

Adding the contribution of the normal pressure eq.(A.51) leads to the total drag or *Stokes drag*,

$$F_{drag} = 6\pi\eta R v_0. \tag{A.56}$$

A buoyant force acts on the solid particle of density $\rho_{part.}$ moving in a fluid of density ρ,

$$F_g = \frac{4\pi}{3}(\rho_p - \rho)R^3 g \tag{A.57}$$

Balancing this with the Stokes drag, $F_{drag} = F_g$, leads to the famous equation for the settling speed of a solid sphere in a fluid,

$$v_0 = \frac{2}{9}\frac{\rho_p - \rho}{\eta} g R^2. \tag{A.58}$$

For the calculation of convective heat and mass transport to a falling sphere in this text we use the stream function Ψ. It was defined in eq.(A.20) as

$$v_r = -\frac{1}{r^2\sin^2\theta}\frac{\partial\Psi}{\partial\theta}. \tag{A.59}$$

Since we have the right hand side of this equation the stream function can be obtained by a simple integration,

$$\Psi = -\int r^2 \sin^2\theta v_r(r,\theta) d\theta \tag{A.60}$$

Performing the integration leads to the equation used in chapter 4,

$$\Psi(r,\theta) = -\frac{v_0}{2}\sin^2\theta\left(r^2 - \frac{3}{2}Rr + \frac{1}{2}\frac{R^3}{r}\right). \tag{A.61}$$

A.3 Droplet moving in a gravity field

The problem of a droplet moving in the gravity field is quite similar to that of a solid sphere moving in the gravity field treated above. The important differences are that there will be fluid flow inside the drop too and that the boundary conditions at the droplet interface will differ from that of a solid sphere. The set-up is similar to the one given above, besides we have a droplet with density ρ_{drop} and viscosity η_{drop} suspended in a matrix of density ρ and viscosity η. The droplet shall move due to gravity along the z-axis as already indicated in Fig. A.2. For the velocity field outside the droplet we take the solutions given above with the undetermined constants which we write here again for convenience,

$$v_r^{mat.}(r,\theta) = \left(-\frac{A_0}{\eta r} + A_1 + \frac{A_2}{r^3}\right)\cos\theta \tag{A.62}$$

$$v_\theta^{mat.}(r,\theta) = \left(\frac{A_0}{2\eta r} - A_1 + \frac{A_2}{2r^3}\right)\sin\theta, \tag{A.63}$$

where we have introduced the superscript *mat.* to indicate that the velocities belong to the matrix fluid. Inside the droplet we again have to solve first for the pressure field Laplace's equation,

$$\nabla^2 P_{drop} = 0, \tag{A.64}$$

where we used the subscript *drop* to indicate that we are dealing with quantities belonging to the droplet interior. Expanding the Laplace operator ∇^2 in spherical polar coordinates leads to

$$\frac{1}{r^2}\frac{\partial}{\partial r}\left(r^2\frac{\partial P_{drop}}{\partial r}\right) + \frac{1}{r^2\sin^2\theta}\frac{\partial}{\partial\theta}\left(\sin\theta\frac{\partial P_{drop}}{\partial\theta}\right) = 0, \tag{A.65}$$

where all derivatives with respect to θ have been set to zero. The general solution of this equation is (see [147])

$$P_{drop}(r,\theta) = \sum_{l=0}^{\infty}\left[A_l r^l + B_l r^{-(l+1)}\right] P_l(\cos\theta) \tag{A.66}$$

where $P_l(\cos\theta)$ are Legendre polynomials (e.g. $P_0 = 1, P_1 = \cos\theta\ldots$). We require that inside the droplet at the origin the pressure is finite, especially it should not tend to infinity. Therefore, the coefficients $B_l = 0$ for all l. We then make the simplest ansatz for the pressure field,

$$P_{drop}(r,\theta) = A_3 + A_4 r\cos\theta, \tag{A.67}$$

taking only $l = 0, 1$. Then, the Navier Stokes equation for the radial velocity component reads

$$\frac{\eta_d}{r^2}\left[\frac{\partial^2(r^2 v_r^{drop})}{\partial r^2} + \frac{1}{\sin\theta}\frac{\partial}{\partial\theta}\left(\sin\theta\frac{\partial v_r^{drop}}{\partial\theta}\right)\right] = \frac{\partial P_{drop}}{\partial r} = A_3\cos\theta. \tag{A.68}$$

As in the section above we try a solution of the form

$$v_r^{drop} = w(r)\cos\theta, \tag{A.69}$$

which leads to the ordinary differential equation for $w(r)$,

$$rw'' + 4w' = \frac{A_3}{\eta_d} r. \tag{A.70}$$

The general solution of this equation is,

$$w(r) = B_d - \frac{C_1}{3r^2} + \frac{A_3}{10\eta_d} r^2 \tag{A.71}$$

Since we require that the flow field shall converge at $r = 0$ we set the coefficient $C_1 = 0$. Thus we have

$$w(r) = B_d + \frac{A_3}{10\eta_{drop}} r^2, \tag{A.72}$$

and the radial velocity inside the drop is given by

$$v_r^{drop}(r,\theta) = \left(B_d + \frac{A_3}{10\eta_{drop}} r^2\right)\cos\theta. \tag{A.73}$$

We again obtain from the continuity equation

$$\frac{1}{r^2}\frac{\partial}{\partial r}\left(r^2 v_r^{drop}\right) + \frac{1}{r\sin\theta}\frac{\partial}{\partial\theta}\left(\sin\theta\, v_\theta^{drop}\right) = 0 \tag{A.74}$$

after inserting eq.(A.73) the polar component of the velocity as

$$v_\theta^{drop}(r,\theta) = -\left(B_d + \frac{A_3}{5\eta_{drop}} r^2\right)\sin\theta. \tag{A.75}$$

To determine the unknown coefficients we have to employ boundary conditions. First, it is obvious that we require

$$v_r^{mat.} = v_0 \cos\theta \quad \text{at} \quad r \to \infty. \tag{A.76}$$

This leads to $A_1 = v_0$. Second, the radial velocities outside and inside the droplet should be equal to zero at the droplet-matrix interface, i.e. at $r = R$. Otherwise the droplet would expand or shrink.

$$v_r^{drop}(R,\theta) = v_r^{drop}(R,\theta) = 0 \tag{A.77}$$

These two equations lead to

$$B_d = -\frac{A_3}{10\eta_{drop}} R^2, \tag{A.78}$$

$$A_0 = \frac{\eta A_2}{R^2} + \eta R v_0. \tag{A.79}$$

The polar components of the velocities shall be equal at the interface, meaning any motion motion along the interface of the matrix induces due to viscous friction an equal motion inside the droplet,

$$v_\theta^{mat.}(r,\theta) = v_\theta^{drop}(r,\theta), \tag{A.80}$$

leading to one new equation relating the coefficients A_3 and A_2,

$$A_2 = \frac{1}{2} v_0 R^3 - \frac{R^5}{10\eta_d} A_3. \tag{A.81}$$

The next condition was not used in the motion of a solid sphere treated in the section above. We require that the shear stress at the interface exerted by the matrix flow, $\tau_{r\theta}^{mat.}$, and that by the flow inside the droplet, $\tau_{r\theta}^{drop}$, are equal:

$$\tau_{r\theta}^{mat.}(R,\theta) = \tau_{r\theta}^{drop}(R,\theta). \tag{A.82}$$

The shear stress is a function of the components of the velocity,

$$\tau_{r\theta}^{drop,mat.}(r,\theta) = \eta^{,d} \left(\frac{1}{r} \frac{\partial v_r^{drop,mat.}}{\partial \theta} + \frac{\partial v_\theta^{drop,mat.}}{\partial r} - \frac{v_\theta^{drop,mat.}}{r} \right). \tag{A.83}$$

Inserting the relations for the velocities at the interface and evaluating all the terms leads after a lengthy calculation to a second equation relating A_3 and A_2, that can be used to calculate the coefficient A_3 using eq.(A.81). The result is

$$A_3 = \frac{\eta\eta_d}{\eta+\eta_d}\frac{5v_0}{R^2}. \tag{A.84}$$

We thus can solve now all equations for the constants, inserting them into the equations for the velocities and obtain finally their expressions

$$v_r^{mat.}(r,\theta) = \frac{v_0\eta}{\eta+\eta_d}\left[1-\frac{R}{r}+\frac{\eta_d}{\eta}\left(1-\frac{3R}{2r}+\frac{R^3}{2r^3}\right)\right]\cos\theta \tag{A.85}$$

$$v_\theta^{mat.}(r,\theta) = -\frac{v_0\eta}{\eta+\eta_d}\left[1-\frac{R}{2r}+\frac{\eta_d}{4\eta}\left(4-\frac{3R}{r}-\frac{R^3}{r^3}\right)\right]\sin\theta \tag{A.86}$$

$$v_r^{drop}(r,\theta) = -\frac{\eta}{2(\eta+\eta_d)}v_0\left(1-\frac{r^2}{R^2}\right)\cos\theta \tag{A.87}$$

$$v_\theta^{drop}(r,\theta) = -\frac{\eta}{2(\eta+\eta_d)}v_0\left(1-2\frac{r^2}{R^2}\right)\sin\theta \tag{A.88}$$

To determine the settling velocity we have, as in the section above, to calculate the pressure and the shear stresses along the surface of the droplet and balance it with the gravity force. We will not go through all the details here, because all basic equations were written down either in the section above or here, and one simply has to go through the tedious task to insert all relations obtained and calculate the drag force. The Stokes drag on a drop as compared to that on a solid spherical particle, given in eq.(A.56), reads,

$$F_{drag} = 2\pi\eta\frac{2\eta+3\eta_d}{\eta+\eta_d}Rv_0. \tag{A.89}$$

Thus the settling speed of a drop is can be calculated as

$$v_0 = \frac{2}{3}\frac{(\rho_d-\rho)(\eta+\eta_d)}{\eta(2\eta+3\eta_d)}gR^2. \tag{A.90}$$

In the same way as outlined above we can calculate the stream functions. The calculation is omitted here. The result is shown in section 4.6.3.

A.4 Thermocapillary convection and motion

Besides Stokes motion there are other possible sources for particle motion in dispersions. Droplets suspended in a fluid can move due to the so-called thermocapillary effect, whose origin is the temperature and/or concentration dependence of the interfacial tension. The interface tension between two liquids generally depends on temperature and concentration like

$$\sigma(x_B, T) = \sigma_0 + \frac{\partial \sigma}{\partial T}(T - T_0) - \alpha_x x_B = \sigma_0 + s_\sigma (T - T_0) \qquad \text{(A.91)}$$

with σ_0 a reference tension at T_0 and $s_\sigma < 0$ is the surface entropy and α_x the variation of surface tension with solute concentration in the bulk (this approximation is only valid in dilute solutions if there is no surface segregation (see Landau [182]; the general form with surface segregation is treated in the textbook of Levich [193]). Therefore, any temperature and/or concentration gradient, which has a component parallel to the interface will induce a variation of the surface tension along the fluid interface (liquid-liquid or liquid-gas). This gives rise to a shear stress. Assuming that the interface tension is isotropic such that it can be described by a scalar function as given in eq.(A.91) the shear force exerted by a temperature gradient is

$$\boldsymbol{\tau} = +\nabla\sigma. \qquad \text{(A.92)}$$

The plus sign in this relation means that the shear force tends to move the surface or interface of the liquid from lower to higher values of surface tension. Inserting the temperature dependent part of eq.(A.91) into eq.(A.92) we obtain

$$\boldsymbol{\tau} = \frac{\partial \sigma}{\partial T}\nabla T + \alpha_x \nabla x_B = -|s_\sigma|\nabla T + \alpha_x \nabla x_B. \qquad \text{(A.93)}$$

In viscous fluids the motion induced by the shear stress at the interface induces a motion inside the bulk fluid as schematically shown in figure A.4. This convection mode is called thermocapillary or Marangoni convection. In dispersions it induces a convection role inside dispersed droplets accompanied by a convection roll in the surrounding matrix. The overall effect on a droplet is a motion towards the hot side, the side where the interface tension would be smallest. For a single droplet moving in a linear temperature field (constant gradient) the problem was solved 1956 by Young, Goldstein and Block [314]. Their now classical formula describing the droplet velocity is

$$U_{Marangoni} = \frac{2}{\eta(2+\hat{\lambda})(2+3\hat{\eta})} \left|\frac{d\sigma}{dT}\right| \nabla T \; R, \qquad \text{(A.94)}$$

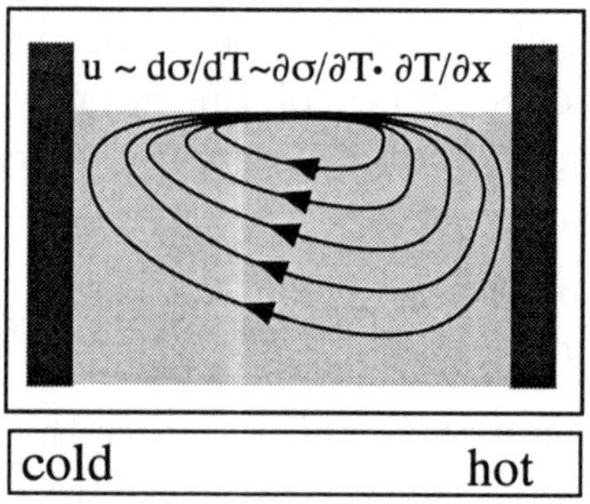

Fig. A.4. Thermocapillary or Marangoni convection in an open boat. The surface tension of a fluid depends on temperature and surface concentration of solute. Thus any temperature gradient along the interface induces a shear stress along the interface and this in viscous fluids a closed convection roll as shown schematically.

where $d\sigma/dT$ is the temperature dependent coefficient of the interfacial tension, $\hat{\lambda}$ the ratio of the thermal conductivities of the droplet to the matrix, $\hat{\eta}$ the ratio of the viscosities and ∇T the temperature gradient. The Marangoni motion is only linearly dependent on the droplet radius (compare to eq.(A.58) for Stokes motion). A short outline of a derivation of this important result is given below.

Consider a droplet of viscosity η_d, radius R and thermal conductivity λ_d being in a fluid of viscosity η and conductivity λ. In the fluid external to the drop a temperature field shall persist with a constant temperature gradient G_0. The situation is illustrated in fig.A.4. The

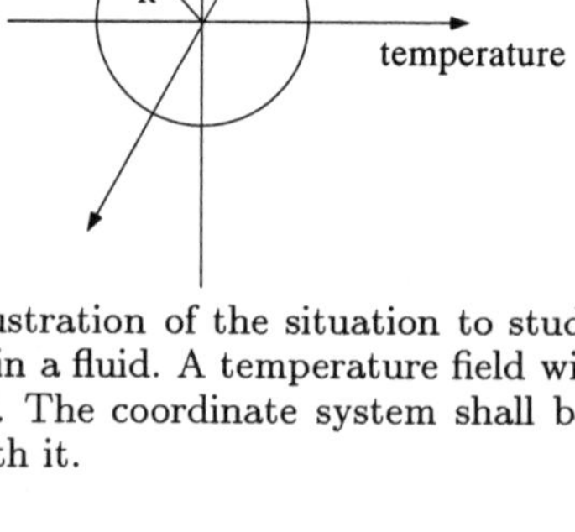

Fig. A.5. Schematic illustration of the situation to study Marangoni motion of a drop which is immersed in a fluid. A temperature field with a constant gradient G_0 shall persist in the fluid. The coordinate system shall be located in the center of the drop and moving with it.

external temperature field far away from the drop can be described by,

$$T(z) = \boldsymbol{G}_0 z \boldsymbol{e}_z = G_0 r \cos\theta \tag{A.95}$$

where $\boldsymbol{e}_z$ is the unit vector along the positive z-axis. Due to the different thermal conductivities of matrix fluid and drop the temperature field around and inside the drop will be disturbed. The temperature will vary along the interface between the drop and the matrix fluid and thus leading according to eq.(A.93) a shear stress along the interface. This shear stress will induce a fluid flow in the drop and the matrix leading finally to a motion of the drop. The calculation of the Marangoni velocity of a drop in a constant temperature gradient proceeds as follows,

1. Calculation of the temperature field in the matrix and the drop given that the far field temperature is described by eq.(A.95).
2. Calculation of the flow field in the drop and the matrix, solving Navier-Stokes equation for creeping flow in fully analogy with the procedures in the sections above on particle or drop motion in a gravity field.
3. The last step is to calculate the velocity of the drop. This is done by using the condition of steady motion and writing that the force exerted by the fluid on the drop is then zero.

We start with a calculation of the temperature field. Assuming that the thermal Peclet number is small the temperature field in the drop T_d and the matrix T_m is obtained from a solution of Laplace's equation utilizing the complete spherical symmetry of the problem

$$\nabla^2 T_{d,m} = \left(\frac{\partial^2 T_{d,m}}{\partial r^2} + \frac{2}{r} \frac{\partial T_{d,m}}{\partial r} \right) = 0 \tag{A.96}$$

The boundary conditions are,

$$T_m(r \to \infty) = G_0 r \cos\theta \tag{A.97}$$

$$T_d(R) = T_m(R) \tag{A.98}$$

$$-\lambda_d \left. \frac{\partial T_d}{\partial r} \right|_{r=R} = -\lambda_m \left. \frac{\partial T_m}{\partial r} \right|_{r=R} \tag{A.99}$$

The general solution of the problem can be written in terms of Legendre polynomials $P_l(\cos\theta)$,

$$T_d(r,\theta) = \sum_{l=0}^{\infty} A_l r^l P_l(\cos\theta) \tag{A.100}$$

$$T_m(r,\theta) = \sum_{l=0}^{\infty} (B_l r^l + C_l r^{-(l+1)}) P_l(\cos\theta) \tag{A.101}$$

The coefficients have to be determined using the boundary conditions. The first boundary condition leads to $B_0 = 0, B_{l>1} = 0$ and $B_1 = G_0$. The second boundary condition leads to $A_0 = C_0 = 0$ and $A_l = C_l = 0$ for all $l \geq 2$, leaving $A_1 = C_1/R^3 + G_0$. The third boundary condition leads to a second equation with the constants A_1 and C_1, namely $-\lambda_d A_1 = -G_0\lambda_m + 2C_1\lambda_m/R^3$. Determining the coefficients yields for the temperature field in the drop,

$$T_d(r,\theta) = G_0 \frac{3\lambda_m}{2\lambda_m + \lambda_d} r\cos\theta \tag{A.102}$$

and in the matrix,

$$T_m(r,\theta) = G_0 r\cos\theta \left[1 + \frac{\lambda_m - \lambda_d}{2\lambda_m + \lambda_d}\frac{R^3}{r^3}\right] \tag{A.103}$$

The far field temperature field of eq.(A.95) thus is only disturbed very close to the drop (see the r^{-3} dependence in eq.(A.103). We later need the tangential component of the gradient of the temperature field, which can be calculated using either eq.(A.103) or eq.(A.102),

$$\frac{1}{R}\frac{\partial T_d}{\partial\theta}\bigg|_{r=R} = \frac{1}{R}\frac{\partial T_m}{\partial\theta}\bigg|_{r=R} = -G_0\frac{3\lambda_m}{2\lambda_m + \lambda_d}\sin\theta \tag{A.104}$$

The calculation of the flow field is performed along the lines described in the preceding section. Since we assume creeping motion, i.e. the Reynolds number is small, the flow fields in the matrix $\boldsymbol{v}^m$ and the droplet $\boldsymbol{v}^d$ are obtained from,

$$\nabla p^{d,m} = \eta_{,d}\nabla^2\boldsymbol{v}^{d,m} \tag{A.105}$$

together with the continuity equation,

$$\nabla\cdot\boldsymbol{v}^{d,m} = 0 \tag{A.106}$$

At the interface the radial flow field in the drop and the matrix shall vanish,

$$v_r^m(R,\theta) = v_r^d(R,\theta) = 0 \tag{A.107}$$

and the tangential components shall be equal,

$$v_\theta^m(R,\theta) = v_\theta^d(R,\theta) \tag{A.108}$$

In contrast to the situation treated in the section above, the interfacial tension and its gradient is of utmost importance in this problem and cannot be neglected. There is in particular first a pressure jump at the interface,

$$p^d - p^m = -\eta \left.\frac{\partial v_r^m}{\partial r}\right|_R + \eta_d \left.\frac{\partial v_r^d}{\partial r}\right|_R + \frac{2\sigma}{R} \tag{A.109}$$

with p^d, p^m the pressure fields inside the drop and the matrix, respectively. This situation is illustrated in fig.A.4 There is, however, another

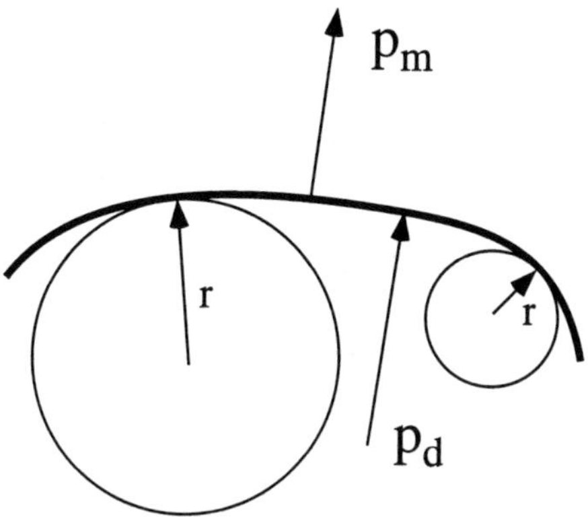

Fig. A.6. Schematic illustration of the pressure jump occurring normal to a curved interface. The pressure inside the drop is increased due to surface tension.

very important new boundary condition. The shear stress exerted on the interface by the fluid flow fields in the drop and the matrix are just balanced by the stress due to the tangential component of the interfacial tension gradient. Thus the condition used before in eq.(A.82) is replaced by,

$$\tau_{r\theta}^m(R,\theta) - \tau_{r\theta}^d(R,\theta) = \frac{1}{R}\left.\frac{\partial T_d}{\partial r}\right|_{r=R}\frac{\partial \sigma}{\partial T} \tag{A.110}$$

The situation is illustrated in fig.A.4. In order to solve the problem we use the solutions of the flow field given in the last section - eqs.(A.62) and (A.63). The flow field in the matrix therefore reads,

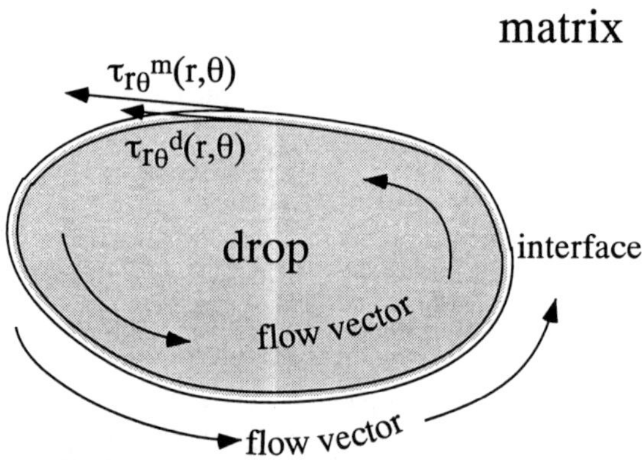

Fig. A.7. Schematic illustration of the jump in shear stress exerted by the fluid flow on both sides of the interface, which is for clarity here drawn with two parallel lines. This jump is balanced by the gradient in the interfacial tension originating from the temperature gradient along the interface.

$$v_r^m(r,\theta) = \left(-\frac{A_0}{\eta r} + v_0\frac{A_2}{r^3}\right)\cos\theta \tag{A.111}$$

$$v_\theta^m(r,\theta) = \left(\frac{A_0}{2\eta r} - v_0 + \frac{A_2}{2r^3}\right)\sin\theta, \tag{A.112}$$

For the flow field inside the drop we obtained the general solution,

$$v_r^d(r,\theta) = \left(B_d + \frac{A_3}{10\eta_d}r^2\right)\cos\theta. \tag{A.113}$$

$$v_\theta^d(r,\theta) = -\left(B_d + \frac{A_3}{5\eta_d}r^2\right)\sin\theta. \tag{A.114}$$

Using the boundary conditions written above one can determine the unknown coefficients A_0, A_2, A_3, B_d and the far field velocity of the fluid v_0, which, however also is the drop velocity. The details of the procedure are omitted here, since there are only minor changes compared to the last section. A lengthy but straight forward calculation yields,

$$v_0 = \frac{2RG_0(\partial\sigma/\partial T)\lambda_m}{(\lambda_d + 2\lambda_m)(2\eta + 3\eta_d)} \tag{A.115}$$

which is identical to the equation given above in eq.(A.94). The flow fields calculated in coordinate system moving with the drop are,

$$v_r^m(r,\theta) = v_0\left(1 - \frac{R^3}{r^3}\right)\cos\theta \tag{A.116}$$

$$v_\theta^m(r,\theta) = -v_0/2\left(2+\frac{R^3}{r^3}\right)\cos\theta \qquad \text{(A.117)}$$

$$v_r^d(r,\theta) = -3v_0/2\left(1-\frac{r^2}{R^2}\right)\cos\theta \qquad \text{(A.118)}$$

$$v_\theta^d(r,\theta) = 3v_0/2\left(1-2\frac{r^2}{R^2}\right)\sin\theta \qquad \text{(A.119)}$$

These equations show an essential difference between Marangoni and Stokes motion: The fluid flow field around a droplet moving by Marangoni motion has the character of a dipole, i.e. it decays with the third power of distance from the droplet surface. Therefore, multi-droplet effects are not so severe and wall effects become only important if the droplets are very close to a solid wall.

B. Nucleation

In this appendix we will briefly review some basic ideas of nucleation. The section is not intended to give a full treatment of all aspects of nucleation, just those results necessary to supplement the material presented in the monograph. The topic is still controversial, although fundamental papers, like that of Gibbs, are more than one hundred years old. The transformation of a single phase material into a multi-phase material is a complicated topic, since the formation of nuclei in a parent phase implies the formation of very small precipitates, droplets or particles, having only a few ten to hundred atoms. Thus, it maybe questioned if conventional continuum thermodynamics should be applied to such a process. A more detailed discussion of this topic can be found in the review papers by Binder and Stauffer [35], Langer [185], Mutaftschiev [211] and others [283, 284, 278, 80, 2, 195, 104, 316]. In the first section we present the classical picture of nucleation that builds upon that given in Gibbs [112], Landau and Lifshitz [183], and the notes of J.E. Hilliard. We shall then discuss the kinetics on the basis of the Fokker-Planck equation in the next section leading to a well known expression for the nucleation rate often used in material science contexts.

B.1 Thermodynamics of nucleation

Imagine a material initially in a stable state, as shown in Fig. B.1. If we quench this system into a metastable state then the energy of the system will be minimized upon the formation of a mixture of two phases.

Since nucleation occurs under conditions of constant T, V and μ_i [1] we must use the Grand Canonical Free energy Ω to calculate the

[1] The critical nucleus is in unstable equilibrium, the μ_i, T are constant, more on this later.

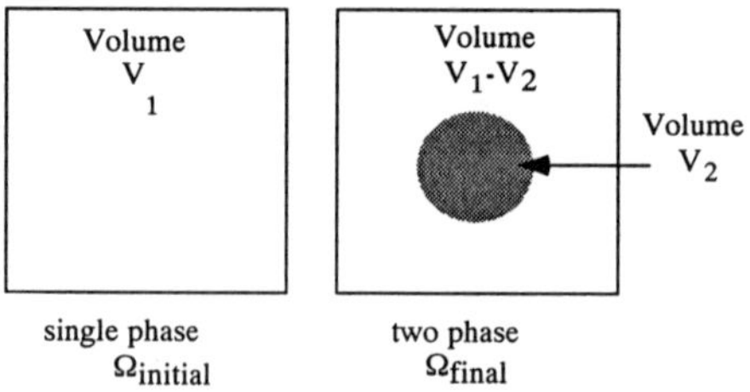

Fig. B.1. Phase change to be a spherical particle The temperature of the system is constant upon the formation of a nucleus as well as the external volume V remains constant.

reversible work for the formation of the critical cluster, W_R^*. The Grand Canonical Free energy is defined as

$$\Omega = E - TS - \sum \mu_i n_i \tag{B.1}$$

and thus the change in the Grand Canonical free energy is

$$dW_R = d\Omega + SdT + PdV + \sum_i n_i d\mu_i . \tag{B.2}$$

Therefore, W_R is equal to the change in Ω at constant T, V and μ_i. From eq.(B.2) we then have

$$dW_R = d\Omega \Rightarrow W_R = \Delta\Omega \tag{B.3}$$

since W_R is not a state function. If we had used the Gibbs Free energy then, the work of formation would have to be calculated from

$$dW_R = dG + SdT - VdP - \sum \mu_i dn_i \tag{B.4}$$

and thus W_R would represent the change in G at constant T, P, n_i. We show below that this is not a proper expression for the work of formation of a nucleus. Now consider the nucleation of β-phase from α-phase. The formation of β-phase within the α-phase results in an increase in energy if $W_R > 0$ for the critical radius. As stated in the thermodynamics section, only processes which result in a decrease in energy are allowed in classical thermodynamics. However, it is still possible to have local increases in energy, fluctuations, as long as the average change in energy is zero or negative. Thus, fluctuations play an important role in the nucleation process. To determine the magnitude of the effects of these fluctuations one must determine W_R for the critical nucleus. The critical nucleus is defined as that cluster which is in unstable equilibrium with the matrix. By unstable equilibrium we mean, that any perturbation, regardless of how small, will make the cluster grow or shrink. As the critical nucleus is in unstable equilibrium the chemical potentials of the matrix and nucleus must be equal, the temperature must be constant, and the system must be in mechanical equilibrium. The first condition can be expressed as

$$\mu_i^\alpha(x_1^0, x_2^0 \ldots x_N^0, P^\alpha) = \mu_i^\beta(x_1^*, x_2^* \ldots x_N^*, P^\beta) \quad i = 1, 2, \ldots N \tag{B.5}$$

and the condition of mechanical equilibrium as

$$P^\beta = P^\alpha + \frac{2\sigma}{R^*}, \tag{B.6}$$

where x_i^0 and x_i^* are the compositions of the matrix and the critical nucleus with radius R^*. i denotes the ith chemical component in the system. Clearly, we cannot use the change in the Gibbs Free energy to measure W_R^*, the reversible work of formation of the critical nucleus, as the pressure is not constant during the nucleation process. In fact the change in Gibbs free energy on forming β-phase can be shown to be zero (see Chapter 2).

The clue as to what energy function to use is provided by the conditions for unstable equilibrium:

1. The chemical potentials, the μ_i's, are constant. We assume that the formation of β-phase does not change significantly the composition of the α-phase.
2. The external volume is fixed.
3. The temperature is constant.

This implies that,

$$dW_R^* = d\Omega \quad \Rightarrow \quad W_R^* = \Delta\Omega \tag{B.7}$$

The reversible work is equal to the change in Grand Canonical Free energy. So, since Ω is a state function, W_R^* is simply found by subtracting $\Omega_{initial}$ from Ω_{final} or $\Delta\Omega = \Omega_{final} - \Omega_{initial}$.

In the initial state we have,

$$\Omega_{initial} = V\Omega_V^\alpha(x_i^0, P^\alpha) \tag{B.8}$$

Where Ω_V is the Grand Canonical Free energy per unit volume. In the final state we also must account for the Grand Canonical Free energy per unit volume on the formation of an α/β interface. This is the interfacial energy, thus

$$\Omega_{final} = (V - V^*)\, \Omega_V^\alpha(x_i^0, P^\alpha) + V^* \Omega_V^\beta(x_i^\beta, P^\beta) + \sigma A^*, \tag{B.9}$$

where A^* is the surface area of the critical nucleus. Thus,

$$W_R^* = \Delta\Omega = V^* \left[\Omega_V^\beta(x_i^\beta, P^\beta) - \Omega_V^\alpha(x_i^0, P^\alpha)\right] + \sigma A^*. \tag{B.10}$$

Since $\Omega = -PV$, see Chapter 2, $\Omega_V = -P$ and

$$W_R^* = V^* \left[P^\alpha - P^\beta\right] + \sigma A^*. \tag{B.11}$$

Since $P^\beta > P^\alpha$ the first term is negative and the second term is positive. Using the Laplace-Young equation, we can relate the pressure difference between both phases to the radius of the critical nucleus,

$$R^* = \frac{2\sigma}{P^\beta - P^\alpha}. \tag{B.12}$$

Since the nucleus is spherical eq.(B.11) can be written as

$$W_R^* = \frac{4}{3}\pi R^{*3}\left(P^\alpha - P^\beta\right) + \sigma 4\pi R^{*2} \tag{B.13}$$

and using eq.(B.12) we obtain

$$W_R^* = \frac{16\pi\sigma^3}{3\left(P^\beta - P^\alpha\right)^2} \tag{B.14}$$

which is a positive quantity. Thus, local increases in energy are required to nucleate a second phase from a metastable matrix.

What determines $P^\beta - P^\alpha$? , or, for that matter $x_i^{\beta *}$ the composition of the critical nucleus? Remember Ω_V^β is a function of both the pressure and composition of the nucleus, $x_i^{\beta *}$ and P^β). Let us consider a binary alloy consisting of A- and B-atoms. In this case the pressure in the nucleus and the composition of the nucleus are simply determined by the equilibrium conditions. For a binary alloy we have

$$\mu_A^\alpha\left(P^\alpha, x_B^0\right) = \mu_A^\beta\left(P^\beta, x_B^*\right) \tag{B.15}$$

$$\mu_B^\alpha\left(P^\alpha, x^0\right) = \mu_B^\beta\left(P^\beta, x^*\right) \tag{B.16}$$

Since P^α and x_B^0 are fixed, the above are two equations in the two unknowns, P^β and x_B^*. Thus, these quantities are uniquely specified. Thus, W_R^* is known and the radius R^*, through eq.(B.12), is also known.

It is also possible to relate the pressure difference to a change in the Gibbs free energy. To do this, however, one must assume that the phases are *incompressible*. This is a reasonable assumption in condensed phases, but a very bad assumption if one of the phases is a gas, since

$$\left.\frac{\partial \mu_i}{\partial P}\right|_{T,n_i,n_j} = \overline{V}_i, \tag{B.17}$$

where $\overline{V}_i$ is the partial molar volume of species i. Thus,

$$\mu_i(P) - \mu_i(P_o) = \int_{P_0}^{P} \overline{V}_i(P')dP', \tag{B.18}$$

where the explicit dependence of μ_i on T and x_i has been dropped for the sake of notational convenience since they are held constant in the above integration. If the partial molar volume is not a function of pressure, then

$$\mu_i(P) - \mu_i(P_0) = \overline{V}_i(P - P_0). \tag{B.19}$$

Thus, for the β-phase (the nucleus) of composition x_B^*

$$\mu_i^\beta(x_B^*, P^\beta) = \mu_i^\beta(x^*, P^\alpha) + \overline{V}_i^*(P^\beta - P^\alpha) \quad i = A, B\ , \tag{B.20}$$

where $\overline{V}_i^*$ is the partial molar volume at x_B^*. Using eq.(B.20) in eq.(B.15) and eq.(B.16) yields

$$\mu_A^\alpha(x_B^0, P^\alpha) = \mu_A^*(x_B^*, P^\alpha) + \overline{V}_A^\beta(P^\beta - P^\alpha) \tag{B.21}$$

$$\mu_B^\alpha(x_B^0, P^\alpha) = \mu_B^\beta x_B^*, P^\alpha) + \overline{V}_B^*(P^\beta - P^\alpha) \tag{B.22}$$

Multiplying eq.(B.21) by n_A^*, the number of moles of A in the critical nucleus, and eq.(B.21) by n_B^*, the number of moles of B in the critical nucleus, and adding them gives

$$\begin{aligned} n_A^*\mu_A^\alpha(x_B^0, P^\alpha) + n_B^*\mu_B^\alpha(x_B^0, P^\alpha) = \\ n_A^*\mu_A^\beta(x_B^*, P^\alpha) + n_B^*\mu_2^\beta(x_B^*, P^\beta) \\ + \left(n_A^*\overline{V}_A^* + n_B^*\overline{V}_B^*\right)\left(P^\beta - P^\alpha\right). \end{aligned} \tag{B.23}$$

Since,

$$V^* = n_A^* \overline{V}_A^* + n_B^* \overline{V}_B^* \tag{B.24}$$

we obtain

$$\begin{aligned} n_A^* &\left[\mu_A^\alpha(x_B^0, P^\alpha) - \mu_A^\beta(x_B^*, P^\alpha)\right] \\ +n_B^* &\left[\mu_B^\alpha(x_B^0, P^\alpha) - \mu_{B'}^\beta(x_B^*, P^\alpha)\right] = \left(P^\beta - P^\alpha\right) V^*. \end{aligned} \tag{B.25}$$

Dividing by $n^* = n_A^* + n_B^*$ gives

$$\begin{aligned} x_A^* &\left[\mu_A^\alpha(x_B^0, P^\alpha) - \mu_A^\beta(x_B^*, P^\alpha)\right] \\ +x_B^* &\left[\mu_2^\alpha(x_B^0, P^\alpha) - \mu_B^\beta(x_B^*, P^\alpha)\right] = V_m^* \left(P^\beta - P^\alpha\right). \end{aligned} \tag{B.26}$$

However, from eq.(2.59) the L.H.S. of eq.(B.26) is simply $-\Delta G_m$ for the formation of a small region of β-phase of composition x_B^* in the α-phase of composition $x_B^)$ and thus

$$P^\beta - P^\alpha = \frac{-1}{V_m^*} \Delta G_m \tag{B.27}$$

evaluated at x_B^*. So, we can represent $P^\beta - P^\alpha$ by ΔG_m. This is quite convenient as the supersaturation dependence of ΔG_m is quite clear, for example by using graphical methods, whereas the supersaturation dependence of $P^\beta - P^\alpha$ is not. Thus,

$$W_R^* = \frac{16\pi\sigma^3}{3\left(\Delta G_m / V_m^*\right)^2} \tag{B.28}$$

and

$$R^* = \frac{2\sigma}{-\Delta G_m / V_m^*}. \tag{B.29}$$

In order to determine the nucleation rate it is necessary to know the reversible work for the formation of a cluster of arbitrary size. These clusters are not in equilibrium, and thus we cannot enforce the conditions for chemical equilibrium, the equality of the chemical potentials at the interface between the cluster and matrix, and mechanical equilibrium, the Laplace-Young equation. We do, however, know the reversible work on the formation of a cluster of size R, W_R, using the same approach as that employed to derive eq.(B.11),

$$W_R = V^{cluster}\left[P^\alpha - P^\beta\right] + \sigma A. \tag{B.30}$$

Assuming that the nucleus is incompressible and using eq.(B.27) gives

$$W_R = V^{cluster}\left(\frac{\Delta G_m}{V_m^{cluster}}\right) + \sigma A. \tag{B.31}$$

For a spherical cluster we have

$$W_R = \frac{4\pi}{3}\left(\frac{\Delta G_m}{V_m^{cluster}}\right) R^3 + 4\pi\sigma R^2. \tag{B.32}$$

The dependence of W_R on the radius is shown in Fig. B.2 for a phase in the metastable state where $\Delta G_m < 0$. The critical radius for nucleation discussed above is at the maximum of this curve, illustrating that the nucleus is in a state of unstable equilibrium.

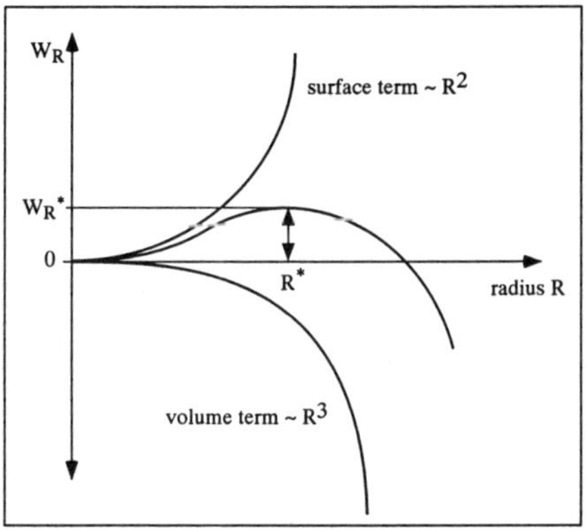

Fig. B.2. The work of formation of a cluster of size R.

These equations show that the larger the gain in volume free energy the smaller the critical radius of a nucleus and the larger the surface or interface energy the larger the critical radius. In a single phase material the gain in volume free enthalpy is directly proportional to the undercooling (see chapter 2). Thus the critical radius in solidification from a melt is inversely proportional to the undercooling and the energy barrier for nucleation proportional to $(\Delta T)^{-2}$.

The number of nuclei and the rate of nuclei production can also be estimated. Assume that all nuclei with just a size of R_c are in thermal equilibrium with the matrix. Then they would distributed according to a Boltzmann distribution:

$$n_c = n_0 \exp\left(-\frac{W_R^*}{k_B T}\right) \tag{B.33}$$

with n_0 the total number of atoms per volume and n_c the number of critical nuclei per volume. The rate of nucleation I_{SS}, i.e. the number of nuclei formed per second and volume, can then be calculated by simply setting:

$$I_{SS} = p n_c \tag{B.34}$$

where p is the number of atoms that enter per unit of time the surface of a nucleus:

$$p = 4\pi R_c^2 \cdot j \tag{B.35}$$

with j the current density of atoms. We could calculate the current density by the diffusional growth in a supersaturated matrix as done in section 4.3 and would have:

$$j \approx D \frac{x_0 - x_\alpha^e}{R_c} \tag{B.36}$$

Using an Arrhenius law for the diffusion coefficient we could write for the nucleation rate:

$$I_{SS} = I_0 \exp\left(-\frac{\Delta G_D}{k_B T}\right) \exp\left(-\frac{W_R^*}{k_B T}\right) \tag{B.37}$$

with ΔG_D the activation energy for diffusion. Using some reasonable values for all quantities in solid metals one can calculate that a typical prefactor I_0 for the nucleation rate is considerable, on the order of 10^{43} and the activation energy is of the order of $65 k_B T_m$. Given this large magnitude, agreement between the predictions of nucleation theory and experiments in metals that differ by as much as 5 orders of magnitude is considered good. At small supersaturations, ΔG_m is small and thus the nucleation rate is very small near the coexistence boundary. If the alloy composition is fixed, upon decreasing the temperature the supersaturation increases and the nucleation rate increases. However, at sufficiently low temperatures the rate for atomic diffusion becomes sufficiently small that the nucleation rate again becomes very small (note the $1/T$ dependence in the exponential of the expression for the diffusion coefficient). Thus it follows that the nucleation rate has a maximum in between.

B.2 Kinetic theory of nucleation

A more realistic treatment of nucleation than the above more or less plausible considerations starts with a concept introduced by Vollmer and Becker and Döring. We treat the nucleation, i.e. the occurrence of heterophase fluctuations in a metastable system in close analogy to a polymerization or polycondensation process: to an existing fluctuation of say (l-1) atoms an addition of 1 atom produces a cluster of l atoms. The dissolution of just one atom from this cluster reduces it to the original (l-1) cluster. Calculating for all cluster size the energy needed to built them and their rate of formation or dissolution we finally arrive at an equation for the number density of clusters with a given size as it depends on time and cluster size. The steady state solution of this equation yields an expression for the nucleation rate. The time dependent solution tells something about the incubation period for nucleation.
The whole chain of clusters that could exist in a supersaturated or supercooled system is schematically depicted in figure B.3. The reactions

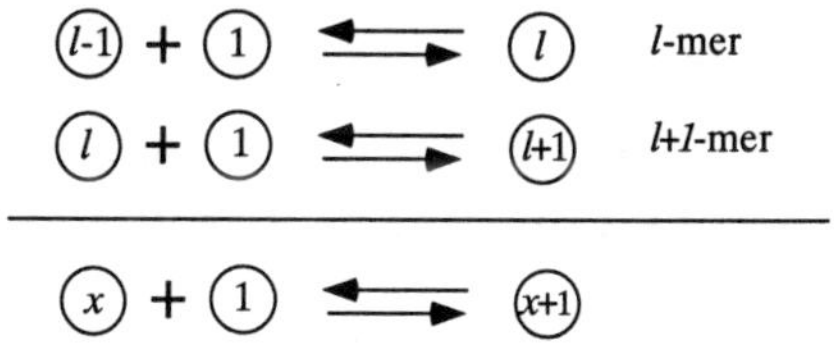

Fig. B.3. If a single phase system is thermodynamically in a metastable state, such that in equilibrium it would be in a two-phase state, there can exist heterophase fluctuations of different size or number of atoms l in it. The occurrence of these clusters could be imagined as a polymerization process: Monomers add to already existing l-mers or an l-mers disappears to to dissolution of a monomer from it.

shown in figure B.3 can be considered as standard chemical reactions. In equilibrium both reaction paths, i.e. $(x) + (1) \Leftrightarrow (x+1)$ have equal probability. Then their concentrations can be described by the law of mass action:

$$\frac{[x+1]}{[x][1]} = \exp(-\Delta G_x / k_B T) \tag{B.38}$$

where the brackets denote concentrations and $\Delta G_x = \epsilon(x+1) - \epsilon(x)$ is the difference of the free enthalpy of clusters ϵ of size $x+1$ and x (size means here exactly the number of atoms in a cluster). In a non-equilibrium situation the rate of formation of a cluster $(x) + (1) \Rightarrow (x+1)$ is different from the rate of decomposition $(x+1) \Rightarrow (x) + (1)$. We will try to calculate the number density of clusters of arbitrary size and therefrom that of critical size leading to a description of the nucleation rate.

Let $n_V^{cluster}(l,t)$ be the number of clusters with l-atoms at time t per unit of volume in the metastable system. The $n_V^{cluster}$ changes with time according to:

$$\frac{dn_V^{cluster}}{dt} = I(l) - I(l+1) \tag{B.39}$$

with $I(l)$ the number of clusters that change their size from $l-1$ to l:

$$I(l) = a(l) n_V^{cluster}(l-1) - b(l) n_V^{cluster}(l) \tag{B.40}$$

where $a(l)$ is the absorption rate and $b(l)$ is the desorption rate. $a(l)$ the denotes the rate with which atoms attach to a cluster of size $l-1$ and $b(l)$ the rate of desorption of an atom from a cluster of size l. In equilibrium we would have $I(l) = 0$, i.e. their appear and disappear the same amount of l-clusters per unit of time:

$$a(l)\overline{n_V^{cluster}(l-1)} = b(l)\overline{n_V^{cluster}(l)} \tag{B.41}$$

where we denoted time averages by a bar indicating also that these are equilibrium values. This equation has the solution:

$$\frac{a(l)}{b(l)} = \frac{\overline{n_V^{cluster}(l)}}{\overline{n_V^{cluster}(l-1)}} = \exp\left(-\frac{\epsilon(l) - \epsilon(l-1)}{k_B T}\right) \tag{B.42}$$

which exactly is the same we would have by using the law of mass action. Using this expression we can rewrite eq.(B.40) to

$$\begin{aligned} I(l) &= a(l)\left[n_V^{cluster}(l-1) - \frac{\overline{n_V^{cluster}(l-1)}}{\overline{n_V^{cluster}(l)}} n_V^{cluster}(l)\right] \\ &= a(l)\left[n_V^{cluster}(l-1) - e^{-\frac{\epsilon(l)-\epsilon(l-1)}{k_B T}} n_V^{cluster}(l)\right] \end{aligned} \tag{B.43}$$

To proceed with this equation we perform a few approximations. First the free enthalpy of a cluster of size l is expanded in a Taylor series:

$$\epsilon(l) = \epsilon(l-1) + \frac{d\epsilon}{dl}[l-(l-1)] + \ldots = \epsilon(l-1) + \frac{d\epsilon}{dl} + \ldots \tag{B.44}$$

We thus obtain:

$$\epsilon(l) - \epsilon(l-1) = \left.\frac{d\epsilon}{dl}\right|_{l-1} \tag{B.45}$$

and thus we can write the exponential in eq.(B.43) as:

$$\exp\left(-\frac{\epsilon(l)-\epsilon(l-1)}{k_B T}\right) \approx 1 + \frac{1}{k_B T}\frac{d\epsilon}{dl} \tag{B.46}$$

Writing:

$$n_V^{cluster}(l) = n_V^{cluster}(l-1) + \frac{\partial n_V^{cluster}}{\partial l} \tag{B.47}$$

we can rewrite eq.(B.43) as:

$$I(l) = a(l)\left[-\frac{\partial n_V^{cluster}}{\partial l} + \frac{1}{k_B T}\frac{\partial \epsilon}{\partial l} n_V^{cluster}(l)\right] \tag{B.48}$$

This expression could be used in eq.(B.39) but we will further approximate:

$$I(l+1) - I(l) = \frac{\partial I}{\partial l} \tag{B.49}$$

Inserting all the expressions we have collected into eq.(B.39) leads to:

$$\frac{\partial n_V^{cluster}}{\partial t} = -\frac{\partial I}{\partial l} = \frac{\partial}{\partial l}\left[a(l)\frac{\partial n_V^{cluster}}{\partial l} + \frac{1}{k_B T}\frac{\partial}{\partial l}\left[a(l) n_V^{cluster}(l)\frac{\partial \epsilon}{\partial l}\right]\right] \tag{B.50}$$

This type of equation is called a Fokker-Planck equation. Its solution would yield the time dependent number density of cluster sizes $n_V^{cluster}(l)$.

B.2.1 Stationary nucleation rate

First we will derive a stationary solution $n_V^{cluster,s}(l)$ giving the time independent distribution of cluster sizes. In the next subsection we will derive an approximate time dependent solution. Stationarity means:

$$\frac{\partial n_V^{cluster}}{\partial t} = 0 \quad \Rightarrow \quad \frac{\partial I}{\partial l} = 0 \quad \Rightarrow \quad I = I_{SS} \tag{B.51}$$

The boundary conditions for a solution of eq.(B.50) are:

1. $n_V^{cluster,s}(l) = \overline{n_V^{cluster}(l)}$ for $l \Rightarrow 0$, meaning extremely small fluctuations shall be in equilibrium.
2. $n_V^{cluster,s}(l) = 0$ for $l \Rightarrow \infty$, meaning extremely big fluctuations do not exist.

We then rewrite eq.(B.50) as:

$$-I_{SS} = a(l)\frac{dn_V^{cluster,s}(l)}{dl} + \frac{1}{k_B T}a(l)\frac{d\epsilon}{dl}nu_s(l) \tag{B.52}$$

This is an ordinary differential equation of first order. Its general solution is:

$$n_V^{cluster,s}(l) = \exp\left(-\frac{\epsilon(l)}{k_B T}\right) \cdot \left[-I_{SS}\int_0^l \frac{1}{a(l')}\exp\left(-\frac{\epsilon(l')}{k_B T}\right) dl' + C\right] \tag{B.53}$$

From the second boundary condition the integration constant is determined:

$$C = I_{SS}\int_0^\infty \frac{\exp\left(-\frac{\epsilon(l')}{k_B T}\right)}{a(l')} dl' \tag{B.54}$$

leading to:

$$n_V^{cluster,s}(l) = I_{SS}\exp\left(-\frac{\epsilon(l)}{k_B T}\right) \cdot \int_l^\infty \frac{\exp\left(-\frac{\epsilon(l')}{k_B T}\right)}{a(l')} dl' \tag{B.55}$$

The first boundary condition leads to:

$$n_V^{cluster,s}(l \Rightarrow 0) = \overline{n_V^{cluster}(l)} = \overline{n_V^{cluster}(1)}\exp\left(-\frac{\epsilon(l)}{k_B T}\right) \tag{B.56}$$

Using eq.(B.55) in the limit $l \Rightarrow 0$ we can determine the prefactor I_{SS} as:

$$I_{SS} = \frac{\overline{n_V^{cluster}(1)}}{\int_0^\infty \frac{\exp\left(-\frac{\epsilon(l')}{k_B T}\right)}{a(l)} dl'} \tag{B.57}$$

We finally obtain the concentration of metastable clusters of size l as:

$$n_V^{cluster,s}(l) = \overline{n_V^{cluster}(1)}\exp\left(-\frac{\epsilon(l)}{k_B T}\right)\frac{\int_l^\infty \frac{\exp\left(-\frac{\epsilon(l')}{k_B T}\right)}{a(l')} dl'}{\int_0^\infty \frac{\exp\left(-\frac{\epsilon(l')}{k_B T}\right)}{a(l')} dl'} \tag{B.58}$$

In equation (B.57) we have the stationary nucleation rate if we would evaluate the expression just at the critical cluster or nucleus size l_c and in eq.(B.58) the concentration of metastable clusters of arbitrary size. Both expressions shall now be evaluated further. We expand the cluster free enthalpy $\epsilon(l)$ in a Taylor series around the critical size l_c:

$$\epsilon(l) = \epsilon(l_c) + \left.\frac{d\epsilon}{dl}\right|_{l=l_c} (l - l_c) + \frac{1}{2} \left.\frac{d^2\epsilon}{dl^2}\right|_{l=l_c} (l - l_c)^2 + \ldots \tag{B.59}$$

The cluster energy has a maximum at the critical size, since it can be written as (see section above):

$$\epsilon(l) = d_1 \Delta G_V l + d_2 \sigma l^{2/3} \tag{B.60}$$

with d_1, d_2 appropriate scaling factors, since ΔG_V is the gain in free enthalpy per volume and the number of atoms in a cluster l is proportional to the volume, i.e. $l \propto R^3$ and in the surface term $l^{2/3} \propto R^2$. The scaling factors adjust this for example to spherical clusters or nuclei. Since $\epsilon(l)$ has a maximum its first derivative vanishes at $l = l_c$ and the second derivative is smaller than zero:

$$b = \left.\frac{d^2\epsilon}{dl^2}\right|_{l=l_c} < 0 \tag{B.61}$$

Calling $\epsilon(l_c) = \epsilon_c$ we can write the Taylor expansion as:

$$\epsilon(l) = \epsilon_c + \frac{1}{2} b (l - l_c)^2 \tag{B.62}$$

neglecting all higher order terms. Inserting this result into both integrals of eq.(B.58) gives for the integral in the denominator:

$$\begin{aligned} &\int_0^\infty \frac{\exp\left(-\frac{\epsilon_c}{k_B T}\right)}{a(l_c)} \exp\left(\frac{b}{2k_B T}(l' - l_c)^2\right) dl' = \\ &\frac{\exp\left(-\frac{\epsilon_c}{k_B T}\right)}{a(l_c)} \int_0^\infty \exp\left(-\frac{|b|}{2k_B T}(l' - l_c)^2\right) dl' \end{aligned} \tag{B.63}$$

The integral can be calculated using the transformation $x = \sqrt{|b|/2k_B T}(l' - l_c)$, leading to an error function:

$$\int_0^\infty \exp\left(-\frac{|b|}{2k_BT}(l'-l_c)^2\right)dl' =$$
$$= \sqrt{\frac{2k_BT}{|b|}}\int_{-l_c\sqrt{|b|/2k_BT}}^\infty \exp(-x^2)dx$$
$$= \frac{2\pi k_BT}{|b|}\frac{1}{2}\left(1+\mathrm{erf}(l_c\sqrt{|b|/2k_BT})\right) \tag{B.64}$$

The integral in the nominator of eq.(B.58) has only the difference in the lower bound of the integral. The result is similar:

$$\int_l^\infty \exp\left(-\frac{|b|}{2k_BT}(l'-l_c)^2\right)dl' =$$
$$= \sqrt{\frac{2k_BT}{|b|}}\int_{-l_c\sqrt{|b|/2k_BT}}^\infty \exp(-x^2)dx$$
$$= \frac{2\pi k_BT}{|b|}\frac{1}{2}\left(1-\mathrm{erf}((l-l_c)\sqrt{|b|/2k_BT})\right) \tag{B.65}$$

Collecting all terms leads to the final expression for the number density of clusters:

$$n_V^{cluster,s}(l) = \overline{n_V^{cluster}(1)}\exp(-\frac{\epsilon(l)}{k_BT})\frac{1-\mathrm{erf}((l-l_c)\sqrt{|b|/2k_BT})}{1+\mathrm{erf}(l_c\sqrt{|b|/2k_BT})} \tag{B.66}$$

One can easily control that the two boundary conditions are fulfilled. If we neglect the factor with the error functions it is just the same expression as given in the above section, since the first term is just a Boltzmann factor. We now can also calculate the stationary nucleation rate using the integrals calculated above:

$$I_{SS} = \frac{\overline{n_V^{cluster}(1)}}{a(l_c)}\exp(-\frac{\epsilon(l_c)}{k_BT})\cdot Z \tag{B.67}$$

with the so-called Zeldovich factor Z:

$$Z = \sqrt{\frac{2|b|}{\pi k_BT}}\frac{1}{1+\mathrm{erf}(l_c\sqrt{|b|/2k_BT})} \approx \sqrt{\frac{2|b|}{\pi k_BT}} \tag{B.68}$$

The error function is usually neglected in this expression. We thus arrive at an expression for the nucleation rate looking quite similar to the simple one given in the preceding section:

$$I_{SS} \propto \overline{n_V^{cluster}(1)}a(l_c)n_V^{cluster}(l_c) \tag{B.69}$$

where $\overline{n_V^{cluster}(1)}$ is the number of atoms per volume, $a(l_c)$ represents for example in diffusional transport of matter the current density of atoms entering the nuclei's surface and the last term is the number of critical nuclei. This expression should be compared with eq.(B.34) to see the similarity.

B.2.2 Time dependent nucleation

We will not solve eq.(B.50) completely, but use a simple approximation first introduced by Kantorowitz 1951:

$$\frac{\partial n_V^{cluster}}{\partial t} = a(l_c)\frac{\partial^2 n_V^{cluster}}{\partial l^2} \tag{B.70}$$

We replaced in eq.(B.50) the absorption function $a(l)$ by a constant value it would have at a critical nucleus site and completely neglect the second term. Thus we have reduced eq.(B.50) to a simple diffusion equation. With the following boundary conditions:

$$n_V^{cluster}(0,t) = n_V^{cluster}(1) \tag{B.71}$$

$$n_V^{cluster}(s,t) = 0 \qquad \text{for} \qquad s \geq l_c \tag{B.72}$$

and the initial condition

$$n_V^{cluster}(l,0) = 0 \tag{B.73}$$

one can solve the equation looking simply into a textbook on heat conduction [55]. The boundary conditions are easily understood: We require that the number density of clusters of size $l = 0$ equals the number of molecules in the system and set with the second condition that no clusters exist being larger than the critical size. The initial conditions is trivial. The solution of equation (B.70) fulfilling the boundary conditions is:

$$\begin{aligned} n_V^{cluster}(l,t) &= n_V^{cluster}(1)(1 - l/s) \\ &- \frac{2n_V^{cluster}(1)}{\pi}\sum_{m=1}^{\infty}\frac{1}{m}\sin(\frac{m\pi l}{s})\exp\left(-\frac{m^2\pi^2 a(l_c)t}{s^2}\right) \end{aligned} \tag{B.74}$$

Approximating the nucleation rate as:

$$I(l,t) \approx -a(l)\frac{\partial n_V^{cluster}}{\partial l} \tag{B.75}$$

$$I(s,t) \approx -a(l_c)\left.\frac{\partial n_V^{cluster}}{\partial l}\right|_s \tag{B.76}$$

The derivative of $n_V^{cluster}$ with respect to l and taking this value at s we obtain:

$$\left.\frac{\partial n_V^{cluster}}{\partial l}\right|_s = -\frac{n_V^{cluster}(1)}{s}\left(1 + 2\sum_{m=1}^{\infty}(-1)^m \exp\left(-\frac{m^2\pi^2 a(l_c)t}{s^2}\right)\right) \tag{B.77}$$

From this we could calculate the nucleation rate. Instead of writing down the final equation we are using another representation for the sum in eq.(B.77) especially useful for short times:

$$\left.\frac{\partial n_V^{cluster}}{\partial l}\right|_s = 4n_V^{cluster}(1)\frac{a(l_c)}{\pi t}\sum_{m=1}^{\infty}\exp\left(-\frac{s^2}{4a(l_c)t(2m-1)^2}\right) \tag{B.78}$$

Leaving only the first term in this sum we obtain for the nucleation rate of critical clusters (see [55]):

$$I(s,t) \approx 4n_V^{cluster}(1)\frac{a(l_c)}{\pi t}\exp\left(-\frac{s^2}{4a(l_c)t}\right) = I_{SS}\exp(-\tau/t) \tag{B.79}$$

with

$$\tau = \frac{s^2}{4a(l_c)} = \frac{l_c^2}{4a(l_c)} \tag{B.80}$$

The nucleation rate is thus initially zero and exponentially increases to the stationary value. This behavior is shown in fig.B.4. The time τ give the incubation time. The larger it is the longer it takes before the nucleation rate increases rapidly. Expression (B.80) implies that a large diffusion coefficient, like that in liquids, leads to a short incubation time and thus the nucleation rate achieves fast the value of the stationary theory. The critical size depends on the energetics of nucleation as written down in eq.(B.60). Calculating the critical cluster size shows that the following relation holds:

$$l_c \propto \left(\frac{\sigma_{12}}{\Delta G_V}\right)^3 \tag{B.81}$$

Thus the incubation time increases with increasing interfacial tension and decreases rapidly if the gain in volume free energy is large.

B.3 Heterogeneous nucleation

Nucleation often does not occur homogeneously in a matrix by for instance concentration fluctuations but most often is provoked and

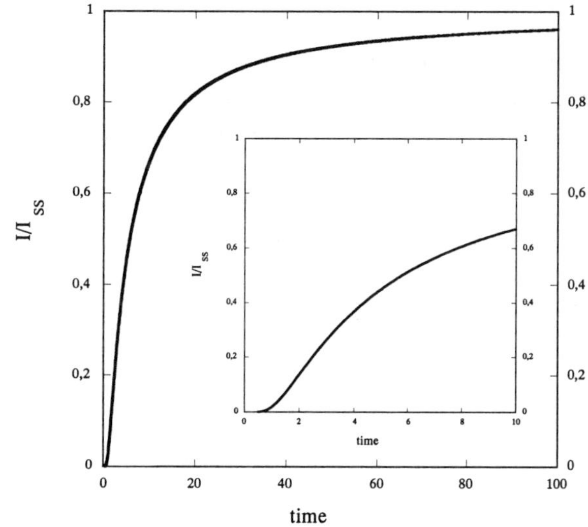

Fig. B.4. Time dependent nucleation rate. Initially the nucleation rate is zero and grows exponentially to the stationary rate.

induced by impurities, dislocations, grain boundaries, immiscible inclusions of all kind, crucible walls etc.. Nucleation of a second phase on generally a substrate occurs easier since the work of formation of a nucleus is reduced. Assume that the substrate is flat as shown in figure B.5 and that the nucleus has the shape of a sphere. The work of

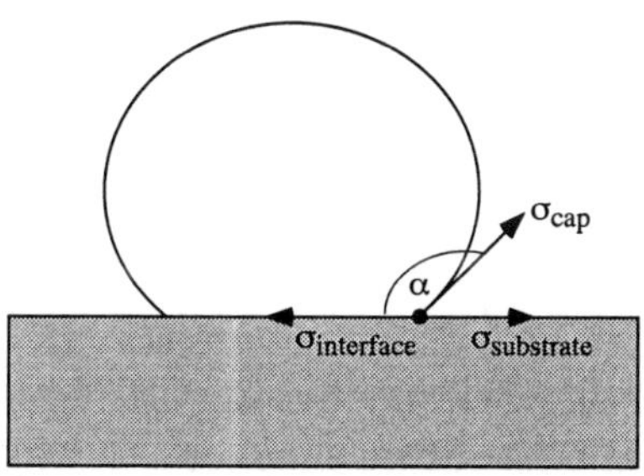

Fig. B.5. Equilibrium shape of a sphere on a flat solid substrate. α is the contact angle between the precipitate and the substrate.

nucleation is calculated as in eq.(B.28), but now the volume and the surface tension change a little bit.

$$W_R = -V_{cap}\Delta G_V + A_{cap}\sigma_c + (\sigma_{interface} - \sigma_{substrate})A_{interface} \quad \text{(B.82)}$$

the first term is as before the gain in free enthalpy per volume, but here the reduced volume has to be calculated. The second term is the work needed to created the surface of the nucleus. The third term describes the work needed to replace the substrate matrix interface by a nucleus interface with a different interface energy. All three interface tensions are related by Youngs equation, stating that the sum of the interface tension must be zero:

$$\sigma_{substrate} = \sigma_{interface} + \sigma_{cap} \cos\alpha \tag{B.83}$$

Using this expression and calculating the volume and the interface

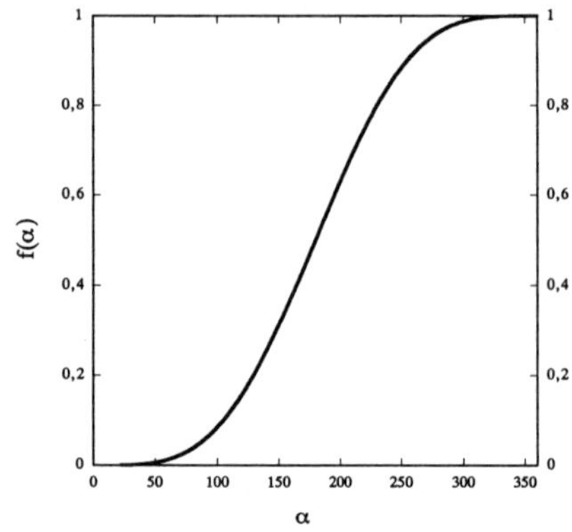

Fig. B.6. Catalytic factor for heterogeneous nucleation; see eq.(B.84).

area one arrives at a new expression for the critical energy barrier of nucleation given by:

$$\Delta W_R^{het} = \Delta G \frac{1}{4}(2 - 3\cos\alpha + \cos^3\alpha) = W_R^* f(\alpha) \tag{B.84}$$

where W_R^* is the work needed for homogeneous nucleation of such a particle in the volume without a substrate (see eq.(B.28)). The function $f(\alpha)$ is shown in fig.B.6. It shows that for a substrate fully wet by the nucleus the energy barrier is zero. Totally incomplete wetting yields the result without substrate.

Bibliography

[**Comments to the bibliography**] The following list of literature on coarsening in solid-solid, solid-liquid and liquid-liquid systems is not a complete survey but is probably a suitable starting point to get an overview on the topic and its development especially in the last decades. For newer literature one is always best advised to look into Acta Materialia, Metallurgical Transactions, Material Science and Engineering A, Physical Review Letters and Physical Review E. A part of the literature is cited in the text but we have added some more references to help people being interested in coarsening to quickly arrange a library of essential papers.

1. H.B. Aaron and G.R. Kotler. Second phase dissolution. *Met. Trans.*, **2** (1971) 393 – 408
2. F.F. Abraham, *Homogeneous Nucleation Theory*, Academic Press, New York 1947
3. N. Akaiwa, S.C. Hardy, and P.W. Voorhees. The effects of convection on Ostwald ripening in solid-liquid mixtures. *Acta metall. mater.*, **39** (1991) 2931 – 2942
4. M. Abramowitz, I.A. Stegun, *Handbook of Mathematical Functions*, Dover Publications, New York, 8th Edition, 1972
5. N. Akaiwa and P.W. Voorhees. Late-stage phase separation: Dynamics, Spatial correlations and Structure Functions. *Phys.Rev.E.* **49** (1994) 3860
6. N.Akaiwa, D.I.Meiron. Numerical simulation of two-dimensional late stage coarsening for nucleation and growth. *Phys.Rev. E* **51** (1995) 5408 – 5421
7. J.Alkemper,V.A.Snyder,N.Akaiwa,P.W.Voorhees. Dynamics of late-stage phase separation: A Test of Theory. *Phys.Rev.Lett.*, **82** (1999) 272 – 278
8. A. Maheshwari and A J Ardell. Morphological evolution of coherent misfitting precipitates in anisotropic elastic media, elastic interactions affect shapes. *Phys. Rev. Lett.*, **70** (1993) 2305
9. A. Maheshwari and A J Ardell. Elastic interaction and their effect on particle shapes in aged dilute NiAl alloys. *Scripta Metall. et Mater.*, **26** (1992) 347.
10. A.J. Ardell. Experimental confirmation of the Lifshitz-Wagner theory of particle coarsening. In *The Mechanism of Phase Transformations in Crystalline Solids*, Monograph and Report Ser. No. 33, pages 111 – 116. The Inst. of Metals, 1969.
11. A.J. Ardell. The effect of volume fraction on particle coarsening: Theoretical considerations. *Acta metall.*, **20** (1972) 61 – 71
12. A.J. Ardell. Isotropic fiber coarsening in unidirectionally solidified eutectic alloys. *Met Trans.*, **3** (1972) 1395 – 1401
13. A.J. Ardell. On the coarsening of grain boundary precipitates. *Acta metall.*, **20** (1972) 601 – 609

14. A.J. Ardell. Precipitate coarsening in solids: modern theories, chronic disagreement with experiment. In G.Lorimer, editor, *Phase Transformations 87*, pages 485 – 494. The Inst. of Metals, Cambridge, 1988.
15. A.J. Ardell. Observations on the effect of volume fraction on the coarsening of γ' precipitates in binary Ni-Al alloys. *Scripta metall. et materialia*, **24** (1990) 343 – 346
16. A.J. Ardell, R.B. Nicholson, and J.D. Eshelby. On the modulated structure of aged Ni-Al alloys (with an appendix on the elastic interaction between inclusions). *Acta metall.*, **14** (1966) 1295 – 1309
17. A.J.Ardell. Late-stage two-dimensional coarsening of circular clusters. *Phys.Rev.B* **41** (1990) 2554 – 2556
18. A.J.Ardell. Temporal behavior of the number density of particles during Ostwald ripening. *Mat.Sci.Eng.A* **238** (1997) 108 – 120
19. George Arfken, *Mathematical Methods for Physicists.* Academic Press, 1970, pp. 761.
20. R. Asimov. Clustering kinetics in binary alloys. *Scripta Metall.*, **11** (1963) 72 – 73
21. G.I. Barenblatt, *Scaling, self-similarity, and intermediate asymptotics*, Cambridge University Press, Cambridge UK 1996
22. G.K.Batchelor, *An introduction to fluid dynamics*, Cambridge University Press, Cambridge 1967
23. H.-J. Diepers, C. Beckermann, I.Steinbach, Simulation of convection and ripening in a binary alloy mush using the phase fiedl method, *Acta mater.* **47** (1999) 3663 – 3678
24. C.W.J. Beenakker. Numerical simulation of diffusion-controlled droplet growth: Dynamical correlation effects. *Phys. Rev. A*, **33** (1986) 4482 – 4485
25. C. W. J. Beenakker and J. Ross. *J. Chem. Phys.*, **84** (1986) 3857
26. W.Bender and L. Ratke. Ostwaldreifung und Koaleszenz in flüssig/-phasengesinterten Co-Cu-Legierungen. *Z. Metallkd.*, **83** (1992) 541 – 547
27. W.Bender and L. Ratke. Ostwald ripening of Cobalt particles in liquid Copper. *Scripta metall. et materialia*, **28** (1993) 737 – 742
28. W. Bender. Ostwaldreifung und Agglomeration beim Flüssigphasensintern. PhD thesis, RWTH-Aachen, Aachen, 1994.
29. W. Bender, L. Ratke, and B. Feuerbacher. Ostwald ripening of cobalt particles in liquid copper. In *Proc. VIIIth European Symposium on Materials and Fluid Sciences in Microgravity*, pages 645 – 649, Brussels, Belgium, 1992. ESA SP-333.
30. W.Bender, L.Ratke. Ostwald ripening of liquid phase sintered Cu-Co dispersions. *Acta Materialia* **46** (1998) 1125 – 1134
31. W.Bender, L.Ratke. Ostwald ripening of the matrix phase surrounding dispersoids. *Z.Metallkde.* **89** (1998) 666 – 671
32. A. Bergman and H. Fredriksson. A study of the coalescence process inside the miscibility gap in Zn-Bi alloys. In G.E. Rindone, editor, *Materials Processing in the Reduced Gravity Environment of Space*, pages 563 – 577, Amsterdam, 1982. Elsevier.
33. K.Binder, *Spinodal decomposition*, in: *Systems Far from Equilibrium*, Lecture Notes in Physics, **132**, Ed.L.Garido, Springer Verlag, Berlin 1980
34. K.Binder, Spinodal decomposition - some recent developments, in: In L. Ratke, editor, *Immiscible Liquid Metals and Organics*, pages 47 – 64. DGM Informationsgesellschaft, Oberursel, 1993.
35. K. Binder and D. Stauffer. Statistical theory of nucleation, condensation and coagulation. *Advances in Physics*, **25** (1976) 343 – 396

36. S. Björklund, L.F. Donaghey, and M. Hillert. The effect of alloying elements on the rate of Ostwald ripening of cementite in steel. *Acta metall.*, **20** (1972) 867 – 874
37. W.J. Böttinger, P.W. Voorhees, R.C. Dobbyn, and H.E. Burdette. A study of the coarsening of liquid-solid mixtures using synchrotron radiation microradiography. *Met. Trans. A*, **18A** (1987) 487 – 490
38. A.D. Brailsford. Diffusion to a random array of identical spherical sinks. *J. Nuclear Materials*, **60** (1976) 257 – 278
39. A.D. Brailsford and P.Wynblatt. The dependence of Ostwald ripening kinetics on particle volume fraction. *Acta Metall.*, **27** (1979) 489 – 497
40. M. Breu, W. Gust, B. Predel, and E. Wachtel. Magnetische Untersuchungen zum Wachstum und Vergröberung von Ausscheidungen in Cu-Co-legierungen - I. Verdünnte binäre Legierungen. *Z.Metallkde.*, **79** (1988) 1 – 7
41. M. Breu, W. Gust, B. Predel, and E. Wachtel. Magnetische Untersuchungen zum Wachstum und Vergröberung von Ausscheidungen in Cu-Co-legierungen - II. ternäre Legierungen. *Z.Metallkde.*, **79** (1988) 8 – 12
42. L.C. Brown. Direct observation of coarsening in Al-Cu alloys. *Acta metall.*, **33** (1985) 1391 – 1398
43. L.C. Brown. A new examination of classical coarsening theory. *Acta metall.*, **37** (1989) 71 – 77
44. L.C. Brown. Reply to comments by Hillert, Hunderi, Ryum and Sätre on "A new examination of classical coarsening theory". *Scripta metall. et materialia*, **24** (1990) 963 – 966
45. L.C. Brown. Reply to comments by Hoyt on "A new examination of classical coarsening theory". *Scripta metall. et materialia*, **24** 2231 – 2234, 1990.
46. L.C. Brown. A new examination of volume fraction effects during particle coarsening. *Acta metall. mater.*, **40** (1992) 1293 – 1303
47. L.C. Brown. Reply to further comments by Hillert, Hunderi and Ryum on "A new examination of classical coarsening theory". *Scripta metall. et materialia*, **26** (1992) 1939 – 1942
48. L.C. Brown. Answer to the rebuttal of Hillert, Hunderi and Ryum. *Scripta metall. et materialia*, **26** (1992) 1939 – 1942
49. J.W.Bullard. Numerical results of transient-stage Ostwald ripening and coalescence in two dimensions. *Mat.Sci.Eng.A* **238** (1997) 128 - 139
50. W.K.Burton, N. Cabrera, F.C.Frank, *Phil.Mag.* **243** (1951) 299 – 358
51. J.W. Cahn, *Metallurgical Soc. AIME* **242** (1968) 166
52. J.W.Cahn, J.E.Hilliard, J.Chem.Phys. **28** (1958) 258; **31** (1959) 668
53. H.A. Calderon, P.W. Voorhees, J.L. Murray, and G. Kostorz. Ostwald ripening in concentrated alloys. *Acta metall. mater.*, **42** (1994) 991 – 1000
54. B. Caroli, C. Caroli, B. Roulet, and P.W. Voorhees. Effect of elastic stresses on the morphological stability of a solid sphere growing from a supersaturated melt. *Acta metall.*, **37** (1989) 257 – 268
55. H.S. Carslaw, J.C.Jaeger, *Conduction of Heat in Solids*, Oxford University Press, 1959
56. P. Casses and M.A. Azouni. Thermal effects on the shape of a solidifying interface near a foreign particle. *J. Crystal Growth*, **130** 13 – 20, 1993.
57. S. Chandrasekhar, *Hydrodynamic and Hydromagnetic Stability*, Clarendon Press, Oxford 1961
58. S. Chandrasekhar, *Stochastic Problems in Physics and Astronomy*, Review of Modern Physics **15** (1943) 2 – 89
59. J.M. Chaix, N. Eustathopoulos, and C.H. Allibert. Ostwald ripening growth rate for nonideal systems with significant mutual solubility - I. binary systems. *Acta metall.*, **34** (1986) 1589 – 1592

60. J.M. Chaix and C.H. Allibert. Ostwald ripening growth rate for nonideal systems with significant mutual solubility - II. ternary systems. application to liquid phase sintering of W-Ni-Cr. *Acta metall.*, **34** (1986) 1593 – 1598
61. D.Z. Che and J.J. Hoyt. Spatial Correlations During Ostwald Ripening – A Simplified Approach, *Acta Metall. Mater.*, **43** (1995) 2551.
62. D.J. Chellman and A.J. Ardell. The coarsening of γ' precipitates at large volume fraction. *Acta Metall.*, **22** (1974) 577 – 588
63. M.K. Chen and P.W. Voorhees. *The Dynamics of Transient Ostwald Ripening.* PhD-thesis, Evanston, Illinois, 1992.
64. J.G. Conley, M.E. Fine, and J.R. Weertman. Effect of lattice disregistry variation on the late stage phase transformation behavior of precipitates in ni-al-mo alloys. *Acta metall.*, **37** (1989) 1251 – 1263
65. S. Coriell and R.L. Parker. Stability of the shape of a solid cylinder growing in a diffusion field. *J.Appl. Phys.*, **36** (1965) 632 – 637
66. S.D. Coughlan and M.A. Fortes. Self similar size distributions in particle coarsening. *Scripta metall. et materialia*, **28** (1993) 1471 – 1476
67. T.H. Courtney. A reanalysis of the kinetics of neck growth during liquid phase sintering. *Met. Trans. A*, **8A** (1977) 671 – 677
68. T.H. Courtney. Microstructural evolution during liquid phase sintering: Part I. development of microstructure. *Met. Trans. A*, **8A** (1977) 679 – 684
69. T.H. Courtney. Microstructural evolution during liquid phase sintering: Part II. microstructural coarsening. *Met. Trans. A*, **8A** (1977) 685 – 689
70. T.H. Courtney and Lee. *Met. Trans A*, **11A** 943, 1988.
71. M.F. Chisholm and D.E. Laughlin. Decomposition in alloys : an overview. In G.Lorimer, editor, *Phase Transformations 87*, pages 1 – 7. The Inst. of Metals, Cambridge, 1988.
72. J. Crank, *Mathematics of Diffusion*, Oxford University Press, London 1950
73. Cummings, A., Wiltzius P. and Bates, F.S. *Phys.Rev.Lett.* **65**, 863 (1990)
74. S.K. Das, A. Biswas, and R.N. Ghosh. Volume fraction dependent particle coarsening in plain carbon steel. *Acta metall. mater.*, **41** (1993) 777 – 781
75. C.K.L. Davies, P. Nash, and R.N. Stevens. Precipitation in Ni-Co-Al alloys - part 1. continuous precipitation. *J. Mat. Sci.*, **15** (1980) 1521 – 1532
76. C.K.L. Davies, P. Nash, and R.N. Stevens. The effect of volume fraction of precipitate on Ostwald ripening. *Acta Metall.*, **28** (1980) 179 – 189
77. R.T. DeHoff. Generalized microstructural evolution by interface controlled coarsening. *Acta metall.*, **32** (1984) 43 – 47
78. R.T. DeHoff. Scale factor coarsening. *Metallography.*, **17** (1984) 203 – 208
79. R.T. DeHoff and C.V. Iswaran. The usefulness of integral mean curvature measurements in the study of the kinetics of coarsening. *Metall. Trans. A*, **13A** (1982) 1389 – 1395
80. W. Döring, *Z. Phys. Chem.* **B36** (1937) 371, **B38** (1938) 292; and R. Becker and W. Döring *Annalen der Physik* **24** (1935) 719
81. M. Doi, T. Miyazaki, and T. Wakatsuki. Precipitate morphology formed under the influence of elastic interaction energies in nickel based alloys. *Mat. Sci. and Engr.*, **54** (1982) 9
82. M. Doi, T. Miyazaki, and T. Wakatsuki. The effect of elastic interaction energy on the morphology of γ' precipitates in nickel-based alloys. *Mater. Sci. Eng.*, **67** (1984) 247 – 253
83. R.L. Drake, *A general mathematical survey of the coagulation equation*, Topics in Current Aerosol Research, Part 2, G.M. Hidy, J.R.Brock, Eds., Pergamon Press, Oxford 1972, 201 - 384

84. L.B. Ekbom and A. Eliasson. Liquid phase sintering of tungsten composites under microgravity. effect of matrix composition. *Microgravity Q.*, **2** (1992) 227 – 232
85. Y. Enomoto. Finite volume fraction effects on coarsening - ii. interface-limited growth. *Acta metall. mater.*, **39** (1991) 2013 – 2016
86. Y. Enomoto and K. Kawasaki. Computer simulation of Ostwald ripening with elastic field interactions. *Acta metall.*, **37** (1989) 1399 – 1406
87. Y. Enomoto, K. Kawasaki, and M. Tokuyama. The time dependent behavior of the Ostwald ripening for the finite volume fraction. *Acta metall.*, **35** (1987) 915 – 922
88. Y. Enomoto, M. Tokuyama, and K. Kawasaki. Finite volume fraction effects on Ostwald ripening. *Acta metall.*, **34** (1986) 2119 – 2128
89. J.D. Eshelby. The determination of the elastic field of an ellipsoidal inclusion, and related problems. *Proc. Roy. Soc.A*, **241** (1957) 376 – 396
90. N. Eustathopoulos. Energetics of solid/liquid interface of metals and alloys. *Intern. Metals Reviews*, **28** (1983) 189 – 210
91. H.E. Exner. Neuere Ergebnisse über die Schrumpfungsmechanismen und die Gefügeausbildung beim Sintern. *Jernkont. Ann.*, **154** (1970) 159 – 171
92. H.E. Exner. Review 159 : Analysis of grain- and particle-size distributions in metallic materials. *Int. Metall. Rev.*, **17** (1972) 25 – 42
93. H.E. Exner. Ostwald-reifung von Übergangsmetallkarbiden in flüssigem Nickel und Kobalt. *Z. Metallkde.*, **64** (1973) 273 – 280
94. E. Exner and H.F. Fischmeister. Theorien des Sinterns, Teil III : Das Sintern mit flüssiger phase. *Metall*, **19** (1965) 941 – 947
95. H.E. Exner and H.L. Lukas. The experimental verification of the stationary Wagner-Lifshitz distribution of coarse particles. *Metallography*, **4** (1971) 325 – 338
96. E. Exner and H.F. Fischmeister. Gefügeausbildung von gesinterten Wolframkarbid-Kobalt Hartlegierungen. *Archiv für das Eisenhüttenwesen*, **37** (1966) 417 – 426
97. H.E.Exner, E.Santa-Marta, G.Petzow, Modern Developments in Powder Metallurgy: Processes, p.315, **4**, Plenum Press, New York, 1971
98. Z. Fang, B.R. Patterson, and M.E. Turner Jr. Growth path envelope analysis of ostwald ripening. *Met. Trans. A*, **22** (1991) 19 – 23
99. Z. Fang, B.R. Patterson, and M.E. Turner. Influence of particle size distribution on coarsening. *Acta metall. mater.*, **40** (1992) 713 – 722
100. P.F.P. Fichtner, H. Schröder, and H. Trinkaus. A simulation study of Ostwald ripening of gas bubbles in metals accouting for real gas behavior. *Acta metall. mater.*, **39** (1991) 1845 – 1852
101. H.F. Fischmeister and E. Exner. Theorien des Sinterns, Teil I : Das Sintern im festen Zustand (Anfangsstadien). *Metall*, **18** (1964) 932 – 940
102. H.F. Fischmeister and E. Exner. Theorien des Sinterns, Teil II : Das Sintern im festen Zustand (Spätstadien). *Metall*, **19** (1965) 113 – 119
103. H. Fischmeister and G. Grimvall. Ostwald ripening - a survey. In G.C. Kuczynski, editor, *Sintering and related Phenomena*, pages 119 – 149, New York, 1973. Plenum.
104. I.J. Frenkel, *Kinetic Theory of Liquids*, Oxford University Press, Oxford 1946
105. A. R. Forouhi and D. de Fontaine. *Acta Met.*, **35** 1863, 1987.
106. S.K. Friedlander and C.S. Wang. The self-preserving particle size distribution for coagulation by Brownian motion. *J. Colloid. Interface Sci.*, **22** (1966) 126 – 132

107. S.K. Friedlander and C.S. Wang. The self-preserving particle size distribution for coagulation by Brownian motion ii. small particle slip correction and simultaneous shear flow. *J. Colloid. Interface Sci.*, **24** (1967) 170 – 179
108. R.L. Fullman. Measurement of particle sizes in opaque bodies. *Transactions AIME J. Metals*, **197** (1953) 447 – 452
109. R.M.German, Liquid Phase Sintering, Plenum Press, New York, 1985
110. G.H. Gessinger, H.F. Fischmeister, and H.L. Lukas. A model for second-stage liquid-phase sintering with a partially wetting liquid. *Acta metall.*, **21** (1973) 715 – 724
111. G. Gessinger, F.V. Lenel, and G.S. Ansell. Rate of rotation of sperical particles under the influence of grain boundary torque. *Scripta Metall.*, **2** (1968) 547 – 550
112. J.W. Gibbs. *The scientific Papers of J. Willard Gibbs Vol 1.* Dover Publications, 1961.
113. U. Glatzel and M. Feller-Kniepmeier. Particle splitting. *Scripta Met.*, **23** (1989) 1839
114. M.E. Glicksman and S.P. Marsh. Microstructural coarsening in 2- and 3-dimensions - Applications of multiparticle diffusion algorithms. In D.J. Srolovitz, editor, *Computer Simulation of Microstructural Evolution*, pages 109 – 124. The Metallurgical Society, Warrendale, Pennsylvania, 1986.
115. M.E. Glicksman and P.W. Voorhees. Ostwald ripening and relaxation in dendritic structures. *Met. Trans. A*, **15** (1984) 995 – 1001
116. A.M. Gokhale, M. Basavaiah, and G.S. Upadhyaya. Kinetics of neck growth during loose stack sintering. *Met. Trans. A*, **19** (1988) 2153 – 2161
117. W.I. Goldburg. Phase separation in confined geometries. In L. Ratke, editor, *Immiscible Liquid Metals and Organics*, pages 37 – 46. DGM Informationsgesellschaft, Oberursel, 1993.
118. S. Goodman and R.W. Smith. Ripening of immiscible liquid droplets in micrography with constrained marangoni drift. In *Proc. VIII European Symp. on Materials and Fluid Sciences in Microgravity*, pages 635 – 640, Brussels, 1992.
119. G.W. Greenwood. The growth of dispersed precipitates in solutions. *Acta metall.*, **4** (1956) 243 – 248
120. J. Gurland. The measurement of distribution, spacing, contact and continuity of particles in a matrix. In F. Benesovsky, editor, *Proceedings of the 4. Plansee Seminar*, pages 507 – 518, Springer Verlag, Wien, 1962.
121. J. Gurland. Spatial distribution of discret particles. In F.N. Rhines R.T. DeHoff, editor, *Quantitative Microscopy*, pages 278 – 290. McGraw-Hill, New York, 1968.
122. M. E. Gurtin and P W Voorhees. The continuum mechanics of coherent two-phase elastic solids with mass transport. *Proc. Roy. Soc. A*, **440** (1993) 323
123. S.C. Hardy and P.W. Voorhees. Ostwald ripening in a system with a high volume fraction of coarsening phase. *Met. Trans. A*, **19** (1988) 2713 – 2721
124. S.C. Hardy, G.B. McFadden, S.R. Coriell, P.W. Voorhees, and R.F. Sekerka. Measurement and analysis of grain boundary grooving by volume diffusion. *J. Crystal Growth*, **114** (1991) 467 – 480
125. R. B. Heady and J.W. Cahn. An analysis of the capillary forces in liquid-phase sintering of sperical particles. *Met. Trans.*, **1** (1970) 185 – 189
126. R.W. Heckel. The growth and shrinkage rates of second-phase particles of various size distributions I. Mathematical growth models. *Trans. Metall. Soc. AIME*, **233** (1965) 1994 – 2000
127. R.W. Heckel and R.L. DeGregorio. The growth and shrinkage rates of second-phase particles of various size distributions II. Spheroidization of a eutectoid steel. *Trans. Metall. Soc. AIME*, **233** (1965) 2001 – 2011

128. C. Herring. Surface tension as a motivation for sintering. In W.E. Kingston, editor, *Physics of Powder Metallurgy.* McGraw-Hill, New York, 1951.
129. M. Hillert, O. Hunderi, N. Ryum, and T.O. Sätre. A comment on the Lifshitz-Slyozov-Wagner (LSW) theory of particle coarsening. *Scripta metall. et materialia*, **23** (1989) 1979 – 1982
130. M. Hillert, O. Hunderi, and N. Ryum. Instability of distribution function in particle coarsening. *Scripta metall. et materialia*, **26** (1992) 1933 – 1938
131. M. Hillert, O. Hunderi, and N. Ryum. Rebuttal to comment by L.C. Brown. *Scripta metall. et materialia*, **26** (1992) 1943 – 1944
132. J.E. Hilliard. Direct determination of the moments of the size distribution of particles in an opaque sample. *Transaction of the Metallurgical Soc. of AIME*, **242** 1373 – 1380, 1968.
133. T. Hirata and D.H. Kirkwood. The prediction and measurement of precipitate number densities in a Nickel 6 .05 wt% Aluminium alloy. *Acta metall.*, **25** (1977) 1425 – 1434
134. A.T. Hjelmfelt, J.F.C. Carney, L.F. Mockros, and S.-L. Lee. Dynamic response of a restrained sphere in a fluid. *J. Eng. Mech. Div., ASCE*, **93** (1967) 41 – 56
135. E. Hornbogen and M. Roth. Die Verteilung kohärenter Teilchen in Nickellegierungen. *Z. Metallkde.*, **58** (1967) 842 – 855
136. J.J. Hoyt. On the steady-state particle size distribution during coarsening. *Scripta metall. et materialia*, **24** (1990) 163 – 166
137. J.J. Hoyt. On the coarsening of precipitates located on grain boundaries and dislocations. *Acta metall. mater.*, **39** (1991) 2091 – 2098
138. J.J. Hoyt and D. de Fontaine. *Acta metall. mater.*, **37** 1611, 1989.
139. O. Hunderi, N. Ryum, and H. Westengen. Computer simulation of grain growth. *Acta metall.*, **27** (1979) 161 – 165
140. W.J. Huppmann. Sintering in the presence of a liquid phase. In G.C. Kuczynski, editor, *Sintering and Catalysis*, pages 359 – 378. Plenum, New York, 1975.
141. W.J. Huppmann and G. Petzow. *The elementary mechanism of liquid phase sintering.* Sintering process. Plenum, New York, 1985.
142. W.J. Huppmann and H. Riegger. Modelling of rearrangement processes in liquid phase sintering. *Acta metall.*, **23** (1975) 965 – 971
143. T. Imaeda and K. Kawasaki. *Physica A*, **186** (1992) 359
144. A.M. Irisarri, J.J. Urcola, and M. Füntes. Kinetics of growth of γ'-precipitates in Ni-6 .75 Al alloy. *Mat. Sci. and Technology*, **1** (1985) 516 – 519
145. C.V. Iswaran and K. Keppler. The study of the kinetics of coarsening - an alternative viewpoint. *Met. Trans. A*, **11** (1980) 1677 – 1681
146. C.V. Iswaran and K. Keppler. A critical appraisal of state of the art coarsening theories. *Res Mechanica*, **12** (1984) 241 – 258
147. J.D. Jackson, *Classical Electrodynamics*, John Wiley & Sons., New York, 1975
148. S.C. Jain and A.E. Hughes. Review - Ostwald ripening and its application to precipitates and colloids in ionic crystals and glasses. *J. Mat. Sci.*, **13** (1978) 1611 – 1631
149. M.A. Jaswon and G.T. Symm. *Integral Equation Methods in Potential Theory and Elastostatics.* Academic Press, London, 1977.
150. C.S. Jayanth and P. Nash. Review - factors affecting particle-coarsening kinetics and size distribution. *J. Mat. Sci.*, **24** (1989) 3041 – 3052
151. John, F., *Partial Differential Equations*, 4th ed., Springer Verlag, New York 1982
152. W.C. Johnson, T.A. Abinandanan, and P.W. Voorhees. The coarsening kinetics of two misfitting particles in an anisotropic crystal. *Acta metall. mater.*, **38** (1990) 1349 – 1367 two particle anisotropic.

153. W.C. Johnson and J.K. Lee. Elastic interaction energy of two sperical precipitates in an anisotropic matrix. *Met. Trans. A*, **10** (1979) 1141 – 1149
154. W.C. Johnson and P.W. Voorhees. Elastic interaction and stability of misfitting cuboidal inhomogeneities. *J.Appl. Phys.*, **61** (1987) 1610 – 1619
155. W C Johnson and J I D Alexander. Coherent interface thermodynamics. *J. Appl. Phys.*, **59** (1986) 2735
156. W C Johnson, M B Berkenpas, and D E Laughlin. Particle shape bifurcations. *Acta Metall.*, **36** (1988) 3149
157. W.C. Johnson, P.W. Voorhees, and D.E. Zupon. The effects of elastic stress on the kinetics of Ostwald ripening: The two-particle problem. *Met. Trans. A*, **20** (1989) 1175 – 1189
158. W.C. Johnson and P.W. Voorhees. On the classification of phase transitions involving changes in composition. *Acta metall.*, **34** (1990) 1183 – 1189
159. W C Johnson and P W Voorhees. Particle shape bifurcations review. *Solid State Phenom.*, **23** (1992) 87
160. W C Johnson and J W Cahn. Particle shape bifurcations. *Acta Metal.*, **23** (1984) 1839
161. W C Johnson. Inverse Ostwald ripening of two particles. *Acta Metall.*, **32** (1984) 465
162. M. Kahlweit, F. Lichterfeld, R. Strey, and J. Jen. Ostwald Ripening of Precipitates. In P.Haasen, V.Gerold, R.Wagner, and M.F. Ashby, editors, *Decomposition of Alloys: the early stages - Proc. of the 2nd Acta-Scripta Metallurgica Conference*, pages 61 – 69, Oxford, 1984. Pergamon Press.
163. R. Kampmann and R. Wagner. Kinetics of precipitation in metastable binary alloys - theory and application to Cu -1.9at% Ti and Ni -14at% Al. In P.Haasen, V.Gerold, R.Wagner, and M.F. Ashby, editors, *Decomposition of Alloys: the early stages - Proc. of the 2nd Acta-Scripta Metallurgica Conference*, pages 91 – 103, Oxford, 1984. Pergamon Press.
164. R. Kampmann, H. Eckerlebe, R. Wagner, Mat. Res. Soc. Proc. **57**, 525 (1987)
165. C.H. Kang and D.N. Yoon. Coarsening of Cobalt grains dispersed in liquid Copper matrix. *Met. Trans. A*, **12** (1981) 65 – 69
166. S.-J.L. Kang, W.A. Kaysser, G. Petzow, and D.N. Yoon. Growth of Mo grains around Al_2O_3 particles during liquid phase sintering. *Acta metall.*, **33** (1985) 1919 – 1926
167. S.S. Kang and D.N. Yoon. Kinetics of grain coarsening during sintering of Co-Cu and Fe-Cu alloys with low liquid contents. *Met. Trans. A*, **13** (1982) 1405 – 1411
168. T.-K. Kang and D.N. Yoon. Coarsening of Tungsten grains in liquid Nickel-Tungsten matrix. *Met. Trans. A*, **9** (1978) 433 – 438
169. M.J. Kaufman, P.W. Voorhees, W.C. Johnson, and F.S. Biancaniello. An elastically induced morphological instability of a misfitting precipitate. *Met. Trans. A*, **20** (1989) 2171 – 2175
170. K. Kawasaki. Aspects of late-stage dynamics of ordering processes. *Physica A*, **163** (1990) 59 – 70
171. K. Kawasaki, Y. Enomoto, and M. Tokuyama. Elementary derivation of kinetic equations for Ostwald ripening. *Physica A*, **135** (1986) 426 – 445
172. W.A. Kaysser, S. Takajo, and G. Petzow. Particle growth by coalescence during liquid phase sintering of Fe-Cu. *Acta metall.*, **32** (1984) 115 – 112
173. A.G. Khachaturyan. *Theory of Structural Phase Transformations in Solids.* John Wiley, New York, 1983.
174. A.G. Khachaturyan, S.V. Semenovskaya, and J.W. Morris. Particle splitting via energy calculations. *Acta Metal.*, **36** (1988) 1563

175. Y. Wang, Long Q. Chen, and A.G. Khachaturyan. Kinetics of strain-induced morphological transformation in cubic alloys with a miscibility gap. *Acta Metal. et Mater.*, **41** (1993) 279–296
176. A.G. Khachaturyan, S.V. Semenovskaya, and J.W. Morris Jr. Theoretical analysis of strain-induced shape changes in cubic precipitates during coarsening. *Acta metall.*, **36** (1988) 1563 – 1572
177. Sung Soo Kim and Duk N. Yoon. Coarsening of Mo grains in the molten Ni-Fe matrix of a low volume fraction. *Acta. metall.*, **33** (1985) 281 – 286
178. A. Kneissl, P. Pfefferkorn, and H. Fischmeister. Ostwald ripening and solidification behavior of near-monotectic Zn-Pb alloys. In *Proceedings of the 4th European Symposium on Materials Sciences under Microgravity*, pages 55 – 61, Madrid. Spain., 1983. ESA SP-191.
179. S. Komura, K. Osamura, H. Fujii, and T. Takeda. *Phys. Rev. B*, **31** (1985) 1278
180. O. Krichevsky and J. Stavans. Correlated Ostwald ripening in two dimensions. *Phys. Rev. Letters*, **70** (1993) 1473 – 1476
181. J.H. Kulwicki and T.H. Sanders Jr. Coarsening of δ'-Al_3Li precipitates in an Al-2 .7 Li-0 .3 Mn alloy. In T.H. Sanders Jr. E.A.. Stark Jr., editor, *Aluminium-Lithium Alloys II - Proc. Sec. Int.Al-Li Conf.*, pages 31 – 51, Warrendale, PA, 1983. AIME.
182. L.D.Landau und E.M.Lifschitz, *Fluid mechanics*, 2nd edition, Pergamon Press, Oxford 1987
183. L.D.Landau und E.M.Lifschitz, *Physikalische Kinetik*, Akademie Verlag, Berlin 1990
184. Langer, J.S. and Schwartz, A.J. *Phys.Rev.A* **21** (1980) 269
185. Langer, J.S., in: *Systems Far from Equilibrium*, Lecture Notes in Physics, **132**, Ed.L.Garido, Springer Verlag, Berlin 1980
186. V.J. Larala, W.C. Johnson, and P.W. Voorhees. The kinetics of Ostwald ripening in stressed solids: The low volume fraction limit. *Scripta metall. et materialia*, **23** (1989) 1749 – 1754
187. Jong K. Lee. Computer simulation of the effect of coherency strain on cluster growth kinetics. *Metall. Trans.*, **2 (1991)2** 1197–1209
188. H.M. Lee, S.M. Allen, and M. Grujicic. Coarsening resistance of MKC carbides in secondary hardening steels: Part 1. theoretical model for multicomponent coarsening kinetics. *Met. Trans. A*, **22** (1991) 2863 – 2868
189. H.M. Lee, S.M. Allen, and M. Grujicic. Coarsening resistance of MKC carbides in secondary hardening steels: Part 2. alloy design aided by a thermochemical database. *Met. Trans. A*, **22** (1991) 2869 – 2876
190. H.M. Lee and S.M. Allen. Coarsening resistance of MKC carbides in secondary hardening steels: Part 3. comparison of theory and experiment. *Met. Trans. A*, **22** (1991) 2877 – 2888
191. P.H. Leo, W.W. Mullins, R.F. Sekerka, and J. Viñals. Effect of elasticity on late stage coarsening. *Acta metall. mater.*, **38** (1990) 1573 – 1580
192. Perry H Leo and Herng-Jeng Jou. Shape evolution of an initially circular precipitate growing by diffusion in an applied stress field. *Acta Metall.*, in press.
193. V.G.Levich, *Physicochemical Hydrodynamics*, Prentice-Hall, Englewood Cliffs, N.J. 1962
194. I.M. Lifshitz and V.V. Slyozov. The kinetics of precipitation from supersaturated solid solutions. *J. Phys. Chem. Solids*, **19** (1961) 35 – 50
195. J. Lothe and G.M. Pound, *J.Chem.Phys.* **45** (1966) 630
196. H.Mandyman, M.E.Glicksman. Late-stage coarsening dynamics in phase separating systems: simulation results. *Mat.Sci.Eng. A* **238** (1997) 121 – 127

197. M. Marder. Correlations and droplet growth. *Phys. Rev. Lett.*, **55** (1985) 2953 – 2956
198. M. Marder. Correlations and Ostwald ripening. *Phys. Rev. A.*, **36** (1987) 858 – 874
199. J.A. Marqusee and J. Ross. Kinetics of phase transitions: Theory of Ostwald ripening. *J. Chem. Phys.*, **79** (1983) 373 – 378
200. J.A. Marqusee and J. Ross. Theory of Ostwald ripening: Competitive growth and its dependence on volume fraction. *J. Chem. Phys.*, **80** (1984) 536 – 543
201. J.A.Marqusee. Dynamics of late stage phase separations in two dimensions. *J.Chem.Phys.* **81** (1984) 976 – 981
202. S.P. Marsh. Kinetics of Diffusion-Limited Microstructural Coarsening. PhD dissertation, Rensselear Polytechnic Institute Troy, New York, Materials Engineering, 1989.
203. S.P.Marsh, M.E.Glicksman. Ostwald ripening in non-spherical morphologies. *Met.Trans. A* **27** (1996) 557 – 567 .
204. S.P.Marsh, M.E.Glicksman. Overview of Geometric Effects on Coarsening of Mushy Zones. *Mat.Sci. Eng. A* **238** (1997) 140 – 147
205. S. P. Marsh and M. E. Glicksman. In S. P. Marsh and C. S. Pande, editors, *Modeling of Coarsening and Grain Growth*, TMS, Warrendale, 1992.
206. S.P.Marsh, M.E.Glicksman. Kinetics of phase coarsening in dense systems. *Acta mater.* **44** (1996) 3761 – 3771
207. J.W.Martin, *Micromechanics in particle hardened alloys*, Cambridge University Press, Cambridge 1980
208. J.W. Martin and R.D. Doherty. Stability of microstructure in metallic systems. Cambridge University Press, Cambridge, UK 1980.
209. M. McCormack, A G Khachaturyan, and J W Morris. *Acta Metall. et. Mater.*, **40** (1992) 325
210. G.B. McFadden, P.W. Voorhees, R.F. Boisvert, and D.I. Meiron. A boundary integral method for the simulation of two-dimensional particle coarsening. *J. of Scientific. Computation*, **1** (1986) 117 –144
211. B. Mutaftschiev, *Nucleation Theory*, in: Handbook of Crystal Growth, Vol.Ia, D.T.J.Hurle, editor, Elsevier, Amsterdam 1993
212. B. Giron, B. Meerson, P. V. Sasorov, Weak selection and stability of localized distributions in Ostwald ripening, Phys.Rev. B **58** (1998) 4213 - 4216
213. B.Meerson, Fluctuations provide strong selection in Ostwald ripening, Phys. Rev. B **60** (1999) 3072 – 3075
214. T. Miyazaki and M. Doi. Shape bifurcation in the coarsening of precipitates in elastically constrained systems. *Mat. Sci. Eng. A*, **110** (1989) 175 – 185
215. J.E. Morral and G.R. Purdy. Particle coarsening in binary and multicomponent alloys. *Scripta Metall. et Mater.*, **30** (1994) 905 – 908
216. P.M.Morse, H.Feshbach, *Methods of Theoretical Physics*, PartI and II, McGraw-Hill, New York,1953
217. H. Müller. Zur allgemeinen Theorie der raschen Koagulation. *Kolloidchemische Beihefte*, Band XXVIII (1928) 223 – 250
218. W.W. Mullins. The statistical self-similarity hypothesis in grain growth and particle coarsening. *J.Appl. Phys.*, **59** (1986) 1341 – 1349
219. W.W. Mullins. The statistical particle growth law in self-similar coarsening. *Acta metall. mater.*, **39** (1991) 2081 – 2090
220. W.W. Mullins. Idealized two dimensional sintering by interface diffusion. *Scripta metall. et mater.*, **29** (1993) 491 – 496
221. W.W. Mullins and J. Viñals. Scaling in linear bubble models of grain growth. *Acta metall. mater.*, **41** (1993) 1359 – 1367

222. P. Nash. Comments on "Coarsening models incorporating both diffusion geometry and volume fraction of particles" (from Tsumuraya/Miyata). *Scripta Metall.*, **18** (1984) 295 – 296
223. A. Nakahara, T. Kawakatsu, and K. Kawasaki. Ostwald ripening in open systems. *J. Chem. Phys.*, **95** (1991) 4407 – 4414
224. A.N. Niemi and T.H. Courtney. Microstructural developement and evolution in liquid-solid sintered Fe-Cu alloys. *J. Mat. Sci.*, **16** (1981) 226 – 236
225. B. Niethammer, R.L. Pego, Non-Self-Similar Behavior in the LSW Theory of Ostwald Ripening, J.Stat.Physics **95** (1999) 867 – 902
226. R.A. Oriani. Ostwald ripening of precipitates in solid matrices. *Acta metall.*, **12** (1964) 1399 – 1409
227. W. Ostwald. Über die vermeintliche Isomerie des roten und gelben Quecksilberoxyds und die Oberflächenspannung fester Körper. *Z.Phys. Chem.*, **34** (1900) 495 – 503
228. C.S. Pande. On a stochastic theory of grain growth. *Acta metall.*, **35** (1987) 2671 – 2678
229. G. Petzow and W.A. Kaysser. Basic mechanisms of liquid phase sintering. In G.S. Upadhyaya, editor, *Sintered Metal-Ceramic Composites*, pages 51 – 70, Amsterdam, 1984. Elsevier.
230. W. Pflügl and U.M. Titulär. The size distribution of liquid droplets during their growth from a vapor. *Physica A*, **198** (1993) 410 – 422
231. A. Pineau. Precipitate evolution in an applied stress field. *Metall. Trans.*, **24** (1976) 559
232. J.K.Platten, J.C. Legros, *Convection in Liquids*, Springer Verlag, Heidelberg 1984
233. D.R.Poirier, S.Ganesan, M.Andrews, P.Ocansey. Isothermal coarsening of dendritic equiaxial grains in Al-15.6wt%Cu alloy. *Mat.Sci.Eng. A* **148** (1991) 289 – 297
234. G.R. Purdy. Widmanstätten precipitation from non-ideal solid solutions: α in β CuZn. *Metal Sci.Journal*, **5** (1987) 81 – 85
235. V. Randle and B. Ralph. *Coarsening behavior of coherent precipitates at grain boundaries*, pages 498 – 500. The Inst. of Metals, Cambridge, 1987.
236. L.Ratke, H.Fischmeister, A.Kneissl, *Coarsening of Liquid Zn-Pb Dispersions - A Spacelab Experiment*, Proc. 6th European Symposium on Material Sciences under Microgravity Conditions, Bordeaux, 1986, Ed. ESA -Paris, 1987, ESA SP 256, p. 161 – 167
237. L.Ratke, H.Fischmeister, A.Kneissl, *Coarsening of liquid Zn-Pb dispersions - final evaluation of a Spacelab 1 experiment*, Proc. VII European Symp. Mat.Sci. Microgravity, Oxford, UK, 1989, ESA SP-295, 1990, pp.135 – 140
238. L.Ratke (Ed.), *Immiscible Liquid Metals and Organics*, DGM-Informationsgesellschaft, Oberursel 1993
239. L. Ratke. Simultaneous coarsening of dispersions by growth and coagulation. *J. Colloid and Interface Sci.*, **119** (1987) 391 – 397
240. L. Ratke. Ostwald ripening in liquids. In J.N. Koster and R.L. Sani, editors, *Low-Gravity Fluid Dynamics and Transport Phenomena*, Vol.130 of Progress in Astronautics and Aeronautics. AIAA, Washington, DC, 1990.
241. L. Ratke and M. Host. Convective contributions to Ostwald ripening in dispersions at low Peclet numbers. *J. Colloid and Interface Sci.*, **141** (1991) 226 – 237
242. L. Ratke and W.K. Thieringer. The influence of particle motion on Ostwald ripening in liquids. *Acta metall.*, **33** (1985) 1793 – 1802

243. L. Ratke, W.K. Thieringer, and H. Fischmeister. Coarsening of immiscible liquid alloys by Ostwald ripening. In M.H. Keller P.R. Sahm, R. Jansen, editor, *Proc. Norderney Symposium on Scientific Results of the German Spacelab Mission D1*, pages 332 – 341, Bonn, 1985. German Ministry of Research and Technology (BMFT).
244. L.Ratke, D.Uffelmann, W.Bender, P.Voorhees. Theory of Ostwald ripening due to a second order reaction. *Scripta Metall. et Materialia* , **33** (1995) 363 – 367
245. L.Ratke. Coarsening of liquid Al-Pb dispersions under reduced gravity conditions. *Mat.Sci.Eng A*, **203** (1995) 399 – 407
246. Rogers, R.R. and Yau M.K. *A Short Course in Cloud Physics*, Pergamon Press, Oxford 1989
247. F. Rosenberger, *Fundamentals of Crystal Growth I*, Springer Verlag, Berlin 1979
248. I. Rubinstein, B. Zaltzmann, Diffusional mechanism of strong selection in Ostwald ripening, Phys.Rev. B **61** (2000) 709 – 717
249. S. Sarian and H.W. Weart. Kinetics of coarsening of spherical particles in a liquid matrix. *J.Appl. Phys.*, **37** (1966) 1675 – 1681
250. G. Sauthoff and M. Kahlweit. Precipitation in Ni-Si alloys. *Acta metall.*, **17** (1969) 1501 – 1509
251. R.J. Schäfer, M.E. Glicksman, and J.D. Ayers. High-confidence measurement of solid/liquid surface energy in pure materials. *Phil. Mag.*, **27** (1975) 725 – 743
252. G.Scharf, J.Eulitz, Metall **29** (1975) 358 – 365
253. W. Schatt, Sintervorgänge, VDI Verlag, Düsseldorf 1992
254. I.Seyhan, L.Ratke, W.Bender, P.Voorhees. Ostwald ripening of solid-liquid Pb-Sn dispersions. *Metall. and Materials Trans. A*, **27** (1996) 2470 – 2478
255. A. Shinozaki and Y. Oono. *Phys. Rev. Lett.*, **66** (1991) 173
256. V.V. Slyozov and V.V. Sagalovich. Theory of diffusive decomposition of supersaturated multicomponent systems. *J. Phys. Chem. Solids*, **38** (1977) 943 – 948
257. A.F. Smith. The isothermal growth of Manganese precipitates in a binary Magnesium alloy. *Acta metall.*, **15** (1967) 1867 – 1873
258. M. Smoluchowski. Drei Vorträge - Zusammenfassende Bearbeitung. *Physik. Zeitschrift*, **17** (1916) 557 – 571 and 585 – 599
259. A. Sommerfeld, *Mechanik der deformierbaren Medien*, Akademische Verlagsgesellschaft, Leipzig 1970
260. W.J. Soppe, G.J.M. Janssen, B.C. Bonekamp, L.A. Correia, and H.J. Veringa. A computer simulation method for sintering in three-dimensional powder compacts. *J. Mat. Sci.*, **29** (1994) 754 – 761
261. G.R. Speich and R.A. Oriani. The rate of coarsening of copper precipitate in an alpha-iron matrix. *Transactions of the Metallurgical Society of AIME*, **233** (1965) 623 – 631
262. M.V. Speight. Growth kinetics of grain-boundary precipitates. *Acta metall.*, **16** (1968) 133 – 135
263. J.P. Stark. Solid state second phase redistribution in an applied field. *Acta Metall.*, **26** (1978) 369 – 378
264. J.P. Stark. Precipitate migration in an applied field by volume diffusion. *Acta Metall.*, **26** (1978) 1139 – 1137
265. J.H. Steele Jr. A nonoverlap model for the dispersion of spherical particles. *Met. Trans. A*, **7** (1976) 1325 – 1332
266. P. Streitenberger and D. Förster. Theory of Ostwald ripening for precipitates with surface fractal dimension. *Mat. Sci. Forum*, **62-64** (1990) 711 – 714

267. S. Takajo, W.A. Kaysser, and G. Petzow. Analysis of particle growth by coalescence during liquid phase sintering. *Acta metall.*, **3 (1984)2** 107–113
268. W.K. Thieringer and L. Ratke. The coarsening of liquid Al-Pb-dispersions. *Acta metall.*, **35** (1987) 1237 - 1244
269. Tiller W..A., The Science of Crystallization - Microscopic Interfacial Phenomena, Cambridge University Press, Cambridge 1991
270. M. Tokuyama and K. Kawasaki. Statistical-mechanical theory of coarsening of sperical droplets. *Physica A*, **123** (1984) 386 - 411
271. M. Tokuyama, K. Kawasaki, and Y. Enomoto. Kinetic equations for Ostwald ripening. *Physica A*, **134** (1986) 323 - 338
272. M. Tokuyama, Y. Enomoto, and K. Kawasaki. *Physica A*, **143** (1987) 183
273. M. Tokuyama and Y. Enomoto. *Phys. Rev. E*, **47** (1993) 1473
274. K. Trinckauf, J. Pesicka, C. Schlesier, and E. Nembach. The effect of the volume fraction on precipitate coarsening in Nickel-base superalloys and Aluminium-Lithium alloys. *Phys. Stat. Sol.(a)*, **131** (1992) 345 - 355
275. R. Triolo, E. Caponetti, and S. Spooner. *Phys. Rev. B*, **39** (1989) 4588
276. K. Tsumuraya and Y. Miyata. Coarsening models incorporating both diffusion geometry and volume fraction of particles. *Acta metall.*, **31** (1983) 437 - 452
277. K. Tsumuraya. Reply to comments on "Coarsening models incorporating both diffusion geometry and volume fraction of particles" (comments by P.Nash). *Scripta metall.*, **18** (1984) 297 - 299
278. D. Turnball, *Metallurgical Trans. AIME* **175** (1949) 774; *J.Appl.Phys.* **21** (1950) 1022
279. A. Umantsev and G.B. Olson. Ostwald ripening in multicomponent alloys. *Scripta metall. et materialia*, **29** (1993) 1135 - 1140
280. D.Uffelmann, W.Bender, L.Ratke, B.Feuerbacher. Ostwald ripening in Lorentz-Force stabilized Cu-Pb Dispersions at Low Volume Fractions, I - Experimental Observations. *Acta Metall. et Mater.*, **43** (1995) 173 - 180
281. R.D. Vengrenovitch. On the Ostwald ripening theory. *Acta metall.*, **30** (1982) 1079 - 1086
282. G. Venzl. Dynamics of first-order phase transition: Theory of coarsening (Ostwald ripening) for open systems. *Phys. Rev. A*, **31** (1985) 3431 - 3440
283. M. Vollmer, *Kinetik der Phasenbildung*, Steinkopf Verlag, Dresden 1939
284. B. Vonnegut, *J.Colloid Sci.* **3** (1948) 563
285. P.W. Voorhees. The theory of Ostwald ripening. *J. Statist. Phys.*, **38** (1985) 231 - 252
286. P.W. Voorhees and M.E. Glicksman. Solution to the multiparticle diffusion problem with applications to ostwald ripening - I. Theory. *Acta metall.*, **32** (1984) 2001 - 2012
287. P.W. Voorhees and M.E. Glicksman. Solution to the multiparticle diffusion problem with applications to ostwald ripening - II. Computer simulations. *Acta metall.*, **32** (1984) 2013 - 2030
288. P.W. Voorhees and M.E. Glicksman. Ostwald ripening during liquid phase sintering: Effect of volume fraction on coarsening kinetics. *Met. Trans. A*, **15** (1984) 1081 - 1088
289. P.W. Voorhees, G.B. McFadden, R.F. Boisvert, and D.I. Meiron. Numerical simulation of morphological development during ostwald ripening. *Acta metall.*, **36** (1988) 207 - 222
290. P.W. Voorhees, G.B. McFadden, and W.C. Johnson. On the morphological development of second-phase particles in elastically-stressed solids. *Acta metall. mater.*, **40** (1992) 2979 - 2992
291. P.W. Voorhees and R.J. Schaefer. In situ observation of particle motion and diffusion interactions during coarsening. *Acta metall.*, **35** (1987) 327 - 339

292. P W Voorhees and W C Johnson. Interfacial equilibrium growth. *J. Chem. Phys.*, **84** (1986) 5108
293. P.W.Voorhees. Coarsening of Binary Solid-Liquid Mixtures. *Met. Trans. A*, **21** (1990) 27 – 37
294. B.J. Spencer, P.W. Voorhees, S.H. Davis and G.B. McFadden. The Effect of Compositionally Generated Elastic Stresses on Morphological Instability During Dirctional Solidification. *Acta Metal. et Mater.*, **40** (1992) 1599-1616
295. C. Wagner. Theorie der Alterung von Niederschlägen durch Umlösen (Ostwald-reifung). *Z.Elektrochemie*, **65** (1961) 581 – 591
296. Gang Wan and P.R. Sahm. Ostwald ripening in the isothermal rheocasting process. *Acta. metall. mater.*, **38** (1990) 967 – 972
297. Gang Wan and P.R. Sahm. Particle growth by coalescence and Ostwald ripening in rheocasting of Pb-Sn. *Acta. metall. mater.*, **38** (1990) 2367 – 2372
298. J.A. Warren, B.T. Murray, *Ostwald ripening and coalescence of a binary alloy in two dimensions using a phase-field approach*, Modelling Simul-Mater.Sci.Eng. **4** (1996) 215 – 229
299. R. Warren. Microstructural development during the liquid-phase sintering of two-phase alloys, with special reference to the NbC/Co system. *J.Mat. Sci.*, **3** (1968) 471 – 485
300. R. Warren. Microstructural development during the liquid-phase sintering of VC-Co alloys. *J.Mat. Sci.*, **7:** (1972) 434 – 442
301. R. Warren and M.B. Waldron. Microstructural development during the liquid-phase sintering of cemented carbides. *Powder Metallurgy*, **15** (1972) 166 – 201
302. R. Watanabe and Y. Masuda. The growth of solid particles in Fe-20wt% Cu alloy during sintering in the presence of a liquid phase. *Trans. JIM*, **14** (1973) 320 – 326
303. Y. Masuda R. Watanabe. The growth of solid particles in some two-phase alloys during sintering in the presence of a liquid phase. In G.C. Kuczynski, editor, *Sintering and Catalysis.* Plenum, New York, 1975.
304. H. Wendt, P. Haasen, *Acta. Met.* **31** (1983) 1649
305. Weibel, E.R. *Stereological Methods Vol.2*, Academic Press, London 1980
306. J.J. Weins and J.W. Chan. The effect of the size and distribution of second phase particles and voids on sintering. In G.C. Kuczynski, editor, *Sintering and related Phenomena*, pages 151 – 163, New York, 1973. Plenum.
307. H. Wendt and P. Haasen. Nucleation and growth of γ'-precipitates in Ni-14at.% Al. *Acta metall.*, **31** (1983) 1649 – 1659
308. J.T. Wetzel, L.D. Roth, and J.K. Tien. Anisotropic growth during the coarsening of Platinum catalyst particles on flat oxide substrates. *Acta metall.*, **32** (1984) 1573 – 1580
309. R.J. White. The particle size distribution in systems evolving from interface-controlled to diffusion-controlled coarsening kinetics. *Mat. Sci. Eng.*, **40** (1979) 15 – 20
310. J. White. Microstructure and grain growth in ceramics in the presence of a liquid phase. In *Proc.3rd Conf. on Sintering and Related Phenomena*, pages 81 – 108, Notre Dame, 1972.
311. R.J. White and S.B. Fisher. The precipitation and growth kinetics of γ' in Nimonic PE16. *Mat. Sci. Eng.*, **33** (1978) 149 – 157
312. D.D. Wicksell. The corpuscle problem. A mathematical study of a biometric problem. *Biometrika*, **17** (1925) 84 – 99
313. S.-C. Yang and P. Nash. Experimental observation of two coalescence mechanisms in liquid phase sintered Fe-70%Cu. *Scripta metall. et materialia*, **24** (1990) 263 – 266
314. N. Young, J. Goldstein, and M. Block, *J. Fluid Mech.* **6** (1959)350

315. J.H. Yao, K.R. Elder, H. Guo, and M. Grant. Theory and simulation of Ostwald ripening. *Phys. Rev. B*, **47** (1993) 14110 – 14125
316. J.B. Zeldovich, Acta Physicochemica USSR **18** (1943) 1
317. Zener, C. *J.Appl.Phys.* **20** (1949) 950
318. J.Z.Zhao, L.Ratke, B.Feuerbacher, Microstructure Evolution of Immiscible Alloys During Cooling Through The Miscibility Gap, *Modelling and Simulation in Materials Science and Engineering* **6** (1998) 123 - 140
319. J.Zhao, L.Ratke, Kinetics of Phase Separation in a Hypermonotectic Alloy, *Z.Metallkde.* **89** (1998) 241 - 246
320. J.Zhao, L.Ratke, Repeated Nucleation of Minority Phase Droplets Induced by Drop Motion, *Scripta Materialia* **39** (1998) 181 - 188
321. Zwillinger. Coarsening of non-sperical particles. *J. Crystal Growth*, **94** (1989) 159 – 165
322. *Zwillinger, Handbook of Differential Equations*, 3rd edition, Academic Press, San Diego 1998

List of Variables

Symbol	Meaning	Definition	Units
$\boldsymbol{a}$	acceleration	-	m/s^2
a_i	constants	-	-
A	area	-	m^2
A^*	surface area of critical nucleus	-	m^2
A_i	fitting parameters	-	-
$\boldsymbol{B}$	magnetic induction	-	gauss
B_i	fitting parameters	-	-
B_i	source/sink strength in multiparticle diffusion analysis	-	-
c_P	specific heat	$\left(\frac{\partial H_m}{\partial T}\right)_P$	J/K mol
c_V	specific heat	$\left(\frac{\partial E_m}{\partial T}\right)_V$	J/K mol
C_P	specific heat	$\frac{1}{V\rho}\left(\frac{\partial H}{\partial T}\right)_P$	J/K kg
C_V	specific heat	$\frac{1}{V\rho}\left(\frac{\partial E}{\partial T}\right)_V$	J/K kg
C	constant in MR theory	-	-
D	diffusion coefficient	$V_m\left[M x_A x_B \frac{\partial^2 G_m}{\partial x_B^2}\right]$	m^2/s
dV	small Volume element	-	m^3
E	total internal energy	-	J
E	constant in MR theory	-	-
E_m	molar internal energy	-	J/mol
F	Helmholtz free energy	$E - TS$	J
F_{drag}	drag on particle in fluid matrix	-	N
F_g	gravitational buoyant force on particle	$(4\pi/3)(\rho_{part.} - \rho)R^3 g$	N
$\boldsymbol{F}_V$	external force on volume element V	-	N
$f(R,t)$, $f(\rho,\tau)$	particle size density distribution	-	$1/m^4$ -
f_s	fraction of solid	-	-
$g(\tau)$	time dependent function for separation ansatz	-	-
$\boldsymbol{g}$	gravity acceleration vector	-	m/s
g	gravity acceleration	-	m/s
G	Gibbs free energy	$H - TS$	J
G_A	activation energy	-	J
G_{fi}	final Gibbs free energy	-	J
G_i	initial Gibbs free energy	-	J
G_m	Gibbs free energy per mole	G/n	J/mol
G_m^a	Gibbs free energy after mixing per mole	-	J/mol

G^b	Gibbs free energy before mixing	-	J
G^b_m	Gibbs free energy before mixing per mole	G^b/n	J/mol
$G^{liq.}_m$	molar Gibbs free energy of the liquid state	-	J/mol
$G^{sol.}_m$	molar Gibbs free energy of the solid state	-	J/mol
G^0_A	Gibbs free energy per mole of pure component A	-	J/mol
G^0_B	Gibbs free energy per mole of pure component B	-	J/mol
$G(r)$	radial distribution function	-	-
$h(\rho)$	length dependent function for separation ansatz	-	-
$h(v_{p.})$	normalized particle volume distribution in MR theory	-	-
H	enthalpy	$E - PV$	J
H_V	enthalpy per unit volume	-	J/m^3
$H^{liq.}_V$	enthalpy per unit volume in liquid	-	J/m^3
$H^{liq.,e}_V$	enthalpy per unit volume in liquid in eqilibrium	-	J/m^3
$H^{sol.}_V$	enthalpy per unit volume in solid	-	J/m^3
$H^{sol.,e}_V$	enthalpy per unit volume in solid in eqilibrium	-	J/m^3
H_m	molar enthalpy	H/n	J/mol
$H_{liq.}$	molar enthalpy of the liquid state	-	J/mol
$H_{sol.}$	molar enthalpy of the solid state	-	J/mol
i, j, k, l	counters	-	-
I	total mass flux of atoms	-	1/s
I_D	total mass flux due to diffusion	-	1/s
I_{solid}	total mass flux towards solid particle	-	1/s
I_{liquid}	total mass flux towards liquid particle	-	1/s
$I_{\alpha\to\beta}, I_{\beta\to\alpha}$	flux of atoms between α- and β-phase	-	mol / m^2
I_{SS}	steady-state nucleation rate	$I_0 \exp(-\frac{\Delta G^*}{k_B T})$	1/sm^3
I_0	nucleation rate prefactor	-	1/sm^3
$I(t)$	nucleation rate	-	1/sm^3
$\boldsymbol{j}_A$	mass flux density of component A	$-M_{AA}\nabla\mu_A - M_{AB}\nabla\mu_B$	mol / m^2
$\boldsymbol{j}_B$	mass flux density of component B	$-M_{BA}\nabla\mu_A - M_{BB}\nabla\mu_B$	mol / m^2
$\boldsymbol{j}_{conv}$	convective mass flux density	$x_B \boldsymbol{v}$	mol m^2 s
$\boldsymbol{j}_{total}$	total mass flux density	-	mol / m^2
j	flux density	-	mol / m^2
$\boldsymbol{J}$	mass flux	jV_m	m/s
$j(R,t)$	particle size density distribution of nucleated particles	-	1/m^{4}s
k_B	Boltzmann coefficient	1.38066×10^{-23} J/K	J/K
k^*	rate constant	-	m/s J^2

k_s^{2nd}	rate constant for concentration driven growth for 2nd order interface kinetics	-	m/s J^2
K_s^{2nd}	rate constant for concentration driven growth for 2nd order interface kinetics	$k_s^{2nd}(\ell^\alpha)^2/x_B^{\beta,e} - x_B^{\alpha,e}$	m/s J^2
k_T^{2nd}	rate constant for temperature driven growth for 2nd order interface kinetics	-	m/s J^2
K_T^{2nd}	rate constant for temperature driven growth for 2nd order interface kinetics	$2k_T^{2nd}\Gamma^2$	m/s J^2
k_s	collection of materials parameters	$\nu \frac{\Omega N_a}{x_B^I} \exp\left(-\frac{G_A}{k_B T}\right)$	m/s
k_T	collection of materials parameters	$\Omega n_a \nu \exp\left(-\frac{G_A}{k_B T}\right) \frac{1}{k_B T} \frac{L}{T_m}$	m/s
k_{total}^2	total sink strength in BW theory	-	1/m^2
k_R^2	sink strength of all particles in the size class R,$R+dR$ in BW theory	-	1/m^2
$K_{diff.}^*$	rate constant for coarsening of spheres controlled by diffusion	-	m^3/s
$K_{int.}^*$	rate constant for coarsening of spheres controlled by interface kinetics	-	m^2/s
K_{sl}^*	rate constant for coarsening of spheres by heat exchange	$\frac{2\kappa\Gamma}{\rho L}$	m^3/s
K_{ss}^*	rate constant for coarsening of spheres by solute exchange	$\frac{D\ell^\alpha}{x_B^\beta - x_B^\alpha}$	m^3/s
K	prticle volume growth rate constant in MR theory	dV^*/dt	m^3/s
$\boldsymbol{k}$	wavevector in reciprocal space	-	1/m
l	length	-	m
L	latent heat of fusion	-	J/kg
L_m	molar latent heat of fusion	$H_{liquid} - H_{solid}$	J/mol
M	mass	-	kg
M	mobility of atoms at interdace	-	mol^2 / m J s
M	moment of a distribution	-	-
M_{kj}	mobilities	-	mol^2 / m J s
m	exponent	-	-
$\boldsymbol{n}$	normal vector	-	-
n	total number of atoms, moles, or particles	-	-
n_a^α	number of atoms per area in α-phase	-	-
n_a^β	number of atoms per area in β-phase	-	-
n_a	number of atoms per area	-	-
n_A	number of atoms or moles of component A	-	-
n_B	number of atoms or moles of component B	-	-

n_A^*	number of atoms or moles of component A in a critical nucleus	-	-
n_B^*	number of atoms or moles of component B in a critical nucleus	-	-
n_A^i	initial number of atoms or moles of component A	-	-
n_B^i	initial number of atoms or moles of component B	-	-
n_A^f	final number of atoms or moles of component A	-	-
n_B^f	final number of atoms or moles of component B	-	-
n_i	number of atoms or moles of component i	-	-
n_V^A	number of A-atoms per unit volume	-	$1/m^3$
n_V^B	number of B-atoms per unit volume	-	$1/m^3$
$n_V^{cluster}$	number of clusters per unit volume	-	$1/m^3$
$n_V^{cluster,s}$	number of clusters per unit volume at steady state	-	$1/m^3$
N_a^α	number of atoms in α-phase in front of an interface	-	$1/m^2$
N_a^β	number of atoms in β-phase in front of an interface	-	$1/m^2$
N_a	number of atoms in front of an interface	-	$1/m^2$
n_a^D	number of defects per interface area	-	$1/m^2$
n_V	number of particles per unit volume	-	$1/m^3$
N	number of chemical components	-	-
N_A	Avogadro's number	6.023×10^{23}	1/mol
P	pressure	-	Pa
P	proportionality constant in MR theory	-	-
P^α	pressure in α-phase	-	Pa
P^β	pressure in β-phase	-	Pa
P^{liquid}	pressure in liquid phase	-	Pa
P^{solid}	pressure in solid phase	-	Pa
P_{total}	total source strength in BW theory	-	1/s
P_R	source strength of all particles in the size class $R, R+dR$ in BW theory	-	1/s
Pe_D	Peclet number for mass transport	$v_c R/D$	-
Pe_T	Peclet number for heat transport	$v_c R/\kappa$	-
$\boldsymbol{q}_{diff}$	diffusive heat current density	$-\lambda \Delta T$	W/m^2 or $J/s\ m^2$
$\boldsymbol{q}_{conv}$	convective heat current density	$Q\boldsymbol{v}$	W/m^2 or $J/s\ m^2$
$\boldsymbol{r}, \boldsymbol{r}_i$	position vector (to center of particle i)	-	m
$\boldsymbol{r}_{ji}$	vector from center of particle j to center of particle i	$\boldsymbol{r}_j - \boldsymbol{r}_i$	m
$\boldsymbol{r}_i^s$	position vector to the surface of particle i	-	m

r	length of position vector, polar coordinate	-	m
r	dimensionless radius variable in MR analysis	-	-
$\hat{r}$	unit vector in radial direction	-	-
r_c	reference length	-	m
R^*	critical radius	-	m
R_c	critical radius of nucleation	-	m
R_{max}	maximum particle radius	-	m
R, R_i	radius (of the ith particle)	-	m
R_{embed}	radius of particle embedding shell in BW theory	-	m
$\boldsymbol{R}_i$	vector pointing to particle surface from particle center	$R_i \hat{r}$	m
R_0	initial particle radius	-	m
R_1, R_2	principal radii of curvature	-	m
s, s_i	surface (of particle i)	-	m^2
s_σ	surface entropy	-	J/m^2 K
S	entropy	-	J/K
S_0	supersaturation	$\frac{x_B - x_B^\alpha}{x_B^\beta - x_B^\alpha}$	-
S_l	molar entropy of the liquid state	-	J/mol K
S_s	molar entropy of the solid state	-	J/mol K
S_m	molar entropy	-	J/mol K
$S(k,t), S(\boldsymbol{k},t)$	structure function	-	-
t	time	-	s
T	temperature	-	K
T_c	critical temperature	-	K
T_m	equilibrium melting temperature	-	K
$T_m(R)$	melting temperature for particle of radius R	-	K
T_0	fixed temperature	-	K
T_I	interface temperature	-	K
T_q, T_{quench}	quench temperature	-	K
T_{anneal}	annealing temperature	-	K
U	internal energy	E	J
V, V_i	volume (of particle i)	-	m^3
V^*	volume of a critical nucleus	-	m^3
V_m^*	molar volume of a critical nucleus	-	m^3
$V^{cluster}$	volume of a cluster of atoms	-	m^3
V_{embed}	volume of particle embedding sphere	$\frac{4\pi}{3} r_c^3$	m^3
$v_{embed}, v_{e.}$	normalized volume of particle embedding sphere	V_{embed}/V_*	-
$V_{particle}, V_{p.}$	volume of particle	$\frac{4\pi}{3} R^3$	m^3
$v_{particle}, v_{p.}$	normalized volume of particle	$V_{particle}/V_*$	-
v_m	maximum normalized volume of particle	-	-
V_*	volume of particle of critical radius	$\frac{4\pi}{3}(R^*)^3$	m^3
V_T	total volume of all particles	$\sum_i^n V_i$	m^3
V_m	molar volume	-	m^3/mol
V_m^{solid}	molar volume of the solid phase	-	m^3/mol

$\overline{V}_i$	partial molar volume of component i	-	m^3/mol
$\mathbf{v}$	velocity vector	-	m/s
v_x	x-coordinate of velocity	-	m/s
v_y	y-coordinate of velocity	-	m/s
v_r	r-coordinate of velocity	-	m/s
v_θ	θ-coordinate of velocity	-	m/s
$v_r^{mat.}$	r-coordinate of velocity of matrix	-	m/s
$v_\theta^{mat.}$	θ-coordinate of velocity of matrix	-	m/s
v	velocity	-	m/s
v_0	fixed velocity	-	m/s
v_c	characteristic velocity	-	m/s
$v_{Marangoni}$	Marangoni velocity	-	m/s
v_{Stokes}	Stokes velocity	-	m/s
W_R	reversible work	-	J
W_R^*	reversible work for the formation of a critical nucleus	-	J
x_A	atomic fraction of component A	n_A/n	-
x_B	atomic fraction of component B	n_B/n	-
x_{B,r_c}	atomic fraction of component B at distance r_c from particle center	-	-
x_B^0	fixed atomic fraction of component B	-	-
x_B^I	atomic fraction of component B at interface	-	-
x_B^∞	atomic fraction of component B in matrix far from the interface	-	-
x_B^m	atomic fraction of component B in matrix	-	-
x_c	cutoff distance for diffusional interactions	-	m
x_i	atomic fraction of component i	n_i/n	-
x_i^*	atomic fraction of component i in a critical nucleus	-	-
x_B^α	atomic fraction of component B in α-phase	-	-
x_B^β	atomic fraction of component B in β-phase	-	-
$x_B^{\alpha,e}$	x_B^α with planar interface	-	-
$x_B^{\beta,e}$	x_B^β with planar interface	-	-
x, y, y_i	exponents in MR-theory	-	-
z	scaled dimensionless radius in MR	-	-
α_x	variation of surface tension with solute concentration	-	J/m^2
β	separation constant	-	-
$\mathcal{R}$	universal gas consant	-	J/K
$\chi(R)$	dimensionless concentration at particle surface	-	-
δ	boundary layer thickness	-	m
$\delta_{s/l}$	boundary layer thickness for solid/liquid interface	-	m
$\delta_{l/l}$	boundary layer thickness for liquid/liquid interface	-	m

ΔG	molar Gibbs free energy of mixing	$G_m^a - G_m^b$	J/mol
ΔG_{atom}	change in Gibbs free energy per atom	$\Delta G/N_A$	J
ΔG^{ex}	excess molar Gibbs free energy of mixing	$G - G^{id}$	J/mol
ΔG^{id}	ideal molar Gibbs free energy of mixing	$-T\Delta S^{id}$	J/mol
ΔG^M	molar Gibbs free energy of mixing	$G_m^a - G_m^b$	J/mol
ΔG^*	activation energy for nucleation	$16\pi\sigma(T)^3/(3\Delta G_V{}^2)$	J
ΔG_V	gain in free energy per volume on nucleation	-	J
ΔH	molar enthalpy of mixing	-	J/mol
ΔH_0	measure of interatomic interaction energy	-	J/mol
ΔS	molar mixing entropy	-	J/mol K
ΔS^{id}	molar mixing entropy in ideal solution	$-\mathcal{R}(x_A \ln x_A + x_B \ln x_B)$	J/mol K
ΔS^{ex}	excess mixing entropy in ideal solution	$\Delta S - \Delta S^{id}$	J/mol K
ΔS_f	entropy of fusion	$S_{solid} - S_{liquid}$	J/mol K
ΔT	temperature difference	-	K
∇T	temperature gradient	-	K/m
ℓ_s	screening length	-	m
ℓ^α	capillary length of the α-phase	-	m
ℓ^β	capillary length of the β-phase	-	m
η	heat source density	-	W/m^3 or J/s m^3
η	vicosity	-	Ns/m^2
η_{drop}	vicosity of droplet	-	Ns/m^2
γ_B	activity coefficient for B atoms	-	-
Γ	Gamma function	-	-
Γ	Gibbs-Thomson coefficient	$(T_m V_m^s \sigma)/L_m$	Km
κ	mean curvature	$\kappa = 1/R_1 + 1/R_2$	- 1/m
κ	thermal diffusivity	$\lambda/(c_V \rho)$	m^2/s
λ	thermal conductivity	-	W/K m or J/s K
λ	scaling variable (varying)	-	-
λ_k	separation distance of kinks	-	m
λ_l	separation distance of ledges	-	m
λ_R	scaling constant for radius of growing particle	$R/\sqrt{t}$	m/s$^{0.5}$
μ_A	chemical potential of component A	$\frac{\partial G}{\partial n_A}\Big\vert_{T,P,n_B}$	J/mol
μ_B	chemical potential of component B	$\frac{\partial G}{\partial n_B}\Big\vert_{T,P,n_A}$	J/mol
μ_i	chemical potential of component i	$\frac{\partial G}{\partial n_i}\Big\vert_{T,P,n_i \neq n_j}$	J/mol
μ_0	chemical potential of pure substance	-	J/mol
ν	jumps frequency of atoms	-	1/s
ν	variable used in LSW analysis	$\frac{K^*_{diff}}{(R^*)^2 \frac{dR^*}{dt}}$	-
ν	kinematic viscosity	η/ρ	Nms/kg
Ω	atomic volume	-	m^3

Ω_k	solid spherical angle in k-space	-	-
Ω	Grand Canonical Free energy	$E - TS - \sum \mu_i n_i$	J
$\Omega_{initial}$	initial Grand Canonical Free energy	-	J
Ω_{final}	final Grand Canonical Free energy	-	J
Ω_V	Grand Canonical Free energy per unit volume	-	J/m^3
ϕ	polar coordinate	-	-
ϕ	potential in fluid flow	-	m^2/s
Φ	volume fraction of particle phase	-	-
Ψ	stream function	-	m^3/s
ψ	-	$d\rho/dt$ -	-
$\Psi(kR_j)$	the Fourier transform of a sphere with uniform concentration	-	-
Ψ	stream function in fluid dynamics	-	m^2/s
ρ	density	-	kg/m^3
$\rho_{part.}$	particle density	-	kg/m^3
ρ_{drop}	droplet density	-	kg/m^3
ρ	scaled particle radius	R/R_c	-
ρ_0	initial scaled particle radius	-	-
$\rho_{min.}$	minimal scaled particle radius	-	-
ρ_m	maximal scaled particle radius	-	-
ρ^*	scaled critical particle radius	R/R_c	-
σ	interfacial energy	-	J/m^2
σ_0	interfacial energy at T_0	-	J/m^2
$\sigma(t)$	dimensionless supersaturation	$(\langle x_B^I \rangle - x_B^{\alpha,e})/x_B^{\alpha,e}$	-
τ	dimensionless time	-	-
τ_0	initial dimensionless time	-	-
τ	viscous shear stress	-	N/m^2
θ	polar coordinate	-	-
Θ	dimensionsless coordinate in MR analysis	-	-

Index

Printing: Saladruck, Berlin
Binding: H. Stürtz AG, Würzburg